Lecture Notes in Physics

Springer
Berlin
Heidelberg
New York
Barcelona
Hong Kong
London
Milan
Paris
Singapore
Tokyo

Physics and Astronomy ONLINE LIBRARY

http://www.springer.de/phys/

Editorial Policy

The series *Lecture Notes in Physics* (LNP), founded in 1969, reports new developments in physics research and teaching -- quickly, informally but with a high quality. Manuscripts to be considered for publication are topical volumes consisting of a limited number of contributions, carefully edited and closely related to each other. Each contribution should contain at least partly original and previously unpublished material, be written in a clear, pedagogical style and aimed at a broader readership, especially graduate students and nonspecialist researchers wishing to familiarize themselves with the topic concerned. For this reason, traditional proceedings cannot be considered for this series though volumes to appear in this series are often based on material presented at conferences, workshops and schools (in exceptional cases the original papers and/or those not included in the printed book may be added on an accompanying CD ROM, together with the abstracts of posters and other material suitable for publication, e.g. large tables, colour pictures, program codes, etc.).

Acceptance

A project can only be accepted tentatively for publication, by both the editorial board and the publisher, following thorough examination of the material submitted. The book proposal sent to the publisher should consist at least of a preliminary table of contents outlining the structure of the book together with abstracts of all contributions to be included.

Final acceptance is issued by the series editor in charge, in consultation with the publisher, only after receiving the complete manuscript. Final acceptance, possibly requiring minor corrections, usually follows the tentative acceptance unless the final manuscript differs significantly from expectations (project outline). In particular, the series editors are entitled to reject individual contributions if they do not meet the high quality standards of this series. The final manuscript must be camera-ready, and should include both an informative introduction and a sufficiently detailed subject index.

Contractual Aspects

Publication in LNP is free of charge. There is no formal contract, no royalties are paid, and no bulk orders are required, although special discounts are offered in this case. The volume editors receive jointly 30 free copies for their personal use and are entitled, as are the contributing authors, to purchase Springer books at a reduced rate. The publisher secures the copyright for each volume. As a rule, no reprints of individual contributions can be supplied.

Manuscript Submission

The manuscript in its final and approved version must be submitted in camera-ready form. The corresponding electronic source files are also required for the production process, in particular the online version. Technical assistance in compiling the final manuscript can be provided by the publisher's production editor(s), especially with regard to the publisher's own Latex macro package which has been specially designed for this series.

Online Version/ LNP Homepage

LNP homepage (list of available titles, aims and scope, editorial contacts etc.):
http://www.springer.de/phys/books/lnpp/

LNP online (abstracts, full-texts, subscriptions etc.):
http://link.springer.de/series/lnpp/

Karl-Ludwig Klein (Ed.)

Energy Conversion and Particle Acceleration in the Solar Corona

 Springer

Editor

Karl-Ludwig Klein
Observatoire de Paris,
Section de Meudon,
LESIA,
5, place Jules Janssen
92195 Meudon Principal Cedex, France

Cover Picture: (see contribution by Bojan Vršnak in this volume)

Library of Congress Cataloging-in-Publication Data.
Cataloging-in-Publication Data applied for

A catalog record for this book is available from the Library of Congress.

Bibliographic information published by Die Deutsche Bibliothek

Die Deutsche Bibliothek lists this publication in the Deutsche Nationalbibliografie;
detailed bibliographic data is available in the Internet at http://dnb.ddb.de

ISSN 0075-8450
ISBN 3-540-00275-8 Springer-Verlag Berlin Heidelberg New York

Springer-Verlag Berlin Heidelberg New York
a member of BertelsmannSpringer Science+Business Media GmbH

http://www.springer.de

© Springer-Verlag Berlin Heidelberg 2003
Printed in Germany

Typesetting: Camera-ready by the authors/editor
Camera-data conversion by Steingraeber Satztechnik GmbH Heidelberg
Cover design: *design & production*, Heidelberg

Printed on acid-free paper
57/3141/du - 5 4 3 2 1 0

Preface

The conversion of energy generated in the Sun's interior creates its hot corona and a wealth of dynamical phenomena such as flares, mass ejections and transient non thermal populations of charged particles. These processes are of general interest in astrophysics. In the case of the Sun they can be probed by a unique combination of imaging, spectrographic and in situ measurements. Radio observations provide important diagnostics, and many instruments are operated by small research groups in Europe. The stimulation of joint investigations using radio diagnostics is a major role of CESRA, the *Community of European Solar Radio Astronomers*. This volume is based on the CESRA Workshop and Euroconference *Energy Conversion and Particle Acceleration in the Solar Corona* held 2–6 July 2001, at Schloss Ringberg near Tegernsee (Germany). It aims to address a broader community of astrophyscists, including graduate students and researchers who want to gain an insight into this subject.

The workshop was organised by a scientific committee composed of C. Alissandrakis (Greece), F. Chiuderi-Drago and G. Einaudi (Italy), M. Karlický (Czech Republic), K.-L. Klein (France), J. Kuijpers (The Netherlands; president of the joint Solar Physics Section of the European Physical Society and the European Astronomical Society), and G. Mann and R. Treumann (Germany). The local organisers were R. Treumann and A. Czaykowska (Max Planck Institut für Extraterrestische Physik, Munich), assisted by the Copernicus Gesellschaft. The participants will remember the pleasant and stimulating atmosphere of Schloss Ringberg, operated by the Max Planck Gesellschaft, and the kind hospitality of Dr. A. Hoermann and his collaborators. Funding by the European Community was essential to enable the participation of young colleagues and keynote speakers. Important financial support by the *Deutsche Forschungsgemeinschaft* and the *Observatoire de Paris* is also gratefully acknowleged. The Editor is indebted to the referees of the contributions. Among them were T. Amari, M. Aschwanden, R. Canfield, D. Delcourt, P. Démoulin, G. Holman, H. Hudson, L. Kocharov, J. Kuijpers, A. Mangeney, S. Pohjolainen, L. Vlahos, and S. White. Last, but not least, he would like to thank Dr. C. Caron (Springer Verlag) for his continued encouragement and helpful advice.

Meudon,
September 2002

Karl-Ludwig Klein
on behalf of the Scientific Organising Committee

Contents

Part III   Aspects of Current Research

List of Contributors

Timothy S. Bastian
National Radio Astronomy
Observatory
Charlottesville, VA 22903, USA
tbastian@nrao.edu

Arnold O. Benz
Institute of Astronomy
ETH Zentrum
8092 Zurich, Switzerland
benz@astro.phys.ethz.ch

Dieter Biskamp
Centre for Interdisciplinary Plasma
Science
Max-Planck-Institut für Plasmaphysik
85748 Garching, Germany
dfb@ipp.mpg.de

Wolfgang Dröge
Bartol Research Institute
University of Delaware
Newark, DE 19716, USA
droege@bartol.udel.edu

Lyndsay Fletcher
Department of Physics and Astronomy
University of Glasgow
Glasgow G12 8QQ, U.K.
lyndsay@astro.gla.ac.uk

Petr Heinzel
Astronomical Institute
Academy of Sciences of the Czech
Republic
25165 Ondřejov, Czech Republic
pheinzel@sunkl.asu.cas.cz

Marian Karlický
Astronomical Institute
Academy of Sciences of the Czech
Republic
25165 Ondřejov, Czech Republic
karlicky@asu.cas.cz

Pierre Kaufmann
CRAAM
Universidade Presbiteriana Mackenzie
São Paulo, SP, Brazil
kaufmann@craam.mackenzie.br

Bernhard Kliem
Astrophysikalisches Institut Potsdam
14482 Potsdam, Germany
bkliem@aip.de

Säm Krucker
Space Sciences Laboratory
University of California
Berkeley, CA 94720-7450, USA
krucker@ssl.berkeley.edu

Yuri E. Litvinenko
Institute for the Study of Earth,
Oceans, and Space
University of New Hampshire
Durham, NH 03824-3525, USA
Yuri.Litvinenko@unh.edu

Alexander L. MacKinnon
Department of Adult and Continuing
Education
University of Glasgow
Glasgow, G3 6LP, UK
a.mackinnon@educ.gla.ac.uk

Reinhard Schlickeiser
Institut für Theoretische Physik
Lehrstuhl IV: Weltraum- und
Astrophysik
Ruhr-Universität Bochum
44780 Bochum, Germany
rsch@tp4.ruhr-uni-bochum.de

Manfred Scholer
Centre for Interdisciplinary Plasma
Science
Max-Planck-Institut für extraterrestrische Physik
85740 Garching, Germany
mbs@mpe.mpg.de

Kiyoto Shibasaki
Nobeyama Radio Observatory
Minamimaki
Minamisaku, Nagano 384-1305, Japan
shibasak@nro.nao.ac.jp

Gérard Trottet
LESIA, Observatoire de Paris
Section d'Astrophysique de Meudon
92195 Meudon Cédex, France
gerard.trottet@obspm.fr

Nicole Vilmer
LESIA, Observatoire de Paris
Section d'Astrophysique de Meudon
92195 Meudon Cédex, France
nicole.vilmer@obspm.fr

Bojan Vršnak
Hvar Observatory
Faculty of Geodesy
Kačićeva 26
10000 Zagreb, Croatia
bvrsnak@geof.hr

Harry P. Warren
Harvard-Smithsonian Center for
Astrophysics
60 Garden Street
Cambridge, MA 02138, USA
hwarren@cfa.harvard.edu

Yihua Yan
National Astronomical Observatories
Chinese Academy of Sciences
Beijing 100012, China
yyh@bao.ac.cn

Introduction

Karl-Ludwig Klein

Solar flares and coronal mass ejections (CMEs) are prominent signatures of the explosive release of energy stored in the coronal magnetic field. At some point during the evolution of a magnetic field configuration the stored energy is converted into heat, kinetic energy of bulk flows (ejecta) and energetic particle populations. Observations with increasing sensitivity and temporal resolution have shown that the corona varies on a broad range of time and length scales. It is now thought that the apparently steadily heated corona itself may owe its existence to similar explosive phenomena of interactions between plasmas and magnetic fields. The manner in which magnetically stored energy is converted to other forms, and prominently to energetic particle populations, is clearly of central importance for coronal physics.

The observational coverage of the Sun improved tremendously during the last decade. The SoHO, TRACE and *Yohkoh* satellites have been providing imaging and spectroscopic data which furthered our current understanding that magnetic reconnection is a key process of energy release in the solar atmosphere. Energetic particle diagnostics from gamma-rays to radio waves have shown the ubiquity of non thermal particle populations in the corona, while in situ measurements of solar energetic particles notably with Ulysses, WIND and ACE demonstrated the variety of these populations and provided new elements for their understanding. The role of radio observations in this field is twofold: radio emissions are an extremely sensitive tracer of non thermal electrons and therefore of energy release processes, and the broad range of frequencies covered by contemporary instruments probes these electrons from the low corona to interplanetary space. Radio observations are therefore crucial to make the connection between the dynamical plasma structures in the corona and in situ measurements of solar energetic particles.

Given the wealth of new observations, and the significant progress in the understanding of magnetic reconnection in collisionless plasmas from numerical studies and from measurements in the Earth's magnetosphere, it appeared timely, at the beginning of the RHESSI mission (Ramaty High Energy Spectroscopic Imager), to critically assess the state of the art, including input from neighbouring fields of solar physics, especially the Earth's magnetosphere. This book is divided into three parts: energy conversion with emphasis on magnetic reconnection, particle acceleration, and a discussion of current research and new diagnostic tools. Extended introductions to these subjects are given in [1, 2, 3].

1 Energy Release and Magnetic Reconnection

In studies of the Earth's magnetosphere and of solar flares it has been realised that the interaction of magnetic flux from different sources creates regions of strong magnetic field gradients (hence intense electric currents – current sheets) where the magnetic field may diffuse and change connectivity. This is the case at the interface between the solar wind and the magnetosphere as well as between magnetic field structures in the solar corona and new flux that emerges from the Sun's interior. One consequence of magnetic reconnection is the detachment and ejection of plasma-magnetic field configurations (plasmoids). Such plasmoid ejections are inferred to exist in the solar corona from imaging observations in visible light, at radio and X-ray wavelengths, and are well studied by in situ measurements in the Earth's magnetotail. Another result of magnetic reconnection is the conversion of energy stored in current systems into heat and bulk motion. The original idea of magnetic reconnection first proposed by Sweet, followed by Parker, in the 1950s is still valid. But their models of relatively long current sheets could not explain the rapidity with which the process is observed to proceed. Petschek's conjecture that reconnection actually involves only a small region where the magnetic field diffuses, whereas the bulk of the plasma flow is diverted outside this region, at two pairs of slow-mode shocks, gave a physical justification of small reconnection regions with a correspondingly high reconnection rate, but did not prove that current sheets actually evolve in this way. The microphysical mechanism of reconnection is different for plasmas that are dominated by collisions and those which are collisionless. Collisionless effects are expected to play a key role in both the solar corona and the Earth's magnetosphere. The reader is referred to [3] for an introductory text on magnetic reconnection.

M. Scholer's introductory lecture gives an overview of phenomena where magnetic reconnection seems to play a role in the solar atmosphere and the Earth's magnetosphere. Imaging and spectroscopic observations of the solar atmosphere, especially at EUV and X-ray wavelengths, show a growing wealth of structures and flows which are reminiscent of scenarios of magnetic reconnection. In situ measurements at the magnetopause and in the magnetic tail of the Earth allow for much more detailed quantitative analyses. They reveal that not all high-speed flows show the relations expected for magnetic reconnection. Slow-mode shocks predicted by Petschek-type reconnection models have been found in the magnetotail, but only in a minority of cases where reconnection seems to occur.

The topologies of magnetic fields and scenarios of magnetic reconnection in the solar atmosphere, as inferred from imaging observations, are discussed by B. Vršnak. He makes use of cartoon scenarios to show how the interaction of magnetic structures may lead to the reconfiguration of the coronal field, plasmoid formation and energy release in 2D and 3D configurations. These scenarios are compared with the morphologies observed during flares and filament eruptions.

While the magnetic field in the corona is the crucial ingredient for the storage and release of energy, direct measurements are difficult because of Doppler broadening of visible emission lines. Infrared measurements recently provided first results, while cyclotron radiation at radio wavelengths is a well-established tool to measure the field strength above sunspots. Other radio techniques hold some promise for the future through multifrequency imaging. For the time being, however, quantitative information on the coronal magnetic fields comes mostly from the extrapolation of measurements in the photosphere. Y. Yan's paper illustrates how the strength and topology of the coronal magnetic field can be inferred through a non-linear force-free field extrapolation. He discusses in particular the 14 July 2000 flare, including evidence for a magnetic flux rope and an estimation of its energy content.

A crucial check of the calculated magnetic field topology in the corona are imaging observations of plasma structures that are supposed to be shaped by the magnetic field. The high-cadence observations at arc-second scale provided by TRACE have set new standards in coronal imaging. L. Fletcher and H.P. Warren illustrate the variety of complex magnetic geometries involved in flares and their build-up phase. Detailed comparison with hard X-ray observations of accelerated electrons underlines the role that these particles play in energy transport from coronal sites of energy release to the chromosphere. The smallness of the flux tubes guiding the particles is a challenge for our understanding of beam propagation in plasmas.

Coherent plasma emission at decimetric and longer wavelengths from the corona is a tracer of electron beam propagation and other processes, which the observations suggest to be related to energy release during flares of various sizes. A. Benz reviews the basics of these processes and discusses the relevance of the observations for understanding energy release and particle acceleration. Combined spectrography and imaging at radio wavelengths, together with X-ray, EUV and visible-light imaging of the corona, show geometries consistent with particle acceleration and energy conversion in common scenarios of magnetic reconnection. However, most of these observations pertain to higher regions in the solar corona than those where radio spectrography suggests the bulk of the energetic particles to be produced during solar flares. The opening of the window between about 500 MHz and several GHz to imaging observations is a necessary step to furthering our understanding of flare energy release.

Radio emission at centimetre and shorter wavelengths is mostly thermal bremsstrahlung (also called free-free emission) from the quiescent atmosphere and active regions, with some contribution of cyclotron (or gyroresonance) emission at low harmonics of the electron cyclotron frequency above sunspots. During flares, gyrosynchrotron radiation from non thermal electrons, sometimes up to relativistic energies (where it is commonly called synchrotron radiation), dominates most often. K. Shibasaki reviews the processes by which electrons generate these radiations and how they can be used to infer information on the radiating particles, and presents flare observations with the Nobeyama Radioheliograph. His interpretation of a flare as a disruption of coronal magnetic structures under the effect of plasma pressure challenges the reconnection scenarios in a magnetically dominated atmosphere, and highlights the ambiguities inherent in present observations of solar flares.

Progress in the theoretical understanding of magnetic reconnection in natural plasmas is discussed by D. Biskamp. A long-standing problem is the slow rate of reconnection in a Sweet-Parker-type current sheet. Using a 2D model of resistive reconnection between two magnetic flux bundles with an arbitrarily imposed localisation of high resistivity, Biskamp shows that reconnection becomes fast, with a rate similar to that predicted by the Petschek scenario. While there is no reason to have an increased collisional resistivity in the reconnection region, non collisional effects related to the dispersion of hydromagnetic waves are shown to provide the required scales. The present status of modelling suggests that energy is mostly converted to ion bulk flows in the reconnecting current sheet, and only a minor fraction into energetic electrons. This may imply that processes which do not occur in the diffusion region, but are the consequence of the ion outflow, like turbulence or shock waves, play the key role in particle acceleration, especially of ions (see, however, Litvinenko's paper).

2 Energetic Particles at and from the Sun

A fundamental consequence of energy release in a plasma is the generation of energetic particle populations. Signatures of energetic particles at and from the Sun range from frequently observed transient enhancements of electron fluxes at typical energies up to a few keV that last between a second and several days to the occasional events where gamma-ray observations and in situ measurements in interplanetary space reveal relativistic electrons and ions. Remote sensing observations and in situ measurements with high time resolution and high sensitivity show that particle acceleration on all scales results from explosive energy release. Flares are still the dominant source of information through remote sensing diagnostics, because the whole range of the electromagnetic spectrum, including nuclear gamma-rays, can be studied. Coronal and interplanetary shocks are another source of energetic particle populations. The respective role of these processes is a subject of ongoing debate which is reflected in this volume.

That a major fraction of energy released during flares is actually transferred to accelerated electrons and nucleons is a consistent result of many years of hard X-ray and gamma-ray analyses of solar flares reviewed by N. Vilmer and A. MacKinnon. The authors describe the basic features of these emissions, as well as radio and visible light diagnostics of energetic particles. They discuss in detail the time scales of the radiation and how they relate to the acceleration and transport of the particles in the coronal magnetic field. A fundamental requirement from the observations is that the acceleration processes act on time scales of seconds or less as shown by the variations of spectra and abundances. The observations are discussed in the light of currently favoured acceleration processes, involving direct electric fields, shock waves and turbulence.

The key role of energetic particles in transporting energy released in the corona to the chromosphere, where the hard X-ray, gamma-ray and optical emissions arise, is demonstrated by the advent of Hα observations with time resolution ≤ 1 s. P. Heinzel and M. Karlický review joint observations in Hα and hard X-rays which show that energy may be transported much faster than by thermal conduction, and discuss how these observations can be used for quantitative analyses. To this end they study the line formation under the effect of pulsed particle beams, using a numerical model which describes both the fluid behaviour of the system and the radiative transfer.

Apart from remote sensing of their characteristic radiations, particles accelerated at the Sun can also be studied through in situ measurements in interplanetary space. One might expect that at coronal acceleration sites where those particles are energised which interact with the solar atmosphere, thereby generating the radiative signatures, particles are also injected into space. This would infer a close correlation between e.g. hard X-ray and radio emission of electrons in the solar atmosphere and electrons detected at 1 AU. S. Krucker's overview of electron and proton measurements with the WIND satellite shows that this simple expectation is sometimes met, but that in a significant number of events the injection of electrons into space is delayed with respect to the coronal radio emission. Suprathermal, but subrelativistic protons often behave in a still different way which may reveal an energy-dependent release at different coronal heights. Krucker argues that besides flares, shock waves driven by large-scale coronal disturbances are significant accelerators of escaping particles at heights >1 $R_\odot$, without injecting electrons in the low corona where they would generate radio emission.

The interaction of energetic particle populations with the turbulent interplanetary magnetic field, especially through pitch angle scattering, blurs the traces of the original acceleration processes. W. Dröge reviews quasi-linear models of particle transport and how they enable one to derive energy or momentum spectra of electrons and protons from in situ measurements. He argues that the spectra can be understood in terms of stochastic acceleration. Again, electrons and protons may have different behaviour. Where the acceleration occurs, and how it is related to small-scale restructuring of the corona as in flares and to large-scale reconfigurations like in coronal mass ejections, is not settled. The question can certainly not be answered by in situ measurements alone, but measurements in the inner heliosphere, where transport effects are minimised because of the shorter travel paths, appear to be an essential tool for further progress.

Theories of particle acceleration processes relevant to natural plasmas are discussed in the articles by Litvinenko and Schlickeiser. R. Schlickeiser presents general aspects showing the necessity of time-varying electric fields. He describes in some detail the acceleration of particles through their interaction with various types of plasma waves which are expected to be abundantly generated in the solar corona, e.g. in the interaction between a reconnection outflow and the ambient plasma or in the vicinity of shock waves in the corona and the interplanetary medium. An interesting feature of stochastic acceleration is that it predicts the selective energisation of particles depending on their charge-to-mass ratio, as is observed in many particle events in interplanetary space, in gamma-ray flares, and more recently in the heavy ion temperatures measured in the quiet corona by SoHO. Wave-particle interactions can also lead to acceleration when a rapid mass flow interacts with an ambient plasma. Schlickeiser shows this for a relativistic flow which may have some relevance in active galaxies. An interesting question in the solar context is if a rapid, though largely non-relativistic, bulk flow such as a coronal mass ejection can accelerate particles through a similar mechanism.

Y. Litvinenko analyses particle trajectories in reconnecting current sheets and evaluates the energy gain and acceleration time scales. The three-dimensional magnetic field structure in the current sheet determines the energy spectra of the accelerated particles. Deka-keV electron beams are expected in large-scale current sheets, and electrons can be accelerated up to MeV energies if a tearing instability creates multiple singular lines. Protons respond differently to the non uniform magnetic field, but acceleration up to GeV energies seems to be suggested by preliminary 2D analyses for relatively strong magnetic fields.

3 Aspects of Current Research

Current research in the field of energetic particles and energy release in the solar atmosphere was discussed by the participants of the CESRA workshop during three half-day working group sessions. The summary by B. Kliem, A. MacKinnon, G. Trottet and T. Bastian gives a flavour of the discussions and puts them into the broader context of research, starting with a brief comparative description of scenarios of magnetic field reconfiguration during flares and mass ejections. Very recent work on the signatures of magnetic reconnection and MHD turbulence, especially at radio wavelengths, of particle acceleration and transport, and of large-scale eruptive phenomena and energetic particles in space is discussed.

A major new diagnostic of the flaring solar atmosphere is observations at wavelengths ≤ 1 mm. Pioneering work using the University of Cologne telescope at Gornergrat (Switzerland) and the newly developed Brazilian *Solar Submillimetre Telescope* (SST) in the Argentinan Andes was reported at the workshop by the groups from Berne and São Paulo. P. Kaufmann reviews the history of solar observations at these wavelengths, presents the setup of SST, and illustrates its diagnostic potential for measuring the synchrotron radiation from relativistic electrons. Sub-millimetre and far infrared observations of solar flares hold promise for providing new constraints on the acceleration processes and also as a diagnostic of the energy transport in the flaring solar atmosphere.

References

1. A.O. Benz: *Plasma Astrophysics - Kinetic Processes in Solar and Stellar Coronae* (Kluwer Academic Publishers, Dordrecht, 1993)
2. T.E. Cravens: *Physics of Solar System Plasmas* (Cambridge University Press, Cambridge, 1997)
3. R.A. Treumann, W. Baumjohann: *Advanced Space Plasma Physics* (Imperial College Press, London, 1997)

Part I

Energy Release and Magnetic Reconnection

Magnetic Reconnection on the Sun and in the Earth's Magnetosphere

Manfred Scholer

Centre for Interdisciplinary Plasma Science,
Max-Planck-Institut für extraterrestrische Physik, 85740 Garching, Germany

Abstract. Observational evidence for magnetic reconnection in the Sun's atmosphere and in the Earth's magnetosphere is reviewed. On the Sun reconnection is inferred from observations of high speed flows derived from motion of X-ray or EUV features, from the identification of structures in the soft X-ray images, and from the observation of energetic particle events in hard X-ray images. In particular the soft X-ray images during flares are supposed to exhibit the large scale magnetic field structure. In the Earth's magnetosphere in situ observations by satellites at a particular location reveal magnetic field and flow variations. To these one can apply conservation laws and verify whether boundaries, like the magnetopause, have a normal magnetic field component, i.e., are rotational discontinuities. In the magnetotail the magnetic field structure can be deduced from the temporal variation of the magnetic field during high speed flow events.

1 Evidence for Reconnection on the Sun

1.1 Long Duration Event Flares

Indication for reconnection being responsible for long duration event flares comes from the observations of cusp-shaped soft X-ray flare loops, from observations of chromospheric evaporation by using extreme-ultraviolet (EUV) spectrohelio-grams, and from observation of inflow into the neutral line by using the motion of structures in EUV images.

Yohkoh soft X-ray observations of cusp-shaped structures in these events have first supported the view that an expansion and restructuring of the magnetic field occurs prior to the flare and that reconnection in a newly created neutral sheet plays a key role in the flare energy release (Tsuneta et al., 1992; Tsuneta, 1996). The tip of the cusp is interpreted to be the remnant of the kink of the reconnected field lines. This tip usually increases in height, at the same time the distance between the footpoints increases. A reconstruction of the temperature structure during the decay phase of a flare by Tsuneta (1996) has shown that the outer shell of the arches has the highest temperature (1.1×10^7 K), whereas the temperature decreases toward the inner arches (7×10^6 K). This is interpreted in terms of energy supply by reconnection near the top of the loop and subsequent cooling of more inner loops by heat conduction and filling with evaporated plasma. Figure 1a shows an X-ray map (negative image) from an east limb flare. North is up and east to the left. Figure 1b shows the plasma temperature; while the

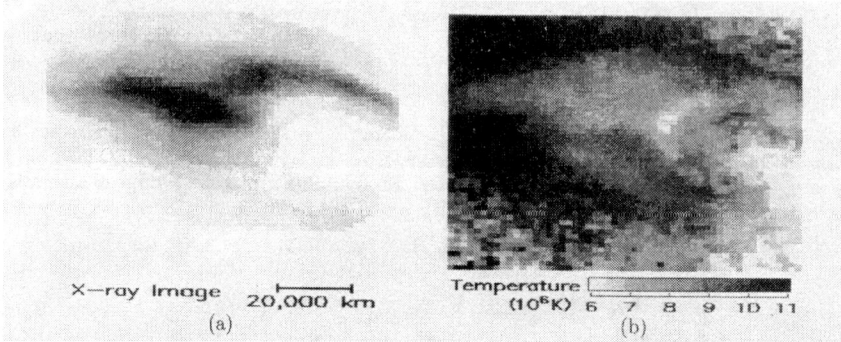

Fig. 1. (a) X-ray and (b) temperature maps of the 1992 February 21 flare. Both the X-ray brightness and temperature are shown by negative image (after Tsuneta, 1996).

X-ray brightness had a peak in the innermost part of the flare region, the plasma temperature was hotter in the outer loops (where darker color corresponds to hotter region). Note that outside the loop the X-ray intensity rapidly drops so that here the temperature map becomes unreliable. Color-coded maps with contours of the X-ray intensity level superimposed can be found in Tsuneta (1996). Although direct evidence for coronal slow MHD shocks has not yet been obtained, the overall morphology and temperature distribution in this type of flares are consistent with the prediction of Petschek-type reconnection with two MHD slow shocks bounding the reconnection outflow region.

Two-dimensional MHD simulations of reconnection including heat conduction predict chromospheric evaporation (i.e. upflows) in the outermost reconnected loops (Yokoyama and Shibata, 1997; 1998). As the loops become disconnected from the reconnection site, they cool and condensation sets in, leading to downflows in the older loop legs. Czaykowska et al. (1999) used data from the Coronal Diagnostic Spectrometer (CDS) on SOHO to demonstrate that in a two-ribbon flare there was indeed gentle upflow near the outer edges of the Hα ribbons. The upflow was observed as blueshifts in the EUV O V, Fe XVI, and Fe XIX lines in low emission regions, and, as time proceeds, the blueshifted regions moved with the Hα ribbons to greater distances away from the magnetic neutral line. In the region between the Hα ribbons and the neutral line bright downflowing plasma is observed. In a later paper, Czaykowska et al. (2001) compared the upflow velocities with those expected from different chromospheric heating models. Since a nonthermal beam of energetic electrons > 15 keV would produce significant hard X-ray emission, which is not observed, they concluded that the most likely energy transport mechnism is thermal conduction, as in the model of Yokoyama and Shibata (1997; 1998).

The final missing piece of evidence for reconnection, that is the inflow into the reconnection region, has recently been reported by Yokoyama et al. (2001). The inflow velocity was derived by tracing the movements of threadlike patterns in images obtained by the EUV Imaging Telescope (EIT) on SOHO. *Yohkoh* soft X-ray images during this flare showed a plasmoid ejection before the inflow sets in and a cusp-shaped loop during the inflow as a result of piling up of reconnected field lines. Yokoyama et al. (2001) derived from the upper limit of the inflow velocity of 5 km/s a reconnection rate of $M_A = 0.001 - 0.03$, where M_A is the inflow Alfvén Mach number.

1.2 Impulsive Flares

For a long time short-lived impulsive flares were believed to be "loop flares" in which the energy release occurs within the loop and not due to reconnection at an X point/line above the loop. This was based on the lack of cusp-shaped structures of impulsive flares in soft X-ray images. However, Masuda et al. (1994) discovered that in some of the impulsive compact loop flares occuring near the solar limb a loop-top hard X-ray (HXR) source appeared well above a soft X-ray bright source during the impulsive phase. Figure 2 shows hard X-ray contour maps obtained during the rising phase of the 13 January 1992 flare near the west solar limb. The soft X-ray loop coincides approximately with the 14-23 keV contour maps (the backbone of the soft X-ray loop is shown by solid lines). The higher energy maps reveal double foot point sources and an additional source well above the apex of the soft X-ray loop. The hard X-ray source above the soft X-ray loop indicates a source of energetic electrons outside of the closed loop. Masuda et al. (1994) proposed that the loop-top hard X-rays are due to nonthermal electrons produced by a fast mode shock when the reconnection flow impinges on the loop. The nonthermal electrons stream along field lines and produce the bright footpoints. This scenario is sketched in Fig. 3.

As pointed out by Shibata et al. (1995) (and indicated in Fig. 3) two ribbon flares should be accompanied by a plasmoid ejection high above the reconnected loops seen as soft X-ray loops. Shibata et al. (1995) actually found an erupting feature in soft X-rays during the 13 January 1992 flare, although the velocity was considerably smaller than the theoretically expected velocity (= Alfvén velocity). Ohyama and Shibata (1998) have analyzed in detail the ejecta during an impulsive flare (Fig. 4). The speed of the ejecta as obtained from the time

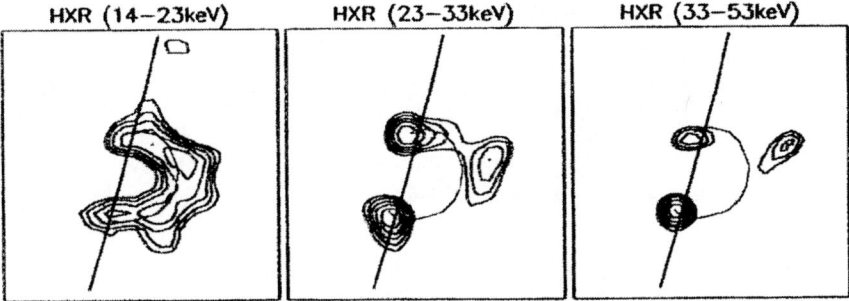

Fig. 2. Hard X-ray images of an impulsive west limb flare in three energy bands obtained from the hard X-ray telescope on *Yohkoh* (after Masuda et al., 1994).

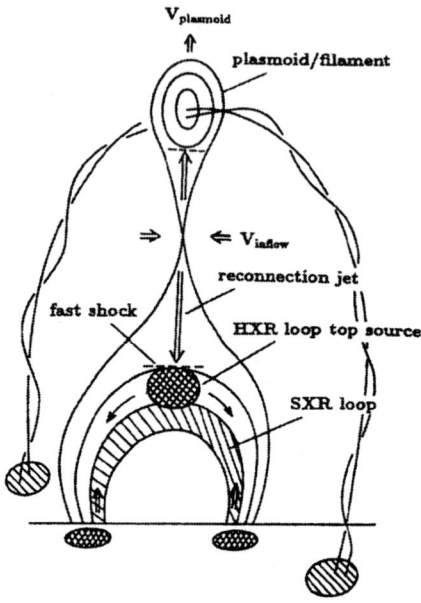

Fig. 3. The reconnection - plasmoid ejection model for compact loop flares (from Shibata et al., 1995).

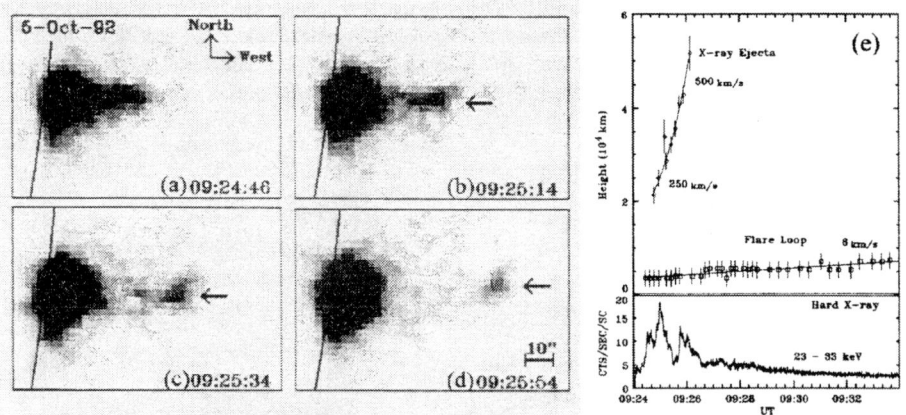

Fig. 4. Left: time sequence of soft X-ray images during the 5 October 1992 flare. The ejected plasma is indicated by an arrow. Right: apparent heights of the plasmoid and flare loop, and hard X-ray intensity versus time (from Ohyama and Shibata, 1998).

sequence analysis shows an outward acceleration (Fig. 4, right hand side) The ejecta is penetrated by a large-scale expanding loop as indicated in the schematic in Fig. 3. The kinetic energy of the ejecta was by an order of magnitude smaller than the thermal energy of the flare loop. This indicates that the plasma ejection is not responsible for stretching the overlying magnetic field into a vertical current sheet, as possibly in long duration event flares, but that plasmoid ejection and reconnection are a consequence of the same driving instability.

1.3 Small-Scale Reconnection in the Corona and in the Chromosphere

Many small-scale X-ray jets are observed to be ejected from emerging flux regions. It has been proposed that separate and opposite magnetic field of emerging flux and of pre-existing coronal field come close together during the rising motion of the emerging flux, which triggers reconnection (e.g., Yokoyama and Shibata, 1995). This leads to heating by Joule dissipation and can produce the so-called Anemone jets if the pre-existing magnetic field is oblique, or a two-sided loop structure if the pre-existing magnetic field is horizontal. MHD simulations have shown that in the oblique case the sling-shot effect of reconnected field lines can induce by a whip-like motion upward flows of cool chromospheric material (Yokoyama and Shibata, 1995). These cool jets may be observed as Hα surges. Canfield et al. (1996) actually studied the relation between Hα surges and X-ray jets and found that all Hα surges in their observations are associated with X-ray jets. Some X-ray jets are associated with type III bursts (Aurass et al., 1994; Kundu et al., 1995). This indicates that electrons with velocities up to ~ 0.3 of the velocity of light are accelerated in these microflares.

Small-scale plasma jets, microflares, and explosive events occur near the boundaries of the supergranulation cells in the chromosphere. Explosive events are short-lived (~ 60 sec), small-scale (1500 km) and exhibit high velocity flows up to 200 km/sec. These velocities are roughly equal to the Alfvén speed in the chromosphere and could be the result of magnetic reconnection. The spatial structure of these flows has been investigated with the SUMER (Solar Ultraviolet Measurements of Emitted Radiation) instrument on SOHO (Innes et al., 1997). In many of these events it has been found that the Doppler-shifts (in the Si IV line profiles) change from red to blue within a few arcseconds during east-west scans. The emission is brightest at the position where the shift changes sign. Furthermore, the two wings of the emission move away from the site of brightest emission across the Sun's surface. The structure of the jets is interpreted as evidence for reconnection, where the reversal of the blue and red shift is due to a reversal of the jet's east-west orientation when the current sheet is inclined in the east-west direction (Fig. 5). Blue-shifted wings have a larger extent in the east-west direction indicating that outflow occupies a larger volume, whereas downflows may be inhibited by the increasing density of the chromosphere.

2 Reconnection in the Magnetosphere

Reconnection in the Earth's magnetosphere is well established by in situ measurements. Whereas in the case of the Sun only indirect measurements, like

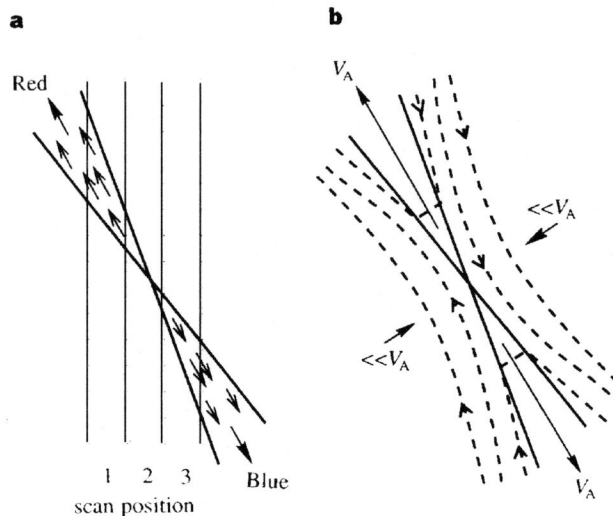

Fig. 5. Left: schematic of the bi-directional jet. Vertical lines are three scan positions of the spectrometer slit during east-west raster. Right: Reconnection model explaining the bi-directional jet measurements (after Innes et al., 1997).

high-speed flows and topological structures of flux tubes, are available, direct in situ measurements of velocities and magnetic fields in the magnetosphere, in particular at the magnetopause, allow a quantitative proof that reconnection is indeed going on. There are two regions in the magnetosphere where reconnection is to be expected: at the magnetopause and in the geomagnetic tail. As first proposed by Dungey (1961), when southward interplanetary magnetic field (IMF) comes into contact at the nose of the magnetopause with the northward Earth's dipolar field, reconnection can occur, and the reconnected IMF field lines are dragged with the anti-sunward moving solar wind into the magnetotail. These field lines are added to the tail lobes, and eventually the antiparallel field lines of the magnetotail can reconnect in the magnetotail current sheet. In the real three-dimensional world the situation is rather complicated: magnetopause reconnection, for instance, can occur for all directions of the IMF; even under northward IMF reconnection has been observed to occur tailward of the polar cusp, where northward IMF field lines and magnetospheric field lines are anti-parallel. The following subsection reviews observations of steady state magnetic reconnection at the magnetopause. We then discuss more bursty magnetopause reconnection events, known as flux transfer events (FTEs). Finally reconnection in the magnetotail and the relation to magnetospheric substorms is briefly reviewed.

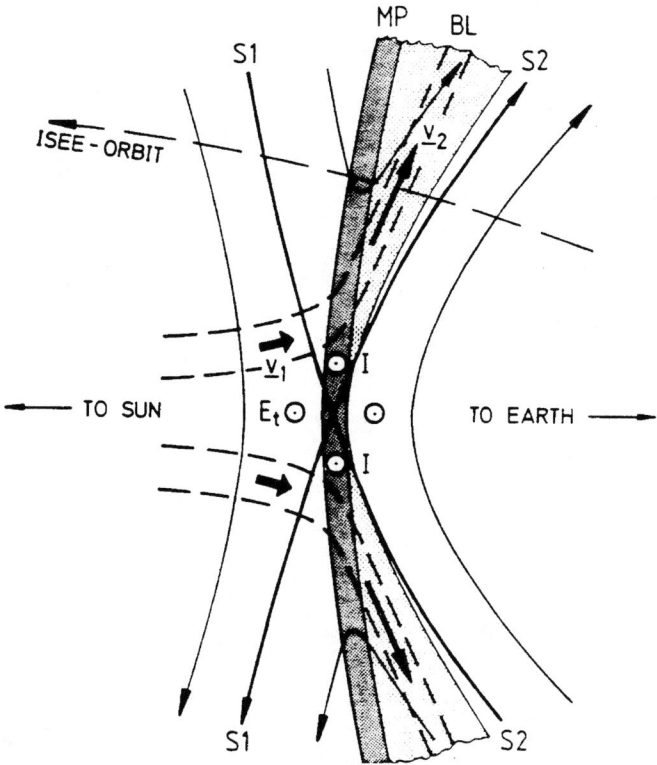

Fig. 6. Meridional cut of the reconnection configuration at the dayside magnetopause under strictly southward IMF. Dotted lines are streamlines, solid lines are magnetic field lines (after Paschmann et al., 1979).

2.1 Steady-State Reconnection at the Magnetopause

Figure 6 is a schematic of a section of the magnetopause in the meridian plane under exactly southward IMF (Paschmann et al., 1979). Three kinds of magnetic field lines are shown: interplanetary field lines, closed magnetospheric field lines, and reconnected field lines which have one foot in the ionosphere and end up in interplanetary space. The magnetosheath plasma enters in such a situation with a normal velocity v_n through the magnetopause and constitutes then a high-speed boundary layer (BL) along the magnetopause within the magnetosphere. S1 and S2 are field lines connected to the neutral line, so-called separatrices. A discontinuity like the magnetopause with a normal magnetic field component is a rotational discontinuity and the plasma and magnetic field data must fulfill the conditions for such a discontinuity. A check whether these conditions are fulfilled is direct proof of steady-state reconnection. We use in the following for tangential components the index t and for the normal component the index n. Mass conservation across the discontinuity can then be written as

$$[F_n] = [\rho(v_n - U_n)] = 0 \qquad (1)$$

where F_n is the mass flow, ρ the density, and U_n the velocity of the discontinuity, and brackets denote the jump of the quantities through the discontinuity. From Maxwell's equations we have constancy of the tangential component of the electric field, $[E_t] = 0$. Finally, the tangential component of the momentum through a discontinuity has to be conserved

$$\left[F_n v_t - \frac{B_n}{\mu_o} B_t\right] = 0 \qquad (2)$$

Using the frozen-in condition $\boldsymbol{E} + \boldsymbol{v} \times \boldsymbol{B} = 0$ one obtains for the normal component of the velocity v_n relative to U_N

$$v_n - U_n = B_n/(\mu_o \rho)^{1/2} \qquad (3)$$

For the tangential component one finds

$$[v_t] = \pm [B_t]/(\mu_o \rho)^{1/2} \qquad (4)$$

Therefore there exists a unique velocity $\boldsymbol{v}_{HT}$ so that the flow velocity $\mathbf{v}$ on both sides of the discontinuity can be written as

$$\boldsymbol{v} = \boldsymbol{v}_{HT} \pm \boldsymbol{B}/(\mu_o \rho)^{1/2} \qquad (5)$$

Consider now an observer moving with velocity $\boldsymbol{v}_{HT}$. In his frame the flow velocity is given by $\boldsymbol{v}' = \boldsymbol{v} - \boldsymbol{v}_{HT} \propto \boldsymbol{B}$, i.e., on both sides of the discontinuity the plasma velocity and the magnetic field are in this particular frame parallel to each other, and thus the $\boldsymbol{v}' \times \boldsymbol{B}$ electric field is zero. This transformation velocity is called the de Hoffmann-Teller (HT) velocity. In order to prove ongoing reconnection one has to show first that there exists a deHoffmann-Teller (HT) velocity. This is a necessary, but not sufficient condition. Step 2 is to verify that in the HT frame $\boldsymbol{v}'$ is given by the Alfvén velocity $\boldsymbol{B}/(\mu_o \rho)^{1/2}$. Since the HT electric field $\boldsymbol{E}_{HT} = \boldsymbol{v}_{HT} \times \boldsymbol{B}$ and the convection electric field $\boldsymbol{E}_c = \boldsymbol{v}_{HT} \times \boldsymbol{B}$ are equal, the HT frame velocity can be obtained by least square fitting of $\boldsymbol{E}_{HT}$ to the convection electric field $\boldsymbol{E}_c$. Figure 7 shows for a magnetopause crossing a scatter plot of all components of $\boldsymbol{E}_c$ versus the corresponding component of $\boldsymbol{E}_{HT}$ (Sonnerup et al., 1990). The high correlation between the HT and the convection electric field demonstrates that there exists for the whole time period a unique HT velocity and a frame in which the flow is parallel to the magnetic field. Figure 8 from Sonnerup et al. (1990) shows component-by-component correlations between $\boldsymbol{v}'$, the velocity in the HT frame, and the Alfvén velocity as the spacecraft travels through the magnetopause. During the event shown in the left panel the velocity in the HT frame is indeed very closely field-aligned and equal to the Alfvén speed. This event fulfills the criteria for a reconnection event. The right panel shows an instance where no agreement is present, even though an excellent HT velocity could be derived. We have gone here in such a detail of the data analysis in order to demonstrate that observations of high

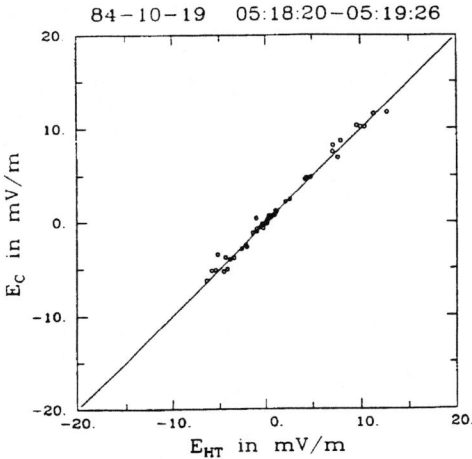

Fig. 7. Scatter plot of the three components of the reconnection electric field versus the components of the de Hoffmann-Teller electric field during a magnetopause crossing (after Sonnerup et al., 1990).

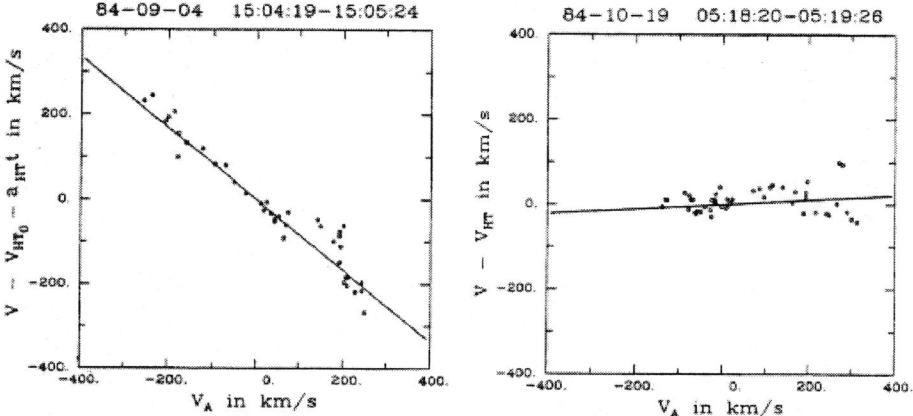

Fig. 8. Scatter plot of the three components of the velocity in the deHoffmann-Teller frame versus the Alfvén velocity during two magnetopause crossings. Event shown in the left panel is a reconnection event, event shown in the right panel is no reconnection event (after Sonnerup et al., 1990).

speed flows are not sufficient to infer reconnection. The choice of $\pm$ in Equation (5) depends on whether the flow is parallel or antiparallel to the magnetic field; for Earthward flow across the magnetopause ($v_n < 0$) the parallel situation ($+$) pertains to a magnetopause region where $B_n < 0$ and the antiparallel situation ($-$) where $B_n > 0$. Determination of the sign allows determination of the position of the observation point relative to the reconnection line, i. e., whether the location of the measurement is north or south of the reconnection line. Statistical analysis has shown that the X-line location varies considerably on the dayside magnetopause and is not hinged at the subsolar or stagnation point.

Single spacecraft observations reveal naturally only single high velocity jets at the magnetopause and the existence of a counter-streaming jet is implicitly assumed. Recently, simultaneous two-spacecraft observations of bi-directional jets at the magnetopause have been reported (Phan et al., 2000). The spacecraft (Geotail and Equator-S) were separated by about 3 R_E (Earth radii) in the east-west direction near the dawn flank, and 4 R_E in the north-south direction. Thus the reconnection line extended at least 3 R_E, but because during southward IMF reconnection is frequently observed near the nose of the magnetopause it is concluded that the X-line may extend well across the subsolar region to the dusk flank. If the length of a dawn-dusk reconnection line along the dayside magnetopause amounts to $\sim 40 R_E$, reconnection is the dominant process of solar wind entry into the magnetosphere when the IMF is southward, with other processes having a minor role at most.

Once a field line is open, magnetosheath ions stream continuously across the magnetopause. These ions have access to the ionosphere along each newly opened field line until it is appended to the tail lobe. A velocity filter effect arises because cusp ions of different field-aligned velocity, injected simultaneously across the magnetopause onto any one field line, have different flight times along that field line. Hence they have different arrival times in the ionosphere, and, as the field line is convecting, are spatially dispersed along the locus of the field line (e.g., Lockwood et al., 1994). An equatorward moving satellite will move onto field lines which have been more recently opened by reconnection so that it will observe an increase in the low-energy cut-off of energetic particles; the lowest energy ions seen at any observation time have the longest flight time and were the first to be injected (and thus were injected close to the X-line). Observation of the low energy cut-off of energetic particles in the ionosphere allows, when the distance to the reconnection line is known, determination of the reconnection rate as a function of time. Lockwood et al. (1994) have shown that, even in cases where reconnection is continuously going on for some length of time, the reconnection rate can be rather bursty on the time scale of minutes.

Under northward IMF reconnection occurs at high latitudes behind the polar cusps. A magnetosheath flux tube draped over the stagnation point on the dayside magnetopause moves relatively slowly with respect to the magnetospheric fields and is likely to reconnect at high latitudes where the magnetosheath and lobe field are antiparallel (Fig. 9). After reconnection the poleward portion of the flux tube convects tailward with the solar wind flow. In the dayside portion of the flux tube the magnetospheric and magnetosheath plasmas mix and the flux tube sinks into the magnetosphere. Such flux tubes could constitute a boundary layer with magnetosheath plasma on the magnetospheric side of the magnetopause, which has been observed frequently under northward IMF. However, as indicated in Fig. 9, there is evidence that reconnection does not occur simultaneously in both hemispheres. High latitude reconnection is inferred from observation of uni-

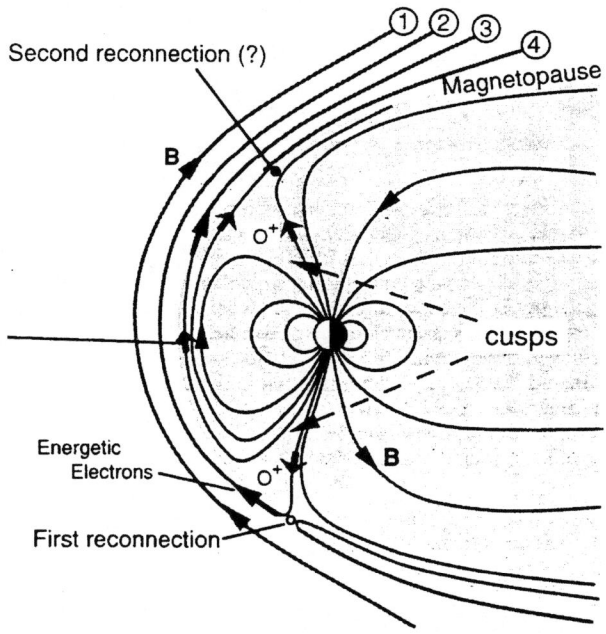

Fig. 9. Schematic of multiple reconnection on the dayside magnetopause under northward IMF. The time progression of a field line labeled 1 through 4 shows a multiple reconnection sequence in the southern and northern cusp, respectively (after Fuselier et al., 2001).

directional streaming of electrons and outflowing ionospheric O^+ beams in the subsolar region: ionospheric O^+ outflow is essentially continuous over the polar caps. Once a field line is reconnected in one hemisphere (southern hemisphere in Fig. 9), a unidirectional O^+ beam together with streaming electrons is observed in the subsolar region. Such events seem to be present for 50% of the time under northward IMF (Fuselier et al., 1997). The field line may reconnect again in the northern high-latitude region. In this case reconnection allows O^+ ions from the southern hemisphere to precipitate into the northern cusp (Fuselier et al., 2001). Such events seem to be rare, indicating that under northward IMF on the dayside field lines remain open long after initial reconnection. Nevertheless, second reconnection may occur again once the open field lines move to the flanks of the magnetopause, generating a boundary layer with magnetosheath plasma on closed field lines.

2.2 Flux Transfer Events

Close to the magnetopause the normal magnetic field component inside the magnetosphere and in the magnetosheath is expected to be approximately zero.

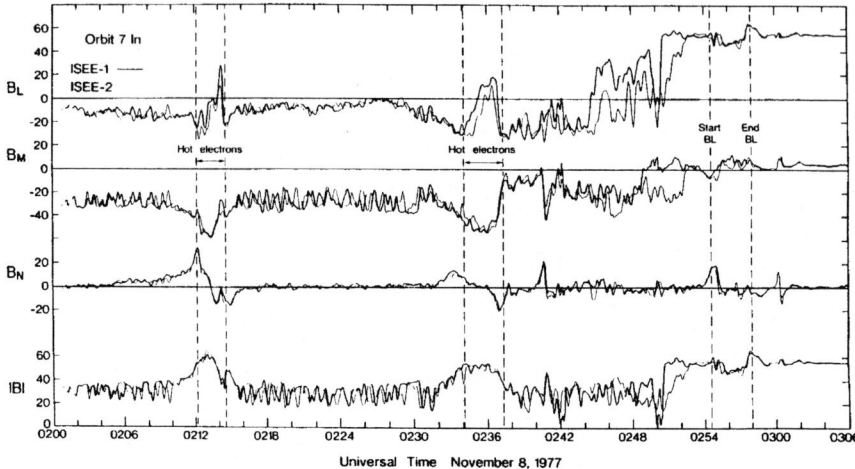

Fig. 10. Magnetic field data during a magnetopause crossing in magnetopause normal coordinates (from Russell and Elphic, 1978).

However, a spacecraft crossing the magnetopause frequently observes typical north-south excursions of the normal magnetic field component on both sides of the magnetopause. Figure 10 shows magnetic field data during a magnetopause crossing in the so-called normal coordinate system: the coordinate L is in the direction of the undisturbed magnetospheric field, N is along the nominal magnetopause normal, and M completes the LMN triad (Russell and Elphic, 1978). The third panel from top shows the normal magnetic field component, the bottom panel shows the magnitude of B. The spacecraft moves from the magnetosheath (lower magnetic field strength) into the magnetosphere (higher field strength). One can see in the magnetosheath one of the principal identifying marks of flux transfer events (FTEs), the bipolar signature in B_N. In the example shown here, the bipolar variation is +/-; in some other FTEs this variation is in the opposite sense, -/+ ("reverse" FTEs). Since the bipolar excursions during this crossing occur in the magnetosheath, one calls the events magnetosheath FTEs. Similar excursions are observed in the magnetosphere. These magnetosheath events are also notable for the large B_M extrema and field strength maxima at the center of the events. Energetic ions are found in the FTEs (Daly and Keppler, 1983; Scholer et al., 1982). These ions stream along the magnetic field away from the Earth, and have energy spectra similar to those of trapped magnetospheric particles. Furthermore, both the plasma ion and electron distributions in the FTEs roughly consist of a mixture of magnetospheric and magnetosheath populations (Thomsen et al., 1987). From single spacecraft data it is not possible to differentiate between spatial and temporal structures, and it is thus not possible to derive the scale size of FTEs. However, a pair of spacecraft, ISEE-1 and ISEE-2, has been separated occasionally by about 5000-7000 km. From simulta-

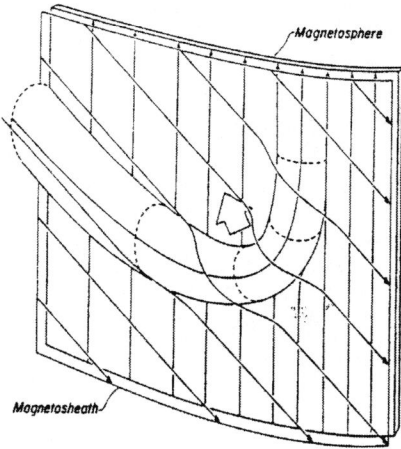

Fig. 11. Schematic of the elbow model for flux transfer events (from Russell and Elphic, 1978).

neous observations at the two spacecraft a FTE size of one Earth radius in the magnetopause normal direction has been deduced (Saunders et al., 1984). Furthermore, it has been found that the magnetic field within the events is twisted, corresponding to a core field-aligned current.

Russell and Elphic (1978) proposed that localized reconnection leads to flux tubes connecting interplanetary and magnetospheric fields through a slanted hole in the magnetopause (Fig. 11). It is implicit in this model that reconnection occurs over a fairly narrow longitude segment, thus creating pairs of elbow-shaped flux tubes, one in the northern hemisphere and a mirror image in the southern hemisphere. As these flux tubes move under the magnetic tension force along the magnetopause, they create at a stationary observer in the northern hemisphere first a positive and then a negative excursion of the interplanetary magnetic field draped around the flux tube. Other models for FTEs are based on multiple neutral line reconnection produced by the tearing mode in the magnetopause current layer (Lee and Fu, 1985) or on time-dependent reconnection at a single X-line (Scholer, 1988; Southwood et al, 1988). In the tearing mode model the tearing islands become in the presence of an IMF B_y component flux tubes with a helical field inside which are embedded in the magnetopause. The field lines at both ends of the flux tubes are connected to both sides of the current sheet, i.e., to the IMF and to the magnetospheric field. In the bursty reconnection model periods of increased reconnection rate or sudden onset of reconnection leads to a pair of bulges in the magnetopause extending in the dawn-dusk direction over a large longitudinal segment. These bulges are then transported northward and southward to the cusp regions essentially with Alfvén speed. Figure 12 shows

(a)   (b)   (c)

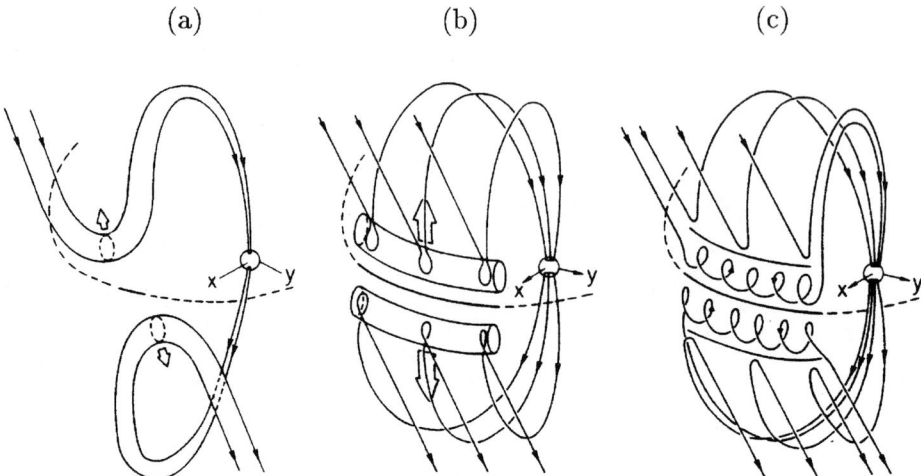

Fig. 12. Schematic of possible FTE models (after Lockwood et al., 1990).

the Russell and Elphic elbow model (a), the Scholer/Southwood et al. bursty reconnection model (b), and the Lee and Fu tearing mode FTE model (c).

2.3 Reconnection in the Earth's Magnetotail

It is widely believed that magnetic reconnection takes place in the magnetotail during substorms. In association with a substorm a neutral line forms in the near-Earth region of the plasma sheet. The plasma sheet is considered to be on closed field lines; the magnetic field lines are highly stretched and pass through the neutral sheet. The last closed field line marks the plasma sheet boundary. Earthward of this neutral line the plasma bulk flow is Earthward, and tailward of the neutral line field lines threading the plasma sheet from north to south are transported with high speed in the tailward direction. Reconnection continues until lobe field lines begin to reconnect and part of the plasma sheet is severed. This part is then free to move as a plasmoid in the tailward direction. Figure 13 (left) shows the temporal development of the magnetotail in the neutral line substorm model. Under southward IMF dayside reconnection adds more and more open flux to the tail lobes. Reconnection then starts at some near-Earth neutral line, and the severed part of the plasma sheet moves as a plasmoid tailward. It expands during this ejection in the north-south and in the dawn-dusk direction. After about 30 min the plasmoid has reached a distance of about 200 Earth radii and can be observed by a deep tail spacecraft. Characteristic signatures are a bipolar north-south excursion of the magnetic field B_z component, combined with a large tailward bulk flow velocity and isotropic energetic electrons of the plasma sheet on the closed plasmoid field lines. Figure 13 (right) is a superposed epoch analysis of observations obtained at 220 R_E in the deep tail. The zero time mark is the first occurrence of the high speed flow. The second panel

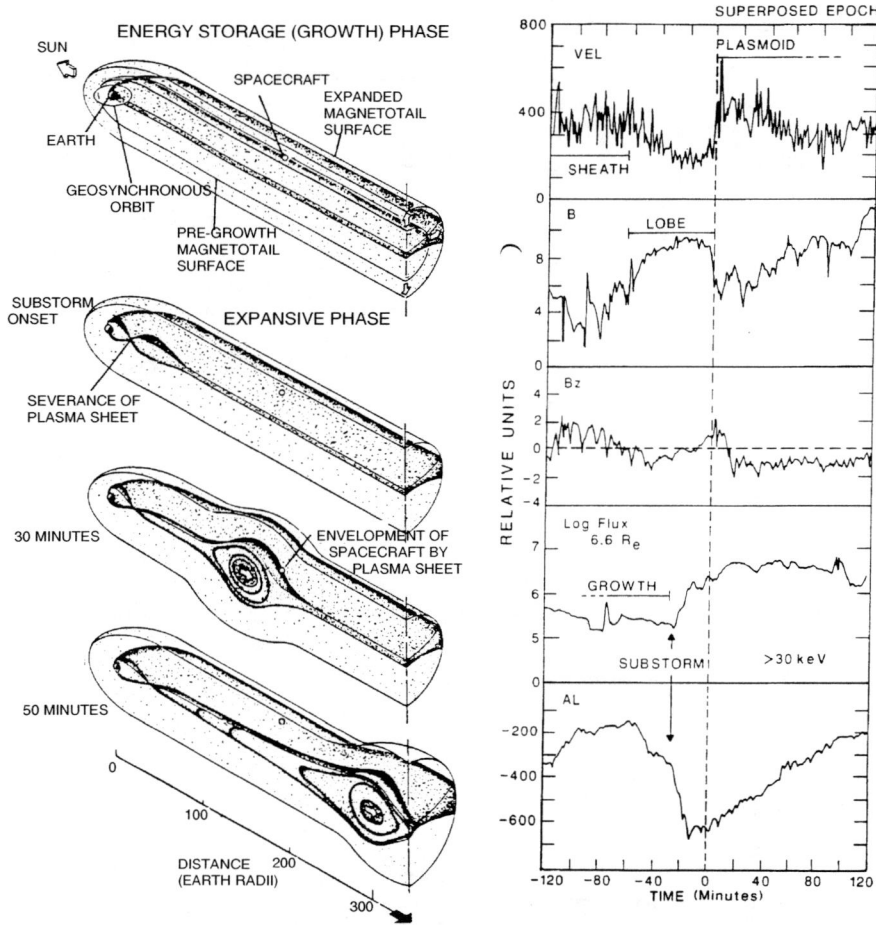

Fig. 13. Left: Schematic of the Hones substorm model. Right: Superposed epoch analysis of tailward velocity, total B, B_z, the flux of > 30 keV electrons at synchronous orbit, and the auroral AL index (after Baker et al., 1987).

from top shows that with the appearance of the high speed flow the spacecraft is engulfed by a region with a lower field strength. The third panel shows the B_z magnetic field component. This component exhibits in the plasmoid first a northward and then southward excursion as the closed flux tubes are convected tailward. The two bottom panels show characterstic signatures of a substorm: the energetic electron flux > 30 keV at synchronous orbit (equatorial spacecraft orbit at a distance of 6.6 Earth radii) increases about 30 min before plasmoid encounter in the deep tail (second panel from bottom). The bottom panel shows the superposed epoch analysis of the the auroral electrojet index AL: a sudden

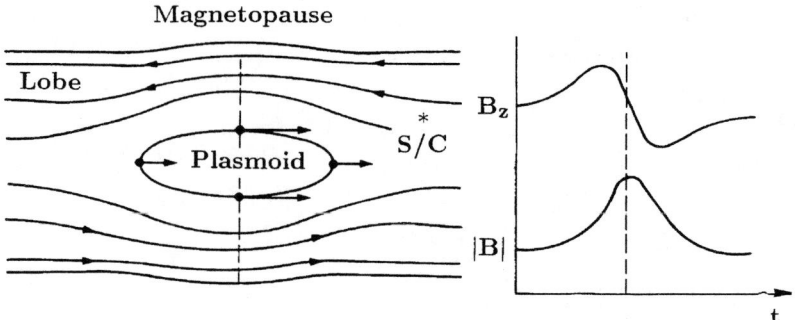

Fig. 14. Topology (left) and characteristic signatures (right) of traveling compression regions (TCR's).

depression of AL is indicative of the substorm onset. As can be seen, substorm onset occurs about 30 min before the passage of a plasmoid in the deep tail.

Quite often a spacecraft in the distant tail will miss the plasmoid, since it is located at relatively high latitudes in the lobes. Nevertheless, a plasmoid passing above or under the spacecraft will lead to characteristic signatures in the magnetic field which have been dubbed traveling compression regions, or TCRs in short. Figure 14 shows a sketch of a part of the tail during the passage of a plasmoid. The plasmoid (1) compresses the field in the lobe, and (2) the field is draped around the plasmoid. This leads at the spacecraft position to bipolar north-south signatures of the magnetic field as a function of time and to an increase in the magnetic field strength (Slavin et al., 1984). Such events are numerous in the deep tail data from the spacecraft ISEE-3 and Geotail and have been used for statistical analysis. Because the ejection of plasmoids at substorm onset is a fundamental prediction of the near-Earth neutral line model, the timing of the TCR, which is taken as a proxy for the direct observation of the plasmoid, relative to substorm onset has received considerable attention. The statistical studies by Slavin et al. (1992) and by Nagai et al. (1994) demonstrate a strong tendency for plasmoids to be released near substorm expansion phase onset.

Determination of the location of the near-Earth neutral line has been made mainly from the indirect statistical study of the plasma flows driven either Earthward or tailward depending on the observer's position with respect to the X-type neutral line. It is concluded that the X-type neutral lines are formed in the magnetotail 20 - 30 R_E away from Earth (Nagai et al., 1998). Recently, Øieroset et al. (2001) have reported on in situ observations of reconnection in the magnetotail at a distance of $60R_E$ behind the Earth. The WIND spacecraft observed at this distance Earthward plasma jetting combined with northward magnetotail field and subsequently tailward plasma jetting combined with southward magnetotail field, indicating that the spacecraft crossed the neutral line region in the tailward direction. In addition, a cross-tail Hall magnetic field was observed on both sides of the neutral line region. The authors conclude from the observation of a Hall field that reconnection was collisionless. It should, however, be noted that in a collisionless plasma the reconnection layer always exhibits a Hall field regardless whether reconnection itself is collisionless or due to an anomalous localized resistivity. The existence of a Hall magnetic field in the reconnection layer can not be taken as evidence that the process of reconnection itself is collisionless.

If magnetic reconnection occurs in the magnetotail, slow mode shocks are predicted to form at the boundaries between plasma sheet and lobe bounding the X-type neutral line. The existence of the slow mode shock was for the first time reported by the ISEE-3 deep tail mission. However, identification was difficult because of the lack of ion plasma data. Recent observations in the distant magnetotail by the Geotail satellite have proved the existence of the slow mode shocks using the magnetic field and velocity moment data (Saito et al., 1995). In order to demonstrate the complexity of in situ data we present here one example of a slow mode shock in the magnetotail at a distance of $\sim 60R_E$ from the Earth. Figure 15 shows magnetic field data and plasma velocity moments during a transition from lobe to plasma sheet. During the entry the plasma density increases by factor 1.6, and the ion and electron temperature increase simultaneously. The bulk flow velocity goes up to 600 km/sec. The upstream and downstream magnetic field and plasma data satisfy the one-dimensional Rankine-Hugoniot relation. This requires determination of the shock normal and conversion of the upstream and downstream parameters into the normal incidence frame. However, only 10% of the plasma sheet lobe boundaries were identified by Saito et al. (1995) as slow mode shocks. Seon et al. (1996) reported that only 3 out of 300 plasma sheet - lobe boundaries could be identified as slow mode shocks. Thus, slow mode shocks are rare in the distant magnetotail.

3 Conclusions

We have concentrated in this brief review on observations which indicate that magnetic reconnection occurs in the solar atmosphere and in the Earth's magnetosphere. As far as the Sun is concerned, the observations are indirect. Reconnection is inferred from observations of high speed flows derived from motion of X-ray or EUV features, from the identification of structures in the soft X-ray images as magnetic flux tubes, and from the observation of energetic events in hard X-rays. In particular the soft X-ray observations during flares show us the large scale magnetic field structure. This structure is rather similar to schematic drawings of the reconnection configuration. This, together with the observation of plasmoids and high speed flows, is rather suggestive that reconnection is indeed going on. In the magnetosphere we are faced with the opposite situation: we have no knowledge about the large-scale structure of the magnetic field and flow, but observe in situ at a particular location the detailed magnetic field and flow variations. To these one can apply conservation laws and verify whether boundaries like the magnetopause have a normal magnetic field component, i.e., are rotational discontinuities. In the magnetotail the large-scale magnetic field structure can be deduced from the temporal variation of the magnetic field during the high speed flow events.

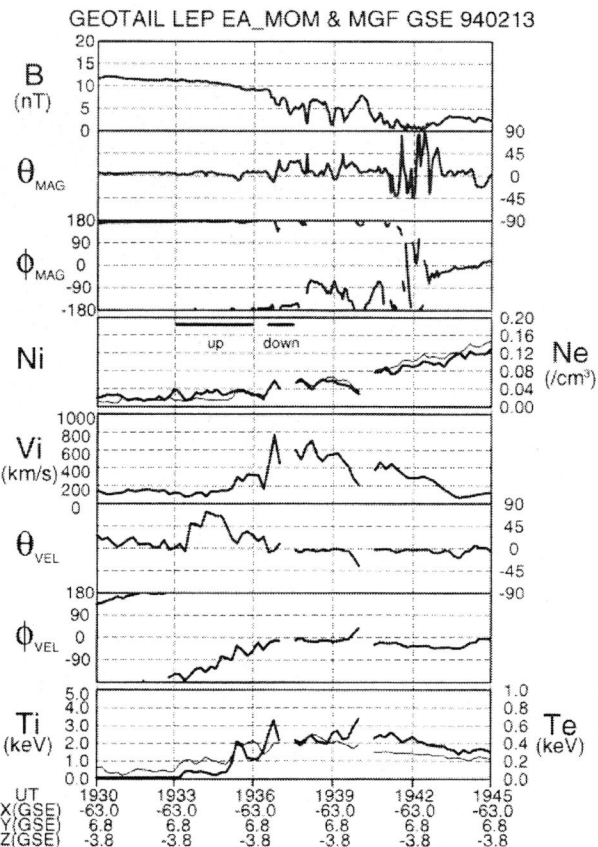

Fig. 15. Magnetic field data and plasma velocity moments during a lobe - plasma sheet crossing by Geotail. From top to bottom: total magnetic field, field polar and azimuthal angles (GSE coordinate system) ion (thick line) and electron (thin line) densities, ion bulk velocity, polar and azimuthal angle of bulk velocity (GSE coordinate system), ion (thick line) and electron (thin line) temperature (from Saito et al. 1995).

Solar and magnetospheric observations are complementary to each other not only because of the different observational approach, but also because of intrinsic differences in physics. For example, in solar flares the characteristic size of the ion gyroradius (or ion inertial length) is of the order of $10^7 - 10^8$, whereas it is only 100, and possibly smaller, in the magnetotail during substorms. Hence the role of non-MHD (kinetic) effects may be rather important in substorms, whereas large-scale configurations in flares may be very well described in terms of MHD. On the other hand, the role of MHD turbulence may be more important in flares than in substorms. A more detailed comparison between flares and substorms can be found in Terasawa et al. (2000).

References

1. H. Aurass, K.-L. Klein, P. C. H. Martens: Solar Phys. Lett. **155**, 203 (1994)
2. D. N. Baker, R. C. Anderson, R. D. Zwickl, J. A. Slavin: J. Geophys. Res. **92**, 71 (1987)
3. R. C. Canfield et al.: Ap. J. **464**, 1016 (1996)
4. A. Czaykowska, B. De Pontieu, D. Alexander, G. Rank: Ap. J. **521**, L75 (1999)
5. A. Czaykowska, D. Alexander, B. De Pontieu: Ap. J. **552**, 849 (2001)
6. P. W. Daly, E. Keppler: J. Geophys. Res. **88**, 3971 (1983)
7. J. W. Dungey: Phys. Rev. Lett. **6**, 47 (1961)
8. S. A. Fuselier, B. J. Anderson, T. G. Onsager: J. Geophys. Res. **102**, 4847 (1997)
9. S. A. Fuselier, S. M. Petrinec, K. J. Trattner, W. K. Petersen: J. Geophys. Res. **106**, 5977 (2001)
10. D. E. Innes, B. Inhester, W. I. Axford, K. Wilhelm: Nature **386**, 811 (1997)
11. M. Kundu et al.: Ap. J. **447**, L135 (1995)
12. L. C. Lee, Z. F. Fu: Geophys. Res. Lett. **12**, 105 (1985)
13. M. Lockwood, S. W. H. Cowley, P. E. Sandholt: EOS Trans. AGU **71**, 709 (1990)
14. M. Lockwood, T. G. Onsager, C. J. Davis, M. F. Smith, W. F. Denig: Geophys. Res. Lett. **21**, 2757 (1994)
15. S. Masuda, T. Kosugi, S. Tsuneta, Y. Ogawara: Nature **371**, 495 (1996)
16. T. Nagai et al.: Geophys. Res. Lett. **21**, 2991 (1994)
17. T. Nagai et al.: J. Geophys. Res. **103**, 4419 (1998)
18. M. Ohyama, K. Shibata: Ap. J. **499**, 934 (1998)
19. M. Øieroset, T. D. Phan, M. Fujimoto, R. P. Lin, R. P. Lepping, Nature **412**, 414 (2001)
20. G. Paschmann, et al.: Nature **282**, 243 (1979)
21. T. D. Phan, et al.: Nature **404**, 848 (2000)
22. C. T. Russell, R. C. Elphic: Space Sci. Rev. **22**, 681 (1978)
23. Y. Saito, et al.: J. Geophys. Res. **100**, 23567 (1995)
24. M. A. Saunders, C. T. Russell, N. Sckopke: Geophys. Res. Lett. **11**, 131 (1984)
25. M. Scholer: Geophys. Res. Lett. **15**, 291 (1988)
26. M. Scholer, D. Hovestadt, F. M. Ipavich, G. Gloeckler: J. Geophys. Res. **87**, 2169 (1982)
27. J. Seon et al.: J. Geophys. Res. **101**, 27383 (1996)
28. K. Shibata, S. Masuda, M. Shimojo, H. Hara, T. Yokoyama, T. Tsuneta, T. Kosugi, Y. Ogawara: Ap. J. **451**, L83 (1995)
29. J. A. Slavin et al.: Geophys. Res. Lett. **11**, 657 (1984)
30. J. A. Slavin et al.: Geophys. Res. Lett. **19**, 825 (1992)
31. B. U. Ö. Sonnerup, I. Papamastorakis, G. Paschmann, H. Lühr: J. Geophys. Res. **95**, 10541 (1990)
32. D. J. Southwood, C. J. Farrugia, M. A. Saunders: Planet. Space Sci. **36**, 503 (1988)
33. T. Terasawa, K. Shibata, M. Scholer: Adv. Space Res. **26**(3), 573 (2000)
34. M. F. Thomsen, J. A. Stansberry, S. J. Bame, S. A. Fuselier, J. T. Gosling: J. Geophys. Res. **92**, 21127 (1987)
35. S. Tsuneta, H. Hara, T. Shimizu, L. W. Acton, K. T. Strong, H. S. Hudson, Y. Ogawara: Publ. Astron. Soc. Japan **44**, L63 (1992)
36. S. Tsuneta: Ap. J. **456**, 840 (1996)
37. T. Yokoyama, K. Shibata: Nature **375**, L61 (1995)
38. T. Yokoyama, K. Shibata: Ap. J. **474**, L61 (1997)
39. T. Yokoyama, K. Shibata: Ap. J. **496**, L113 (1998)
40. T. Yokoyama, K. Akita, T. Morimoto, K. Inoue, J. Newmark: Ap. J. **546**, L69 (2001)

Magnetic 3-D Configurations of Energy Release in Solar Flares

Bojan Vršnak

Hvar Observatory, Faculty of Geodesy, Kačićeva 26, 10000 Zagreb, Croatia

Abstract. Basic concepts and principles usually used to interpret the solar flare phenomenon are summarized, and traditional classification schemes based on 2-D magnetic field representations are briefly reviewed. The extension to 3-D opens new aspects, some of which are sketched in this paper: In particular two-loop interactions, two-ribbon flare geometry, and the plasmoid formation in a current sheet are considered. It is shown that alternative onset processes exist beside the standard two-ribbon flare scenario in which the pre-flare arcade evolves slowly through a series of equilibrium states until the eruption. For example, the arcade can become unstable after an abrupt reformation of its core through a sequence of loop interactions. The restructuration results in an impulsive compact flare and the formation of an unstable 'sigmoid' whose eruption provides the two-ribbon phase aftermath. Possible modalities of main phase are emphasized. Especially the secondary plasmoid formation is considered and its fate discussed stressing the 3-D aspect of the process and the effect of line-tying. Finally, complex events composed of several distinct, but causally related energy release processes are described.

1 Introduction

Solar flare is a process of an abrupt energy release in the solar atmosphere. Coronal plasma can be heated up to $\approx 4 \times 10^7$ K, particles are accelerated to high energies, violent mass motions are ignited, and MHD blast waves launched. Total energy liberated varies from, say, 10^{22} J in subflares to several 10^{25} J in the largest events. The energy release can last from seconds to hours. Flares, especially those of long durations, are closely related to huge eruptions called coronal mass ejections (CMEs) that propagate into interplanetary (IP) space and drive IP shock waves.

Radio observations reveal that the primary energy release takes place in the low corona [9]. During the flare build-up phase sub-photospheric convective motions twist, drag, and bring out the magnetic field, inducing the electric currents in the corona. In this way free energy is deposited into coronal magnetic structures. Speaking in general terms, convective motions drive the magneto-hydrodynamic (MHD) dynamo that converts the mechanical energy into an electromagnetic one [80].

A part of the energy transported into the corona is instantaneously spent for coronal heating, whereas a part accumulates and stays stored in non-potential

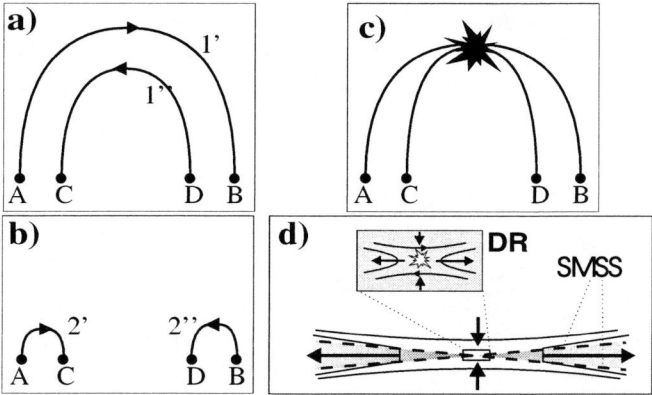

Fig. 1. Schematic 2-D presentation of the topological change of the coronal magnetic field. a) Initial non-potential configuration ($\mathbf{j} \propto \nabla \times \mathbf{B} \neq 0$): Field lines 1' and 1" connect region A with B and region C with D, respectively. b) Final (lower energy) configuration: Field lines 2' and 2" connect region A with C and B with D, respectively. c) The rearrangement a→b requires *reconnection*. d) Basic features of fast reconnection.

magnetic fields. Field lines are rooted in the inert photosphere (so called *line-tying condition*). The only way to release the excess energy, i.e. to achieve a lower energy configuration, is by readjusting the coronal field line system and changing its topology. If the rearrangement is fast enough, providing a powerful energy release, a flare is created.

The *topological change* of magnetic field under the line-tying condition means that field lines must re-connect (Fig. 1). The coronal plasma is characterized by a very high electric conductivity, i.e. very low *magnetic diffusivity* [80]. Under such conditions the fields can meet and interact only within extremely thin layers where the diffusion is still effective – the magnetic flux inflow must be balanced by the diffusion [49]. On the other hand, the plasma inflow into the layer must be balanced by the outflow along the layer. This implies that, in order to have a relatively fast inflow (high *reconnection rate*) i.e. to achieve an efficient energy release, the layer length cannot be much larger than its thickness. So, the reconnection can be fast only if it occurs within an extremely small volume, called the *diffusion region* (DR).

Such a mechanism of *fast reconnection* [13] (sometimes also referred as *Petschek regime* after [76]), including its modalities (see, e.g., [81, 82]), is essential for flares [94]. A simple concept of the magnetic field annihilation in a long current sheet, usually called *Sweet–Parker regime* after [103] and [73], is too slow to account for the energy release in flares.

Besides a small DR, the fast reconnection mechanism anticipates formation of two pairs of *slow magnetosonic standing shocks* (SMSSs) extending from DR (Fig. 1). The plasma inflow is almost perpendicular to the inflowing magnetic field implying that the flow is faster than the corresponding slow mode waves[1] and the shocks appear in the region where merging flows "collide". SMSSs are sometimes called switch-off shocks [107] since the downstream magnetic field is almost perpendicular to the shock. It should be noted that the majority of the energy is released at SMSSs [13, 107, 108, 109] which after completion [72] extend all along the contact region between the merging magnetic systems. The role of small DR is only to turn-on the reconnection.

In SMSSs plasma is heated, whereas the inflow is deflected and accelerated [13] to form two fast *outflow jets* of hot plasma (Fig. 1d). Electric fields associated with DR and SMSSs accelerate particles [42, 14] and trigger plasma kinetic instabilities [11, 23, 24, 46, 57, 93, 102]. These nonthermal processes excite emission in the decimeter/meter wavelength range [11, 12], providing the most immediate signature of the *primary energy release* [9].

Various magnetic field configurations and amounts of stored energy can be involved in a flare. A broad variety of physical processes becomes feasible, yielding diversity of evolutionary scenarios and modes of energy release. In this paper characteristical features and processes are systematically ordered and some new results are presented.

2 General Properties of Flares

2.1 Energy Build-Up Phase

The flare occurrence rate in some region and the corresponding energy release rate is governed by the rate at which the energy is transported through the photosphere. It is limited by the upward (z-direction) component of the Poynting flux $\mathbf{P} = \mathbf{E} \times \mathbf{B}/\mu$ integrated over the area of the region. The electric field is induced by the motion perpendicular to the magnetic field $\mathbf{E} = -\mathbf{v} \times \mathbf{B}$. There are two contributions to P_z: the *magnetic field emergence* provides $P_z^e = v_z B_x^2/\mu$ and the *shearing motion* contributes with $P_z^s = v_x B_z B_x/\mu$. Obviously, the energy is stored faster in strong magnetic field regions. Consequently, a large majority of flares take place in sunspot groups and only occasionally in spotless regions [20, 90, 91].

Using the order of magnitude values for the velocity and magnetic field $v = 0.1$ - 1 km s^{-1} and $B = 0.01$ - 0.1 T, respectively, one finds $P_z \approx 10^5$ - 10^6 W m^{-2}. An active region (AR) covering 10^5 km $\times$ 10^5 km ($A = 10^{16}$ m^2) can store in one day 10^{25}-10^{26} J. This accounts for one large flare per day, a dozen of small-to-medium flares or hundreds of small flare-like brightenings. The largest flares can appear only in big AR complexes.

Since pressure gradients and gravity under normal coronal conditions are much smaller than the Lorentz force, the pre-flare field is practically force-free [80]. The electric current $\mathbf{j}$ is parallel to the field $\mathbf{B}$, meaning $\mu \mathbf{j} = \nabla \times \mathbf{B} =$

---

[1] slow mode magnetosonic waves have a very low speed when propagating almost perpendicular to the field (see, e.g. [80])

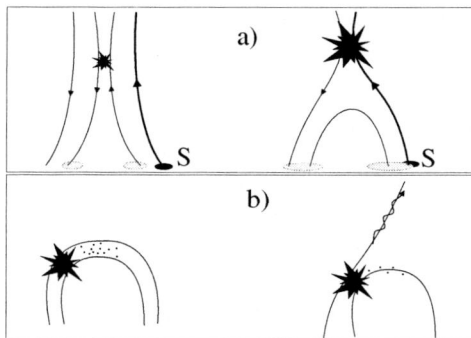

Fig. 2. a) Flare intensification by inclusion of strong field (the field line drawn *bold*). The release enhances after the strong fields are embraced – indicated by the protrusion of chromospheric emission (*gray*) over the sunspot (S). b) Electron trapping and/or escape: In small loops (*left*) the nonthermal particles are "trapped" (*dots*), whereas if large scale loops or "open" field lines are included electron beams (*wavy arrow*) can escape away from the flaring region (*right*)

$\alpha \mathbf{B}$, where α is a degree by which the field is sheared. For a given magnetic field, the regions more stressed contain stronger currents and thus more of free energy[2]. Analogously, for a given α currents are concentrated in stronger fields [39, 41, 92].

In reality, flares occur preferably in close vicinity of magnetic *inversion lines*[3] at locations where the field is strongly sheared [33, 34, 38, 39, 41, 65, 92]. Furthermore, the energy release in a particular flare significantly intensifies when the process embraces strong magnetic fields rooted in sunspots (Fig. 2a; see also [21, 25, 37, 64, 89, 91, 117]). Consistently, flares in spotless regions are much weaker than AR flares [20, 90, 91].

Flares in highly stressed strong fields are generally also more impulsive and show more prominent nonthermal signatures [6]. There are two basic reasons for this. Since the development of an MHD process depends on the *Alfvén travel time*[4], the events embracing strong fields with steep gradients (small length scales) will progress faster [101]. On the other hand, since higher current densities and stronger electric fields are involved the thresholds for *kinetic plasma instabilities* [11] and particle acceleration are more easily reached.

Significant differences in flare characteristics can also appear depending on the nature of field lines included in the energy release process (Fig. 2b). If only small scale loops are involved, the accelerated particles stay trapped in the flaring region. If field lines extend over large distances they can propagate to other ARs (see, e.g., [5]) possibly triggering a *sympathetic flare* [104]. If "open" field lines

---

[2] The energy excess above the energy of potential field ($\mathbf{j}=0$) can be expressed heuristically as LI^2 [44], where L is the self-inductivity of the current system and $I = jA$ is the current associated with the considered photospheric area A

[3] Lines where the photospheric magnetic field changes sign; also called *neutral lines*

[4] $t_A = d/v_A$; d is the length scale involved and $v_A = B/\sqrt{\mu\rho}$ is the Alfvén velocity

are included, beams of accelerated particles can protrude to the high corona (see, e.g., [79, 117, 122]), or eventually escape to IP space [22, 55, 85].

2.2 Flare Classification

Most generally, flares are divided into confined and dynamical [22, 55, 85]. In *confined flares* the overall magnetic structure involved remains preserved. *Dynamical flares* are associated with a disruption of pre-flare magnetic structure.

When the morphology is considered flares are described either in terms of loop assemblies, or distinct magnetic loops. For example, flares associated with emerging/merging magnetic configurations can be described as an interaction of two arcade-like field line systems (Fig. 3a) or as two interacting loops (Fig. 3b).

Apparently, flares show a large variety of morphological and evolutionary characteristics. Yet, majority are traditionally sorted into three classes (see, e.g. [80]):

– **Two-Ribbon Flares**, being by definition dynamical, represent a majority of large, long duration events. In the standard scenario (see, e.g., [80]) a sheared arcade with twisted neutral line filament becomes unstable and erupts. Field lines embedding the filament's magnetic rope stretch and below the rope a current sheet is formed (Fig. 3c). When the sheet becomes long enough the *tearing instability* [32] sets in and fast reconnection starts [29, 60, 72, 107, 121]. The energy released at DR and SMSSs is transported down the field

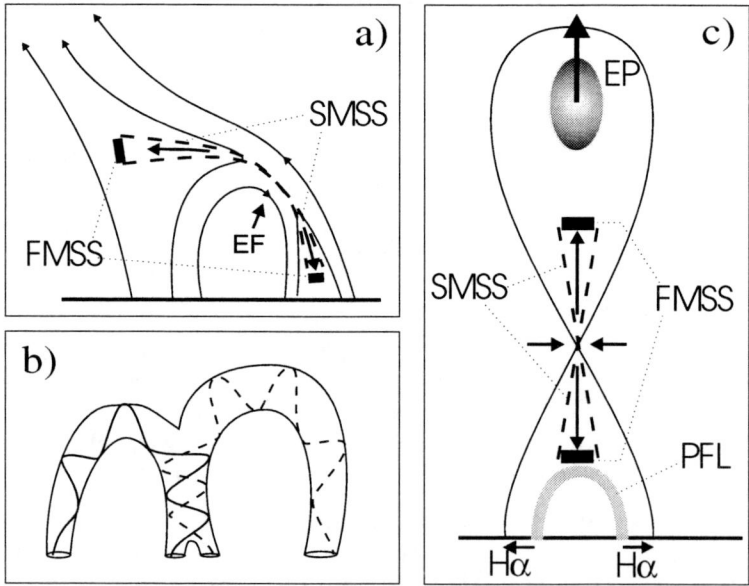

Fig. 3. a) Emerging flux (EF) flare. b) Interacting loops. c) Two-ribbon flare; postflare loops (PFL) and Hα ribbons are indicated. The anticipated locations of slow and fast mode standing shocks (SMSS and FMSS, respectively) are indicated in a) and c)

lines by electron beams and thermal conduction. The transition region and chromospheric layers are impulsively heated and bright ribbons are formed at both sides of neutral line. As the reconnection proceeds field lines anchored at successively larger distances from the neutral line enter into DR. The ribbons expand outwards from the inversion line and the postflare loop system grows (Fig. 3c).

- **Interacting-Structure Flares** include *interacting-loop flares, merging* or *emerging-flux flares*, etc. The reconnection between interacting magnetic systems (Fig. 3a,b) is governed by emerging/merging motions, or is driven by *coalescence instability* [26, 50, 106]. A frequent feature associated with flares of this class is an opposite polarity "intrusion" within the dominant photospheric field, often created by a newly emerging flux. The tension of overlying field prevents the expansion of emerging flux, providing the current sheet formation (Fig. 3a). An analogous, yet distinct class of events is driven by arcade eruption that rushes into the overlying field [115]. In contrast to previous examples, these are not confined flares and can be called the *erupting-flux flares*. Another specific class are flares caused by a sequence of loop interactions within an arcade[5] [114].

- **Simple-Loop Flares** are presumably caused by an energy release confined within a single loop [80, 81]. The energy release mechanism could be: (*i*) *cylindrical tearing* [118, 10] but then a rather specific magnetic field configuration is required; (*ii*) coalescence instability of fine structure current filaments within a loop [42, 43, 56]; (*iii*) energy release based on the *double layer mechanism* [1, 68]. On the other hand, it is well possible that in the system of two interacting loops one loop is not resolved. Further option is suggested by [98, 71] who found soft X-ray ejecta in all analyzed simple-loop flares and revealed that the primary energy release is frequently above the flaring loop (see also [4]). This indicates that at least in some cases apparently simple-loop-flares are unresolved small dynamical flares.

This traditional scheme, based on 2-D representation, can be in fact simplified (see, e.g., [3] and references therein). The first and third class can be attributed to the reconnection process in a bipolar magnetic structure, leading to a postflare bipolar arcade or a single loop, respectively. The second class, embracing various forms of interactions between different magnetic structures, can be attributed to a quadrupolar reconnection, leading to interacting flare loop pairs or quadrupolar arcades (see, e.g. [69] and [47], respectively)

Finally, let us stress a complementary classification proposed by [59] which sorts the flares according to their role in the restructuration of coronal magnetic fields of an active region. The smallest and most numerous are nanoflares. They are a prompt coronal response to field changes that are governed by subphotospheric convection (e.g. flux emergence). Presumably, the energy released by nanoflares provides coronal heating and can be related to gradual readjustments of coronal fields. Yet, a part of the deposited energy remains stored. It is released

---

[5] The Hα morphology in this type of events might resemble the two-ribbon flare, but since the overall arcade structure is preserved these are confined flares, see, e.g. [6].

by various forms of confined flares, which intermittently relax the coronal fields towards simpler configurations, finally forming a large scale arcade. Next "generation" of flares serves to create a flux rope in the arcade. Eventually, when a sufficient fraction of the arcade field is "detached" from the photosphere, i.e. when the flux rope is sufficiently long, the arcade erupts. A large scale coronal mass ejection is launched and as a byproduct a major two-ribbon flare occurs.

2.3 Flare-Associated and Flare-Like Processes

Processes similar to those in flares can occur also on much smaller and much larger scales. For example, interacting loop analogies are observed within the EUV bright points at scales of few 10^3 km [74, 75, 15]. Furthermore, the plasmoids presumably formed within the current sheet should coalesce causing the flare short-time (<1 s) modulations [40, 51, 52, 86, 95, 96].

The coalescence of twisted flux tubes might be important also in processes occurring in the high corona and IP space. Interactions of CMEs, sometimes called *CME canibalism* [35], are accompanied by a strong nonthermal radio emission at appropriate plasma frequencies [35]. Thus flare-like energy release processes are observed at distances beyond ten solar radii, involving interacting magnetic ropes that have diameters comparable with the solar radius.

Furthermore, it can be expected that CMEs interact with the ambient coronal or IP field (Fig. 4a) in a similar way as in erupting flux flares [115] mentioned in Sect. 2.2. The interaction enables the escape of particles accelerated in the two-ribbon flare below the CME, as illustrated by Fig. 4. Indeed, the CME lift-off time is often closely associated with the onset of IP type III bursts [84].

The interaction with ambient fields (drawn *gray* in Fig. 4) affects also the CME dynamics. If the magnetic field in front of the CME cannot be pushed aside, the overlying field tension acts as a restoring force, and the CME must "eat its way" throughout [2]. Once the overlying field is reconnected away, the CME is suddenly free to accelerate much faster. If the CME (or IP *magnetic cloud* [16]) moves *along* the ambient magnetic field and the field is perpendicular to its axis (Fig. 4a), the reconnection between the ambient field and the azimuthal field of the CME takes place at one side of the CME [17], deflecting it [110] from the original direction (Fig. 4a-right).

3 3-D Aspect Emphasized

The examples shown in Fig. 4 illustrate how some important features of the magnetic field restructuration cannot be fully represented by 2-D or $2\frac{1}{2}$-D models[6]. In this Section some intrinsically 3-D features of energy release are stressed.

---

[6] 2-D model means that magnetic field has only two components and that all quantities are invariant in the third direction. In $2\frac{1}{2}$-D models the third component of the magnetic field is added, but all quantities are still invariant in the third direction

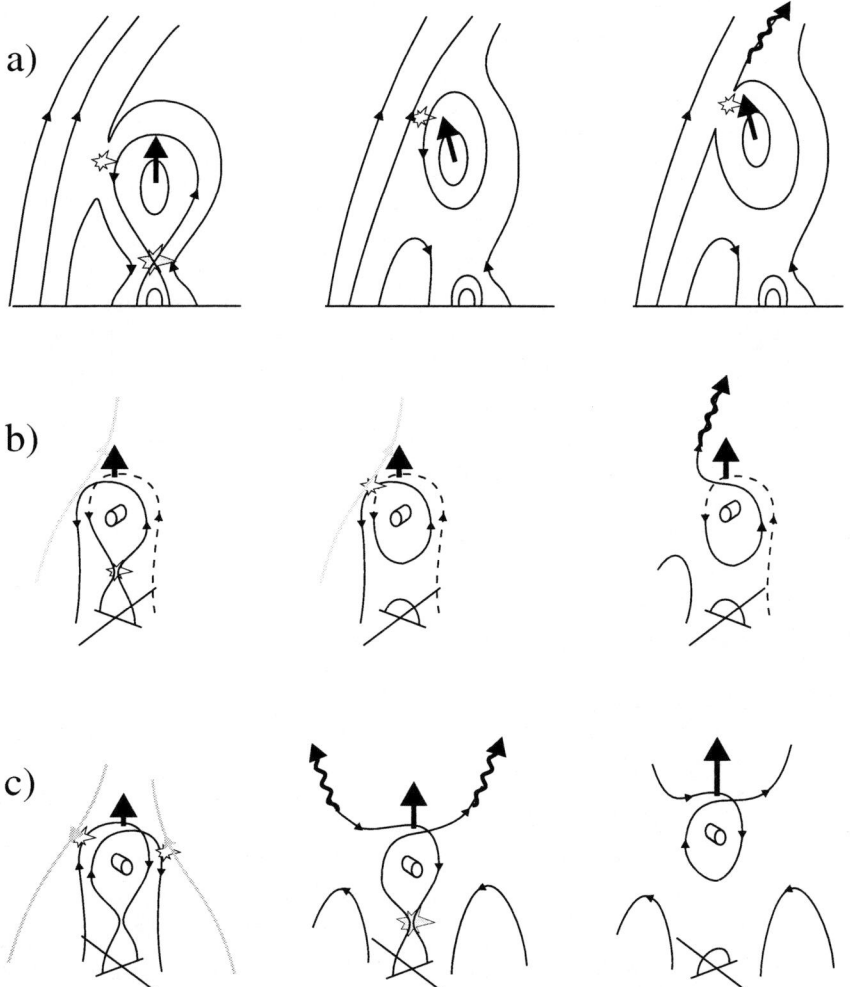

Fig. 4. a) 2-D sketch of an interaction of dynamical flare with the overlying field. In 3-D various options become feasible – two examples are shown in b) and c) where only the part of a structure that is embraced by the interaction is drawn. The accelerated particles can stay trapped in the plasmoid, can escape (*wavy arrow*) after some time, or can escape immediately after being accelerated.

3.1 Two-Loop Interactions

Aligned loops can interact in two different manners[7]. The one involving the coalescence instability is caused by the attraction of longitudinal currents driving the reconnection of the azimuthal fields (Fig. 5a). Another possibility is the reconnection of longitudinal fields (Fig. 5b).

---

[7] For numerical simulations including some other mutual orientations see [58]

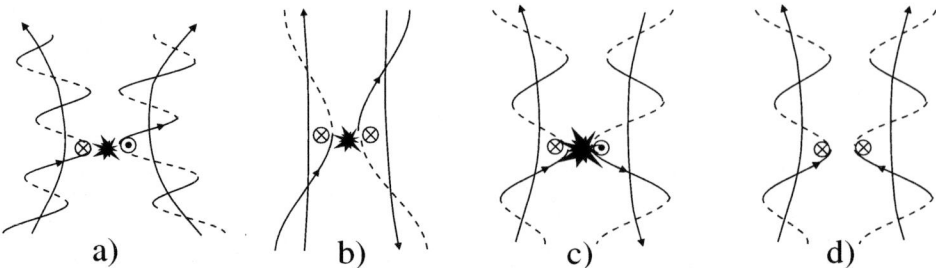

Fig. 5. Two-loop interaction: a) coalescence-mode; b) axial field reconnection; c) optimum case – the axial and azimuthal field components are both anti-parallel; d) no interaction – both components are parallel

The first option requires highly twisted loops (large pitch angles of field lines). Since the pitch angle decreases towards the loop axis [80] the energy release can last only until the magnetic field component perpendicular to the plane of reconnection becomes too large slowing down the reconnection [30, 100]. The second possibility requires anti-parallel axial magnetic fields in the contact region and small field line pitch angles.

It can be presumed that coronal loops probably have small pitch angles, since otherwise they would be unstable and erupt [112, 113, 116]. A further drawback of the coalescence mechanism is that in the final state the loops should stay "glued" in the contact region (see Fig. 3c and [58]) which is not clearly demonstrated by observations.

The most favourable situation for the interaction is when two loops have anti-parallel axial fields and opposite helicities (sign of α) since then also the azimuthal fields are anti-parallel (Fig. 5c). Such configurations are found in ARs substructures [77], although they are probably rare at larger scales due to a *helicity segregation rule* [87, 88]. Finally, note that when loops have opposite helicities and parallel axial fields the reconnection is not possible (Fig. 5d).

3.2 Early Stage of Two-Ribbon Flares

The standard two-ribbon flare scenario begins with the arcade/filament eruption. It is presumed that the pre-flare structure slowly evolves through a series of equilibrium states, comes to the state when the equilibrium is lost, and erupts thereafter [8, 80, 82, 112, 113, 116]. The flare starts with the onset of fast reconnection below the rising filament.

However, sometimes basically different onset scenarios are observed [70]. One example is the early phase of the 25 October 1994 flare described in [6]. The pre-flare arcade was strongly sheared, dominated by four closely spaced kinked SXR loops. These loops shaped an interrupted sigmoidal pattern (see also [70, 78]). The flare started by a sequence of neighbouring loop interactions (Fig. 6). The final result of interactions were three relaxed postflare SXR loops and an overlying sigmoid anchored at the outermost edges of the pre-eruption arcade as

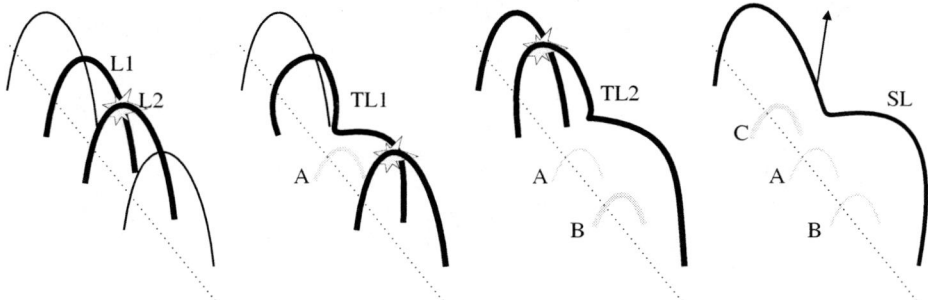

Fig. 6. Sequence of loop interactions as observed in the 25 Oct. 1994 flare. Postflare loops are drawn *gray*, whereas interacting features are drawn *bold*. Process started by the interaction of inner loops L1 and L2. Transient kinked loops are denoted as TL1 and TL2, and the final one (the unstable sigmoid) as SL

shown in Fig. 6. It should be stressed that although two flare ribbons appeared at the loop footpoints it was not a true two-ribbon flare at this stage since the ribbons did not expand laterally and the arcade eruption did not begin yet.

The phase of loop interactions was by all means a confined flare event. Only after the sigmoid was completed and erupted, the true two-ribbon (dynamical) flare phase started. In contrast to the filament in the standard scenario, the sigmoid was born unstable. Such an event can be in a way characterized as a *complex flare* since it consists of two distinct, but causally related flare events – the compact interacting loop flare that created the unstable sigmoid, and a dynamical two-ribbon flare aftermath.

The second event (6 February 1992) described by [6] also developed into a two-ribbon flare after an unstable sigmoid formation. However, the event took place in a less stressed (although larger) arcade and the sigmoid formation was less violent. In the pre-flare phase, lasting for hours, the sheared arcade consisting of a number of kinked loops was *gradually* reconnecting into a sigmoid. Loop interactions were accompanied only by minor soft X-ray brightenings. As in the previous example the sigmoid erupted after being completed, and the two ribbon flare began.

To conclude, the energy release in a sheared arcade can follow various scenarios. Beside the standard one, at least three alternative options are possible:

1. series of loop interactions resulting in the relaxed postflare loop system and a stable sigmoid (purely compact, *impulsive* flare; see [70])
2. gradual formation of an unstable sigmoid followed by a two-ribbon flare (purely dynamical flare, predominantly *gradual*; e.g. flare of 6 February 1992)
3. abrupt formation of unstable sigmoid (confined flare) followed by two-ribbon dynamical flare (*impulsive & gradual* phase; e.g. flare of 25 October 1994)[8]

---

[8] Compare with [70] who proposed that the confined and dynamical phase are not distinct stages

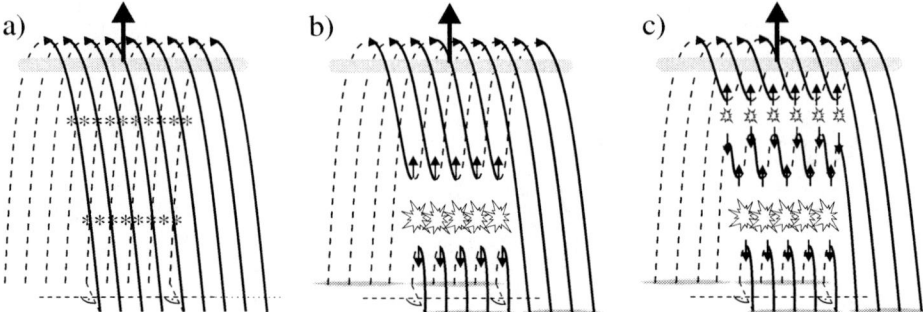

Fig. 7. Reconnection in a segment of an idealized two-ribbon flare configuration. a) Field lines are stretched by the arcade eruption. *Asterisks* indicate the locations of the two X-lines where the reconnection drawn in b) and c) takes place. b) Reconnection (*explosion symbols*) at one X-line below the arcade core (*gray*). c) Reconnection at two X-lines resulting in the secondary plasmoid formation. Reconnection outflows are indicated in b) and c) by straight *arrows*.

3.3 Main Phase of Two-Ribbon Flares

Sheared Arcade with Uniform Field. In Fig. 7a an idealized magnetic configuration presumably corresponding to the main phase of two-ribbon flares is shown. The sheared arcade field lines are stretched by the eruption and electric currents are redistributed, forming the current sheet below the arcade core [63]. In front of the erupting structure plasma is compressed creating flows towards the trailing edge of eruption [17]. The vortices formed behind the arcade core drive the plasma inwards [17] pressing the current sheet. The resulting increase of current density excites kinetic plasma instabilities [11, 46, 102] causing a localized anomalous resistivity enhancement[9]. The increased magnetic diffusivity provides initiation of reconnection and subsequent triggering of tearing instability [107, 108].

Since the formation of vortices is governed by the eruption kinematics, the onset of reconnection should be expected during its acceleration phase. Indeed, the flare onset is often contemporaneous with a fast acceleration phase of filament eruption [45]. Note also the feedback: the reconnection "supplies" the erupting arcade core with a "fresh" azimuthal field, enhancing the upward component of the Lorentz force and increasing the eruption acceleration [30, 100].

The fast reconnection regime cannot set-in if the current sheet is not long enough [107], i.e. the flare cannot start before the eruption attains some critical height. In the simple two ribbon spotless flare of September 12, 2000 (Fig. 8) the erupting Hα filament was still visible at the time of flare onset (Fig. 8a). Measuring the height of the lower edge of the filament and the initial separation of flare ribbons it is possible to estimate the current sheet width to length ratio

---

[9] The necessary merging velocity for a transition to the anomalous resistivity regime is in the order of only $10\text{-}100 \text{ m s}^{-1}$ [111], however to get a significant level of resistivity it must be in the order of 10 km s^{-1} [111]

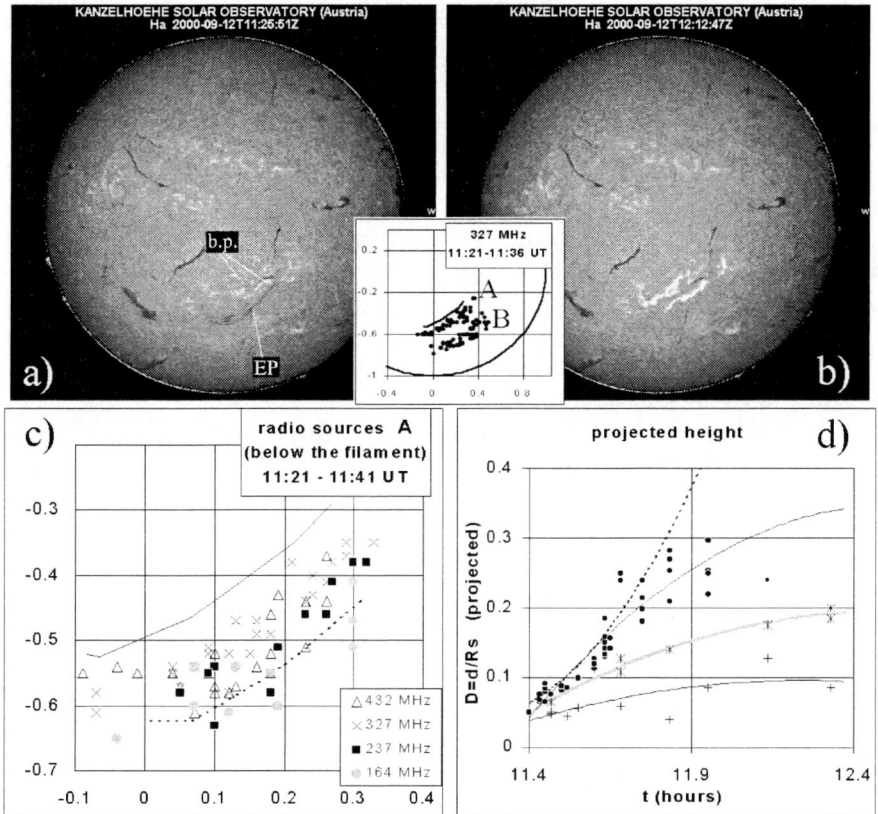

Fig. 8. The spotless two-ribbon flare of September 12, 2000. a) Early phase. b) Flare maximum. Radio sources are clustered in two groups: scattered sources below the filament (A) and moving type IV sources in front of the filament (B). 327 MHz sources are shown in the inserted small map. c) The A-sources observed between the neutral line (*full*) and the lower edge of filament (drawn for 11:39 UT by *dotted* line) in the early flare phase. d) The projected distance of sources from the neutral line shown as a function of time. The filament's lower edge and the second degree polynomial fits for three source sub-groups are shown by the *dotted*, *full*, *gray*, and *thin* line, respectively.

as $\delta/\lambda \approx 1/10$-$1/20$. A consistent value of $1/15$ is anticipated by numerical simulations [107].

The simplest situation where the reconnection occurs at only one X-line is shown in Fig. 7b and it corresponds to the 2-D representation shown in Fig. 3c. Note that there is a non-zero horizontal magnetic field present in the current sheet. The reconnection takes place between pairs of field lines whose footpoints are located on the line that is perpendicular to the inversion line. In $2\frac{1}{2}$-D and 3-D situation these footpoints are not magnetically connected before the reconnection. After the reconnection, loops below the X-line are lying in planes perpendicular to the inversion line, whereas the upper "loops" form helical field lines (see, e.g., [18, 19, 36]).

If too strong, the horizontal field component prevents the reconnection [30, 100]. This is another reason why the reconnection cannot start before the eruption attains an appropriate height – the field lines have to stretch enough to decrease sufficiently the horizontal to vertical field ratio.

In a long current sheet it is likely that two (or more) X-lines form, and in-between the secondary plasmoid(s) is created [96]. This is a common feature of 2-D numerical simulations of two-ribbon flare reconnection. Depending on the procedure applied, the plasmoid is ejected upwards [108] or downwards [31]. In Fig. 7c the secondary plasmoid formation is sketched in 3-D. Likewise in Fig. 7b the "nearest" field lines reconnect: above the perpendicular loops two sets of helical field lines form (see also [3]).

The 3-D aspect of secondary plasmoid formation introduces new important moments (Fig. 7c). The plasmoid is in fact a twisted flux tube anchored in the photosphere. Its length depends on the length of X-lines. The twist of helical field lines depends on the tube length and the initial arcade shear. The tube radius and the mass embraced depend on the duration of its formation which is determined by the distance between the two X-lines [51]. These parameters determine further behaviour of the plasmoid: It can be unstable and erupts, or is stable and the restoring force drives it towards an equilibrium position [30, 100]. Anyhow, the plasmoid speed is not determined by the speed of reconnection out-flow [94], but by its internal structure and shape. This is a possible explanation for unexpectedly slow downflows observed in *cusp structured flares* [66, 67] and flows in coronal streamer *disconnection events* [99, 119, 120] – the velocities observed are at least several times slower than expected from the fast reconnection theory.

If the plasmoid moves downwards it can reconnect with the postflare loops [31]. If however an equilibrium state is met before reaching postflare loop region the flux tube becomes a bay where the neutral line filament can be formed again. Since the tube is heated during its formation, the additional plasma is supplied from below by the evaporation flows. After radiative cooling the plasma becomes dense but it does not leak downwards since the twisted field provides dips where it condenses [19]. The decreased pressure in the magnetically dominated tube can eventually initiate *siphon-flows* [48] of the cold plasma, sucked-in from the chromosphere.

Sheared Arcade with Irregularities. The coronal magnetic field is highly structured and the real situation is obviously far from the idealized one shown in Fig. 7. Furthermore, the powerful energy release of impulsive phase is associated with violent and turbulent motions additionally distorting the magnetic field.

It can be presumed that in such an intricate and turbulent environment localized reconnection sites are scattered all over a distorted current sheet. In Fig. 8c positions of impulsive radio bursts observed by the Nançay Radio Heliograph during the 12 September 2000 flare are shown. The map reveals an apparently

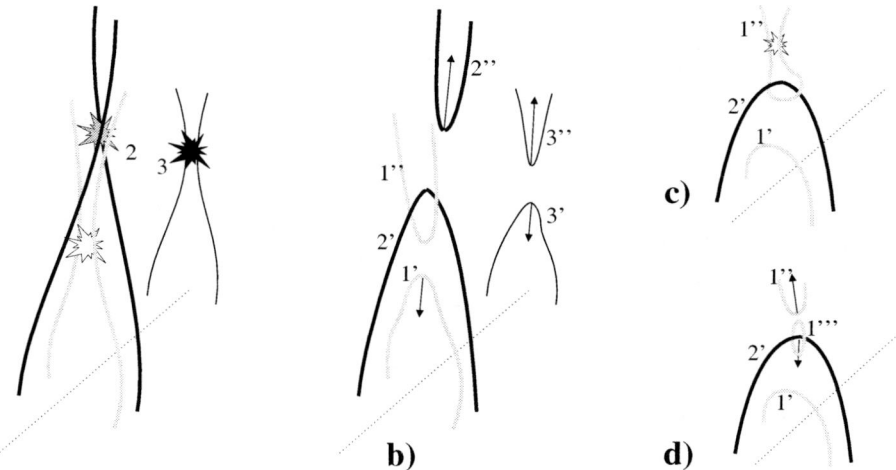

Fig. 9. Some features of "random" reconnection in a distorted current sheet. Reconnection outflows are indicated by *arrows*. a) The reconnection of intertwined field lines occurs at locations 1 and 2, whereas for a comparison, the site denoted by 3 represents the "regular" two-ribbon flare pattern. b) The reconnected, downward closing field line 1' bridges over the upward disconnecting line 2'. c) and d) Formation of small scale features (1''')

random "flickering" across the region below the erupting filament. Yet, preferable sites are found to be close to the lower edge of the filament and above the neutral line (shown in Fig. 8d by dots and crosses, respectively), i.e. at the two locations where intermittent FMSSs are expected (Fig. 3c).

Such a behaviour can be (and usually is) interpreted by the intermittent nature of unsteady reconnection, which produces temporary magnetic islands by the tearing instability, which then merge by the coalescence instability. In this way a range of temporal fine structures seen in radio bursts can be produced (see, e.g., [51]). Most of interactions could be expected (as observed) at the edges of the current sheet, where moving islands meet obstacles (postflare loops below, and erupting flux rope above the current sheet).

However, bearing in mind the complexity of the 3-D aspect of the island formation, geometry, and evolution, and some serious disagreements between the mentioned interpretation and observations stressed by [66], it might be instructive to recall a complementary option. In Fig. 9, a cartoon based on the concept proposed by [54] is worked out to emphasize some details. The most important feature is that in real situations field line bundles might be intertwined (Fig. 9a), so some of the downward closing field lines bridge over some of those disconnecting upwards (Fig. 9b) and local reconnection outflows are slowed down. Eventually small scale features are formed (Fig. 9c,d) that might be observed as various types of inhomogeneities. Such a scenario was proposed in [66] as a tentative explanation for the observed comparatively slow reconnection outflows. It should be stressed that when the "localized reconnection" is considered the $2\frac{1}{2}$-D approach is definitely inappropriate – a full 3-D treatment of reconnection is necessary (see, e.g. [97], and for observations [27, 28]).

The "smooth" reconnection in a "homogeneous" arcade (Fig. 7b) could be established in a late flare phase when more ordered large scale flows might form. The reconnection outflow is expected to be super-magnetosonic [29, 72, 60, 121] and more persistant FMSSs should appear (Fig. 3c). A possible signature, the "non-drifting type II-like" radio burst late in the 7. April 1997 flare, was reported by [7] (presented also at this workshop by H. Aurass, cf. [53]). In analogy with ordinary type II bursts which are caused by traveling fast mode MHD shocks the event was interpreted in [7] as the radio signature of FMSS formed in the downward reconnection jet.

3.4 Complex Flares

Large ARs frequently show complex magnetic configurations consisting of various subsystems. The stored energy is often released by flares of different types and sizes, appearing more or less independently at favourable locations. Yet, sometimes conditions are met for a large complex event, where all subsystems release their energy in a sequence of causally related processes [105]. In the following it will be illustrated how such an apparently intricate event can be decomposed and represented in terms of few basic processes described in previous sections.

In Sect. 3.2 one type of composite energy release process (series of interacting loop processes followed by a two-ribbon flare in the 25 October 1994 event) was already described. However, the same flare was even more complex – the two-ribbon phase showed some additional distinct stages.

First, the energy release considerably intensified when one of the ribbons protruded over a part of the largest sunspot in the AR. The energy release enhancement associated with the ribbon-spot contact activated the emerging flux region near the main spot. The magnetic field lines anchored in the spot were pushed towards the oppositely oriented emerging flux, starting the reconnection between the two systems (see Fig. 8 in [6]). This episode was very much like the spine-reconnection[10] event described in [27] (see Figs. 9 and 11 in [27] and compare with Fig. 8 in [6]). It was an independent, sympathetic energy release process, triggered when the two-ribbon flare embraced a part of the sunspot field flux.

On the other hand, during the "emerging flux episode" an interaction of the erupting arcade/sigmoid with the medium-scale field line system anchored in the main sunspot occurred (see Fig. 9 in [6]). A new reconnection point was temporarily formed providing a branching in the field line system analogous to that shown in Fig. 4b. The nonthermal electrons coming from the emerging flux region split at this point, being directed to two remote regions at opposite sides of the active region.

Finally, in the latest stage of the event a large scale coronal restructuration took place, which is a frequent feature of eruptive events (see, e.g. [62]. It was

---

[10] for the terminology in 3-D reconnection see, e.g., [3]

caused by an interaction of the erupting field and global magnetic structures. The restructuration was revealed by the appearance of two transient coronal holes at AR boundaries [6, 61].

4 Summary

The corona is built of magnetic loop assemblies forming intricate and permanently changing structures. Driven by the sub-photospheric convection the magnetic fields imbedded in a highly conductive coronal plasma are stressed, electric currents are generated, and free energy is deposited in coronal magnetic structures. The relaxation of stored energy causes a broad variety of phenomena, each of which plays its role in the coronal response to the solar magnetic cycle [59]. One of the most intriguing forms of the energy release are flares.

The morphology and development of a particular flare is prescribed by the pre-flare magnetic field environment. Since pre-flare coronal structures are generally complex, flares show a large diversity of appearances. Yet, some characteristic patterns are recognizable, and consequently some basic flare classes can be distinguished. The basic processes involved are also identified as constitutive elements of complex events where the energy release takes place in composite magnetic configurations consisting of a number of subsystems. Furthermore, some of the basic processes observed in flares are involved also in phenomena occurring at entirely different scales, from transition region fine structures to interplanetary space.

In this paper an attempt is made to disclose basically different processes in some apparently similar types of events. In particular different options of the two loop interaction process are illustrated, being the most basic element of any flare type. The type of interaction and the efficiency of energy release are determined by mutual orientation of axial field, the amount of twist, and the sense of twist. The most powerful option is the interaction of loops that have antiparallel axial fields and opposite senses of twist.

Similarly, different onset scenarios are revealed in the case of two-ribbon flares. Usually it is considered that the arcade eruption associated with two-ribbon flares starts after a phase of slow rise of the arcade filament which can be well described by the evolution through a sequence of equilibrium states. However, besides such a scenario, some other options are observed. Some events start by an abrupt transformation of highly sheared arcade core causing an impulsive confined flare phase. In this process an unstable sigmoid is formed, leading to the arcade eruption and the two-ribbon dynamical flare phase aftermath. Sometimes the arcade core transformation preceding the eruption is more gradual, being provided by a number of small subflares, flare-like brightenings, and microflares. In all of these options the arcade erupts only after the flux rope is created in its core, supporting the line of arguments presented by [59]. Note that the creation of the rope is not a sufficient condition since sometimes after the restucturation the arcade remains stable.

The examples shown indicate that complex events can be interpreted in terms of basic processes. Series of quadrupolar loop interactions can be a prelude to a two-ribbon flare, an erupting arcade can interact with overlying fields, the expansion of a heated region can drive or trigger loop interactions in its surroundings, etc. To reveal and identify these processes a detailed analysis of morphology in all available wavelength ranges is necessary. And even then, the unique and unambiguous interpretation quite often can not be reached.

Acknowledgements

The author is grateful to the organizers of CESRA Workshop for the invitation and sponsorship of his attendance at the Tegernsee meeting. A significant part of new results presented in the paper is realized in a collaboration with H. Aurass and K.-L. Klein during the author's stay at Observatoire de Meudon and Astrophysikalisches Institut Potsdam. Fruitful and intriguing discussions with H. Aurass, H.S. Hudson, A. Klassen, K.-L. Klein, and G. Mann substantially advanced some of ideas incorporated in the paper. The author is especially thankful to the referee (M. Aschwanden) for a exceptionally careful reading of the manuscript and very constructive comments and suggestions.

References

1. H. Alfvén, P. Carlquist: Solar Phys. **1**, 220 (1967)
2. S.K. Antiochos, C.R. Devore, J.A. Klimchuk: Astrophys. J. **510**, 485 (1999)
3. M.J. Aschwanden: Space Sci. Rev. (2002, in press)
4. M.J. Aschwanden, T. Kosugi, H.S. Hudson, M.J. Wills, R.A. Schwartz: Astrophys. J. **470**, 1198 (1996)
5. H. Aurass, K.-L Klein: Astron. Astrophys. Suppl. **123**, 279 (1997)
6. H. Aurass, B. Vršnak, A. Hofmann, V. Ruždjak: Solar Phys. **190**, 267 (1999)
7. H. Aurass, B. Vršnak, G. Mann: Astron. Astrophys. **384**, 273 (2002)
8. A.A. van Ballegooijen, P.C.H. Martens: Astrophys. J. **343**, 971 (1989)
9. T.S. Bastian, A.O. Benz, D.E. Gary: Ann. Rev. Astron. Astrophys. **36**, 131 (1998)
10. H. Baty, J. Heyvaerts: Astron. Astrophys. **308**, 935 (1996)
11. A.O. Benz: *Plasma Astrophysics* (Kluwer Acad. Publ., Dordrecht 1993)
12. A.O. Benz: *this volume*
13. D. Biskamp: *this volume*
14. E.G. Blackman: Astrophys. J. **484**, L79 (1997)
15. D.S. Brown, C. Parnell, E. Deluca, R. McMullen, L. Golub: *Observed Magnetic Structure of X-Ray Bright Points from TRACE and MDI* in 'Magnetic Fields and Oscillations' ed. by B. Schmieder, A. Hofmann, J. Staude (ASP Conf. Series **184**, Astron. Soc. Pacific, San Francisco 1999) pp. 81-85
16. L.F. Burlaga: J. Geophys. Res. **93**, 7217 (1988)
17. P.J. Cargill, J. Chen, D.S. Spicer, S.T. Zalesak: J. Geophys. Res. **101**, 4855 (1996)
18. P. Démoulin, E.R. Priest: Astron. Astrophys. **214**, 360 (1989)
19. P. Démoulin, M.A. Raadu: Solar Phys. **142**, 291 (1992)
20. H.W. Dodson, E.R. Hedeman: Astron. J. **65**, 51 (1960)
21. H.W. Dodson, E.R. Hedeman: Solar Phys. **13**, 401 (1970)

22. W. Dröge: *this volume*
23. A. Duijveman, P. Hoyng, J.A. Ionson: Astrophys. J. **245**, 721 (1981)
24. C.T. Dum: Space Sci. Rev. **42**, 467 (1985)
25. B.N. Dwivedi, H.S. Hudson, S.R. Kane, Z. Švestka: Solar Phys. **90**, 331 (1984)
26. J.M. Finn, P.K. Kaw: Phys. Fluids, **20**, 72 (1977)
27. L. Fletcher, T.R. Metcalf, D. Alexander, D.S. Brown, L.A. Ryder: Astrophys. J. **554**, 451 (2001)
28. L. Fletcher, H.P. Warren: *this volume*
29. T.G. Forbes: Astrophys. J. **305**, 553 (2001)
30. T.G. Forbes, J.M. Malherbe: Astrophys. J. **302**, L67 (1986)
31. T.G. Forbes, E.R. Priest: Solar Phys. **84**, 169 (1983)
32. H.P. Furth, J. Killeen, M.N. Rosenbluth: Phys. Fluids **6**, 459 (1963)
33. V. Gaizauskas, Z. Švestka: Solar Phys. **114**, 389 (1987)
34. G.A. Gary, R.L. Moore, M.J. Hagyard, B.M. Haisch: Astrophys. J. **314**, 782 (1987)
35. N. Gopalswamy, S. Yashiro, M.L. Kaiser, R.A. Howard, J.-L. Bougeret: Astrophys. J. **548**, L91 (2001)
36. J.T. Gosling: J. Geophys. Res. **98**, 18937 (1993)
37. J.P. Hagen, D.F. Neidig: Solar Phys. **18**, 305 (1971)
38. M.J. Hagyard, D. Teuber, E.A. West, J.B. Smith: Solar Phys. **91**, 115 (1984)
39. M.J. Hagyard: Solar Phys. **115**, 107 (1988)
40. T. Haruki, J.-I. Sakai: Astrophys. J. **552**, L175 (2001)
41. A. Hofmann, V. Ruždjak, B. Vršnak: Hvar Obs. Bull. **16**, 29 (1992)
42. P. Hoyng: Astron. Astrophys. **55**, 23 (1977)
43. P. Hoyng, A. Duijveman, T.F.J. van Grunsven, D.R. Nicholson: Astron. Astrophys. **91**, 17 (1980)
44. J.D. Jackson: *Classical Electrodynamics* 2nd edn. (John Wiley & Sons, New York 1975)
45. S.W. Kahler, R.L. Moore, S.R. Kane, H. Zirin: Astrophys. J. **328**, 824 (1988)
46. S.A. Kaplan, V.N. Tsytovich: *Plasma Astrophysics* (Pergamon, Oxford 1973)
47. J.T. Karpen, S.K. Antiochos, C.R. Devore, L. Golub: Astrophys. J. **495**, 491 (1998)
48. J.T. Karpen, S.K. Antiochos, M. Hohensee, J.A. Klimchuk, P.J. MacNeice: Astrophys. J. **553**, L85 (2001)
49. E.R. Priest: 'Magnetohydrodynamics'. In *Plasma Astrophysics* ed. by A.O. Benz, T.J.-L. Courvoisier (Springer, Berlin, Heidelberg 1994) pp.1-112
50. B. Kliem: Astrophys. J. Suppl. Ser. **90**, 719 (1994)
51. B. Kliem: *Coupled Magnetohydrodynamic and Kinetic Development of Current Sheets in the Solar Corona* in 'Coronal Magnetic Energy Releases' ed. by A.O. Benz, A. Krüger (Springer, Heidelberg 1995) pp. 93-114
52. B. Kliem, M. Karlický, A.O. Benz: Astron. Astrophys. **360**, 715 (2000)
53. B. Kliem, A. MacKinnon, G. Trottet, T. Bastian: *this volume*
54. J.A. Klimchuk: *Post-Eruption Arcades and 3-D Magnetic Reconnection* in 'Magnetic Reconnection in the Solar Atmosphere' ed. by R.D. Bentley, J.T. Mariska, (ASP Conf. Series **111**, Astron. Soc. Pacific, San Francisco 1996), pp. 319-330
55. S. Krucker: *this volume*
56. J. Kuijpers, P. van der Post, C. Slottje: Astron. Astrophys. **103**, 331 (1981)
57. M. Kuperus: Solar Phys. **47**, 79 (1976)
58. M.G. Linton, R.B. Dahlburg, S.K. Antiochos: Astrophys. J. **553**, 905 (2001)
59. B.C. Low: Solar Phys. **167**, 217 (1996)

60. T. Magara, S. Mineshige, T. Yokoyama, K. Shibata: Astrophys. J. **466**, 1054 (1996)
61. P.K. Manoharan, L. van Driel-Gesztelyi, M. Pick, P. Démoulin: Astrophys. J. **468**, L73 (1996)
62. Ch. Marqué, P. Lantos, K.-L. Klein, J.M. Delouis: Astron. Astrophys. **374**, 316 (2001)
63. P.C.H. Martens, N.P.M. Kuin: Solar Phys. **122**, 263 (1989)
64. M.J. Martres, M. Pick: Ann. Astrophys. **25**, 293 (1962)
65. E.B. Mayfield, J. Lawrence: Solar Phys. **96**, 293 (1985)
66. D.E. McKenzie: Solar Phys. **195**, 381 (2000)
67. D.E. McKenzie, H.S. Hudson: Astrophys. J. **519**, L93 (1999)
68. D.B. Melrose: Astrophys. J. **387**, 403 (1992)
69. D.B. Melrose: Astrophys. J. **486**, 521 (1997)
70. R.L. Moore, A.C. Sterling, H.S. Hudson, J.R. Lemen: Astrophys. J. **552**, 833 (2001)
71. N.V. Nitta, J. Sato, H.S. Hudson: Astrophys. J. **552**, 821 (2001)
72. S. Nitta, S. Tanuma, K. Shibata, K. Maezawa: Astrophys. J. **550**, 1119 (2001)
73. E.N. Parker: J. Geophys. Res. **62**, 509 (1957)
74. C.E. Parnell: *Discussion and Application of X-Ray Bright Point Models*, in 'JOSO Annual Report 95' ed. by M. Saniga (Astron. Inst. Tatranska Lomnica, 1995) pp. 121-123.
75. C.E. Parnell, E.R. Priest, L. Golub: Solar Phys. **151**, 57 (1994)
76. H.E. Petschek: AAS-NASA Symp. on Phys. of Solar Flares, NASA SP-50 (1964) pp. 425
77. A.A. Pevtsov, R.C. Canfield, T.R. Metcalf: Astrophys. J. **425**, L117 (1994)
78. A.A. Pevtsov, R.C. Canfield, H. Zirin: Astrophys. J. **473**, 533 (1996)
79. M. Poquérusse, P.S. McIntosh: Solar Phys. **159**, 301 (1995)
80. E.R. Priest: *Solar Magneto-hydrodynamics* (D. Reidel Publ. Co., Dordrecht 1982)
81. E.R. Priest: Rep. Prog. Phys. **48**, 955 (1985)
82. E.R. Priest, T.G. Forbes: J. Geophys. Res. **91**, 5579 (1986)
83. E.R. Priest, T.G. Forbes: Solar Phys. **126**, 319 (1990)
84. M.J. Reiner, M.L. Kaiser, J. Fainberg, R.G. Stone: J. Geophys. Res. **103**, 29651 (1998)
85. M.J. Reiner, M. Karlický, K. Jiřička, H. Aurass, G. Mann, M.L. Kaiser: Astrophys. J. **530**, 1049 (2000)
86. J.A Robertson, E.R. Priest: Solar Phys. **114**, 311 (1987)
87. D.M. Rust: *Magnetic Helicity, MHD Kink Instabilities and Reconnection in the Corona* in 'Magnetic Reconnection in the Solar atmosphere' ed. by R.D. Bentley, J.T. Mariska (ASP Conf. Series **111**, Astron. Soc. Pacific, San Francisco 1996) pp. 353-358
88. D.M. Rust, A. Kumar: Astrophys. J. **464**, L199 (1996)
89. V. Ruždjak, B. Vršnak, P. Zlobec, A. Schroll: Solar Phys. **104**, 169 (1986)
90. V. Ruždjak, M. Messerotti, M.Nonino, A. Schroll, B. Vršnak, P. Zlobec: Solar Phys. **111**, 103 (1987)
91. V. Ruždjak, B. Vršnak, A. Schroll, R. Brajša: Solar Phys. **123**, 309 (1989)
92. T. Sakurai: Adv. Space. Res. **13**, 109 (1993)
93. R. Schlickeiser: *this volume*
94. M. Scholer: *this volume*
95. J. Schumacher, B. Kliem: Phys. Plasmas **3**, 4703 (1996)
96. J. Schumacher, B. Kliem: Phys. Plasmas **4**, 3533 (1997)

97. J. Schumacher, B. Kliem, N. Seehafer: Phys. Plasmas **7**, 108 (2000)
98. K. Shibata, S. Masuda, M. Shimojo, H. Hara, T. Yokoyama, S. Tsuneta, T. Kosugi, Y. Ogawara: Astrophys. J. **451**, L83 (1995)
99. G.M. Simnett, S.J. Tappin, S.P. Plunkett, and 16 co-authors: Solar Phys. **175**, 685 (1997)
100. A.M. Soward: J. Plasma Phys. **28**, 415 (1982)
101. D.S. Spicer: Solar Phys. **53**, 305 (1977)
102. D.S. Spicer, J.C. Brown: 'Solar Flare Theory'. In *The Sun as a Star*, NASA SP-450, ed. by S. Jordan (NASA, Washington, 1981) pp.413-470
103. P.A. Sweet: *The Neutral Point Theory of Solar Flares* in 'Electromagnetic Phenomena in Cosmical Physics', ed. by B. Lehnert, IAU Symp. **6** (Cambridge University Press) pp.123
104. Z. Švestka: *Solar Flares* (D. Reidel Publ. Co. Dordrecht, 1976)
105. Z. Švestka: Solar Phys. **121**, 399 (1989)
106. T. Tajima, J. Sakai, H. Nakajima, T. Kosugi, F. Brunel, M.R. Kundu: Astrophys. J. **321**, 1031 (1987)
107. M. Ugai: Geophys. Res. Lett. **14**, 103 (1987)
108. M. Ugai: Phys. Fluids, B **4**, 2953 (1992)
109. M. Ugai: J. Geophys. Res. **104**, 6929 (1999)
110. M. Vandas, S. Fisher, M. Dryer, Z. Smith, T. Detman: J. Geophys. Res. **101**, 2505 (1996)
111. B. Vršnak: Solar Phys. **120**, 79 (1989)
112. B. Vršnak: Solar Phys. **129**, 295 (1990)
113. B. Vršnak: Ann. Geophys. **10**, 344 (1992)
114. B. Vršnak, V. Ruždjak, M. Messerotti, P. Zlobec: Solar Phys. **111**, 23 (1987)
115. B. Vršnak, V. Ruždjak, M. Messerotti, Z. Mouradian, H. Urbarz, P. Zlobec: Solar Phys. **114**, 289 (1987)
116. B. Vršnak, V. Ruždjak, B. Rompolt: Solar Phys. **136**, 151 (1991)
117. B. Vršnak, V. Ruždjak, R. Brajša, P. Zlobec, L. Altaş, A. Ozguç, H. Aurass, A. Schroll: Solar Phys. **194**, 285 (2000)
118. B.V. Waddell, M.N. Rosenbluth, D.A. Monticello, R.B. White: Phys. Rev. Lett. **41**, 1386 (1978)
119. Y.-M. Wang, N.R. Sheeley, J.H. Walters, and 6 co-authors: Astrophys. J. **498**, L165 (1998)
120. Y.-M. Wang, N.R. Sheeley, R.A. Howard, O.C. St.Cyr, G.M. Simnett: Geophys. Res. Lett. **26**, 1203 (1999)
121. T. Yokoyama, K. Shibata: Astrophys. J. **549**, 1160 (2001)
122. P. Zlobec, V. Ruždjak, B. Vršnak, M. Karlický, M. Messerotti: Solar Phys. **130**, 31 (1990)

Theoretical Models of Magnetic Configurations Relevant to Energy Release

Yihua Yan

National Astronomical Observatories, Chinese Academy of Sciences, Beijing 100012, China

Abstract. In this paper, we mainly discuss the reconstruction of 3D coronal magnetic fields from observed boundary magnetogram data under the general (non-constant-α) force-free assumption for flare events. These methods can perhaps be divided into 4 classes according to their mathematical bases for practical implementations: direct numerical discretization of the force-free field equations by the finite difference method (FDM); numerical treatment of the variational problem of the force-free field, or the finite element method (FEM); numerical treatment of the boundary integral equation of the force-free field, or the boundary element method (BEM); and the quasi-physical evolution from MHD equations to the force-free state. We will, however, present the results mostly obtained by the BEM and compare the reconstructed 3D coronal magnetic structures with Hα, soft X-ray, UV/EUV images and/or radio observations so as to understand the triggering and energy release processes in these flare events.

1 Introduction

It is believed that the magnetic field plays a central role in the solar activities. In order to understand different solar phenomena controlled by the magnetic field, the magnetic field computations in the solar corona is presently still a dynamic research area [5, 15].

In the past 4 decades, various computational methods for reconstructing the non-potential coronal field from boundary data have been proposed. All of them assume the magnetic field to be force-free, which is generally believed to be a very good approximation in the low solar corona [16] (cf. also [26]). With the available magnetograph data of vector magnetic fields in the photosphere and the presently increasing computer capability, it is necessary to develop reliable and robust methods for the general non-linear force-free field computation. There have been several reviews on the computations of coronal magnetic fields based on boundary data from complementary aspects [25, 7, 15, 32, 29]. In [15, 32] the topics are focused on nonlinear problems. Some mathematical properties of nonlinear force-free fields were discussed in [1]. According to the mathematical bases for practical implementations, here we discuss the numerical reconstruction of 3D coronal magnetic fields from observed boundary magnetogram data under non-linear force-free condition. These practical methods, available at present, can perhaps be classified into 4 classes: a direct numerical discretization of the force-free field equations (e.g., [30]) by the finite difference method (FDM); a numerical treatment of the variational problem of the force-free field [23], or the finite element method (FEM); a numerical treatment of the boundary integral equation of the force-free field [34, 35], or the boundary element method (BEM); and the quasi-physical evolution from MHD equations to the force-free state (e.g., [17, 22]).

In the next section, we discuss the modelling of boundary value problems for the 3D non-linear force-free fields. Then in §3 the reconstruction of coronal fields by BEM from boundary data is introduced and the results by the BEM are demonstrated. Finally, we summarize our conclusions in §4.

2 Modelling Coronal Magnetic Fields

To understand the energy release and conversion processes in the solar corona we need a detailed knowledge of solar magnetic fields since most phenomena we observe are produced directly by a subtle nonlinear interaction between the solar atmospheric plasma and the magnetic field [19, 21]. However, we can only measure the magnetic field reliably at the solar surface, and so a knowledge of the overlying fields depends on an extrapolation from those values upwards. The classical way of doing this was to assume the magnetic field is potential, but this has been shown to be completely inadequate in many cases since the equilibria are instead force-free when the magnetic configuration contains excess energy that can be released to heat the corona or drive solar flares or great eruptions from the solar surface.

Understanding the origin of coronal heating or of eruptions from the Sun (which have profound effects on the Earth) then depends directly on understanding of the nature of force-free fields in the solar atmosphere. The first attempts to do so assumed the fields were linear, but such fields have many undesired properties and are usually not realistic at all [19, 20]. The present situation is therefore that we need to develop reliable and robust methods for modelling non-linear force-free fields. There are also some attempts to consider non-force-free fields but the technique is presently not practical [13, 18, 8]. Throughout this paper we focus our attention to the non-constant-α force-free field problems.

In modelling the solar coronal magnetic field, it is generally assumed that $\Gamma = \{(x, y, z)|z = 0\}$ plane corresponds to the photosphere and semi-space $\Omega = \{(x, y, z)|z > 0\}$ to the solar atmosphere. Under the force-free condition, the magnetic field yields the following equations,

$$\nabla \times \mathbf{B} = \alpha \mathbf{B} \ , \ \ \nabla \cdot \mathbf{B} = 0 \quad \text{in} \quad \Omega, \tag{1}$$

where α is a function of spatial location and is determined consistently from the boundary conditions since it is constant along field lines.

The boundary condition is

$$\mathbf{B} = \mathbf{B}_o \quad \text{on} \quad \Gamma. \tag{2}$$

$\mathbf{B}_o$ here denotes known boundary values which can be supplied from vector magnetograph measurements. Since there are three independent variables involved in the general force-free field equations (1), specifying both the vector field on the boundary and the connectivity of field lines will over-determine the boundary value problem. Therefore, different methods have different choices of the boundary condition in order to avoid this over-specification. In general, the normal component of the magnetic field, B_n, is always prescribed over the boundary. In addition, normal current component, J_n, (or curl of the tangential components of the field) over one polarity [24] is employed, or J_n is specified over the whole boundary [17]. Others [30, 22] used all three components of the field. In [34, 35] the vector field is also used as boundary conditions and it is assumed that the boundary values are prescribed consistently with the force-free field in space.

Physically the magnetic field should tend to zero identically at infinity, and mathematically the boundary value problem (1,2) has a well-known ill-posed feature as mentioned in many papers [25, 15, 32]. Therefore an asymptotic condition at infinity must be added in order to avoid this ill-posed property [2, 3].

$$\boldsymbol{B} \to 0, \quad \text{when} \quad R \to \infty. \tag{3}$$

For a magnetic field with finite energy content in open space above the Sun, the asymptotic condition needed is actually as follows [34, 35],

$$\boldsymbol{B} = O(R^{-2}) \ , \quad \text{when} \quad R \to \infty. \tag{4}$$

Physically, it is also accepted that no current flows at infinity [2, 3]. The point is how to formulate the boundary value problem (1,2,3) or (1,2,4) because the ill-posedness will appear if the asymptotic condition is not incorporated into the formulation properly.

It should be noted that the 180 degree ambiguity in the transverse field is intrinsic and cannot be resolved from measurements. It remains a key problem in the research on the observational study of solar magnetic field. Therefore it has significant influence on the results of 3D coronal fields based on the reconstruction from boundary data, as will be disucssed further when the numerical result is demonstrated.

3 Reconstructions of Coronal Magnetic Fields from Boundary Data

It should be noted that at present there is no mathematical proof on the existence of the solution of the boundary value problem (1,2,4) but the non-linear force-free field exists in the solar corona physically. Even if we ignore the above issues, there still exists a stability problem, i.e., the ill-posed feature may occur under certain situations as mentioned above [25, 12, 15, 32]. In this section we will briefly introduce the BEM method.

3.1 Boundary Integral Equation Representation – BEM

More recently, Yan and Sakurai [34, 35] have for the first time proposed a boundary integral equation representation of the general force-free field with the help of Green's theorem so that the asymptotic condition (4) can be taken into account. Therefore the solution of the exterior boundary value problem (1,2, 4) can be represented by the following boundary integral equation,

$$c_i \mathbf{B}_i(\mathbf{r}_i) = \int_\Gamma \left[Y(\mathbf{r}_i; \mathbf{r}) \frac{\partial \mathbf{B}}{\partial n}(\mathbf{r}) - \frac{\partial Y}{\partial n}(\mathbf{r}_i; \mathbf{r}) \mathbf{B}(\mathbf{r}) \right] d\Gamma, \qquad (5)$$

where c_i is a constant depending upon the location of the point $\mathbf{r}_i$. If $\mathbf{r}_i$ is in Ω, $c_i = 1$, and if $\mathbf{r}_i$ is on Γ, $c_i = 1/2$. Y is the proposed fundamental solution,

$$Y(\mathbf{r}_i; \mathbf{r}) = \frac{\cos(\lambda |\mathbf{r} - \mathbf{r}_i|)}{4\pi |\mathbf{r} - \mathbf{r}_i|}, \qquad (6)$$

where $\mathbf{r}_i$ is the field point in Ω, $\mathbf{r}$ is the source point on Γ, and λ is a parameter to be found by

$$\mathbf{E}(\lambda, \mathbf{r}_i) = \int_\Omega Y(\mathbf{r}_i; \mathbf{r})[\lambda^2(\mathbf{r}_i)\mathbf{B}(\mathbf{r}) - \alpha^2(\mathbf{r})\mathbf{B}(\mathbf{r}) - \nabla\alpha(\mathbf{r}) \times \mathbf{B}(\mathbf{r})] d\Omega$$
$$= 0. \qquad (7)$$

Thus the parameter λ, which is an implicit function of $\mathbf{r}_i$, will be determined. Actually (7) is a vector equation, and for each component we introduce different λ and Y functions. The boundary integral equation (5) indicates that if the values of $\mathbf{B}$ and $\partial \mathbf{B}/\partial n$ over Γ are known, the field value $\mathbf{B}$ at any point in Ω is determined by the integration of products of the proposed fundamental solution Y and the values of $\mathbf{B}$ and $\partial \mathbf{B}/\partial n$ over Γ. The quantity $\partial \mathbf{B}/\partial n$ is not supplied from boundary data. When (5) is applied to the point $\mathbf{r}_i$ on Γ, we obtain a set of linear algebraic equations for $\partial \mathbf{B}/\partial n$. By solving this equation, we obtain a consistent relation between the boundary values and their normal derivatives. This is a standard procedure in the boundary element method (BEM [4]). Yan and his colleagues [38, 39, 31] first applied this technique to the solar linear force-free magnetic field computations.

Yan and Sakurai's method for nonlinear fields [34, 35] can deal with noisy boundary data properly because it is an integral method which is advantageous in finding an approximate solution: the errors in some elements may be compensated by errors from other elements [32]. Although the value of λ has to be selected to satisfy the constraint (7), we have chosen the average value of the boundary α distribution as $\lambda(= \bar{\alpha})$ for the numerical analysis, and still we assume that the constraint (7) holds. The discrepancy in this assumption can be checked later when the solution is evaluated. The good coincidence between the extrapolated field lines and observations have been obtained in many applications by using this approximation method [34, 10, 11, 40, 27, 33, 36]. Quantitative analyses of errors, including the comparison with an analytical solution [12], were also demonstrated in [35, 28], showing that this extrapolation method is very effective.

3.2 Reconstructed Magnetic Fields and Disucssions

As mentioned above, the BEM has been applied to many event analyses and the good coincidence between the extrapolated field lines and observations has been obtained [34, 10, 11, 40, 27, 33, 36]. Here we show the analysis of the great flare/CME event in NOAA 9077 on 14 July 2000 (cf. also [6, 9]).

The X5.7/3B flare at position N22W07 in active region NOAA 9077 at 10:24 UT on 14 July 2000 (so called Bastille-day event) was well-observed with ground- and space-based instruments in hard and soft X-rays, extreme ultraviolet wavelengths (EUV), Hα, radio and magnetic fields, etc. Thus it provides us a good chance to examine the mechanism of triggering, energy release, and related dynamic effects in the flare process.

The data employed here were Huairou vector magnetogram and Fig. 1 shows the calculated magnetic field lines projected onto the photospheric magnetogram. It can be seen that the main feature of the magnetic fields is the multi-layer magnetic field lines forming arcades with different orientations. Inside this magnetic arcade, there is a magnetic rope suspended in the corona above the stretched neutral line. The calculated field lines of the rope rotate around its axis for more than 3 turns. Therefore the presence of a magnetic rope is, for the first time in [33], revealed from the extrapolation of the 3D magnetic field structure.

As shown in Fig. 1, the right branch of the rope was co-spatial with the Hα filament and the EUV bright lane [33]. The left branch of the rope was co-spatial with the Hα flare bulb at 10:20:25 UT even with the same shape. The rope was located at the center of the flare patch as shown in Fig. 1(d). The front and side views are shown in Figs. 1(e,f) as well. From these figures we see that the reconstructed rope in the chromosphere and low corona has a total projected length of about $100''$ and the height ranging from about $2''$ to $30''$. The rope thickness is about $20''$, as shown in Fig. 1(f). The estimated free magnetic energy in this rope system is about 1.6×10^{25} J [33]. This result is consistent with observations in interplanetary space [14].

Recently, however, we found that if the transverse field is calibrated with a potential model, there is no flux rope in the extrapolated 3D field ! On the other hand, flux ropes were obtained as the arrows of transverse field were mostly directed from S polarity to N polarity when we reconstruct the 3D field in AR 9077 [33, 37]. This indicates that the 180 degree ambiguity in the observed transverse field is a very important problem for vector field analyses because it cannot be resolved from measurements. Therefore the pre-assumption to resolve the 180 degree ambiguity in the transverse boundary field will have a significant influence on the reconstructed 3D field in higher solar atmosphere.

In general, it is believed that the free energy released in solar flares, coronal mass ejections, filament eruptions, etc., is stored originally as magnetic energy in the stressed magnetic fields [19, 21]. This free energy is mainly determined by the transverse field from vector magnetograph observations and the non-potentiality may be evaluated by the discrimination from a potential state. Therefore many theoretical models that are favourable for the energy release have been suggested to explain the various phenomena, like solar flares, filament eruptions, coronal

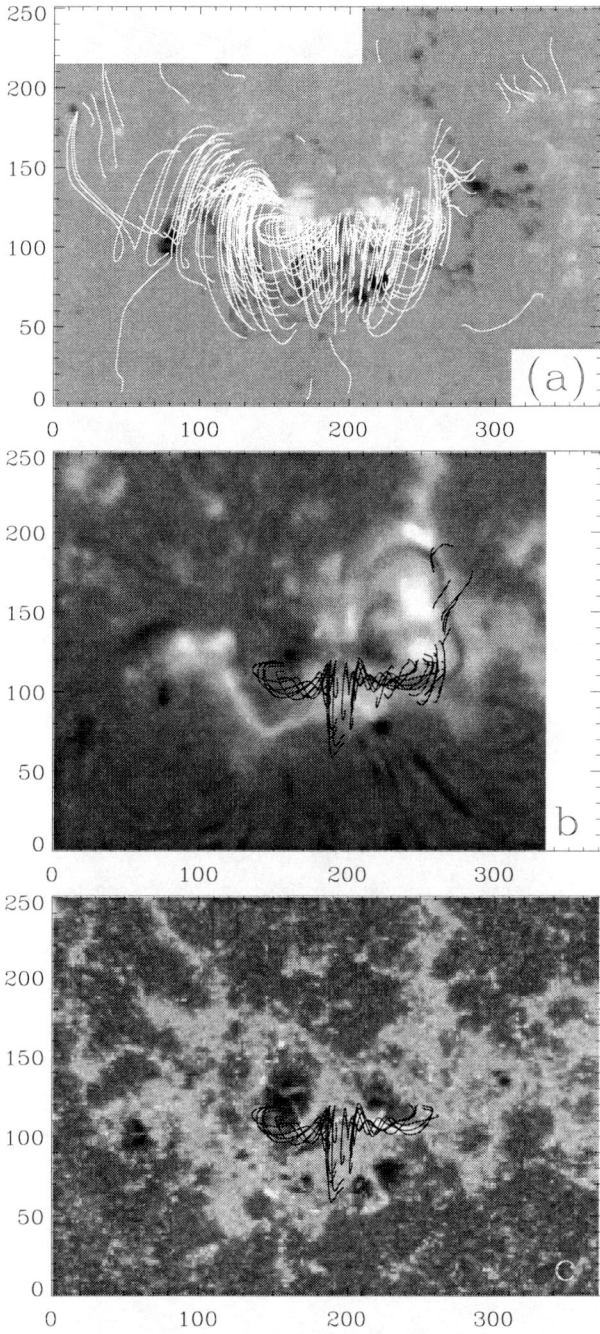

Fig. 1. (a) The reconstructed magnetic field lines projected onto the photospheri magnetogram at 01:19 UT showing a magnetic rope along the neutral line embraced by overlying arcades. The length unit is arc sec. The alignment of the flux rope with (b) Hα filament at 04:42:53 UT; (c) TRACE UV 1600 Å bright lane at 09:28:10 UT.

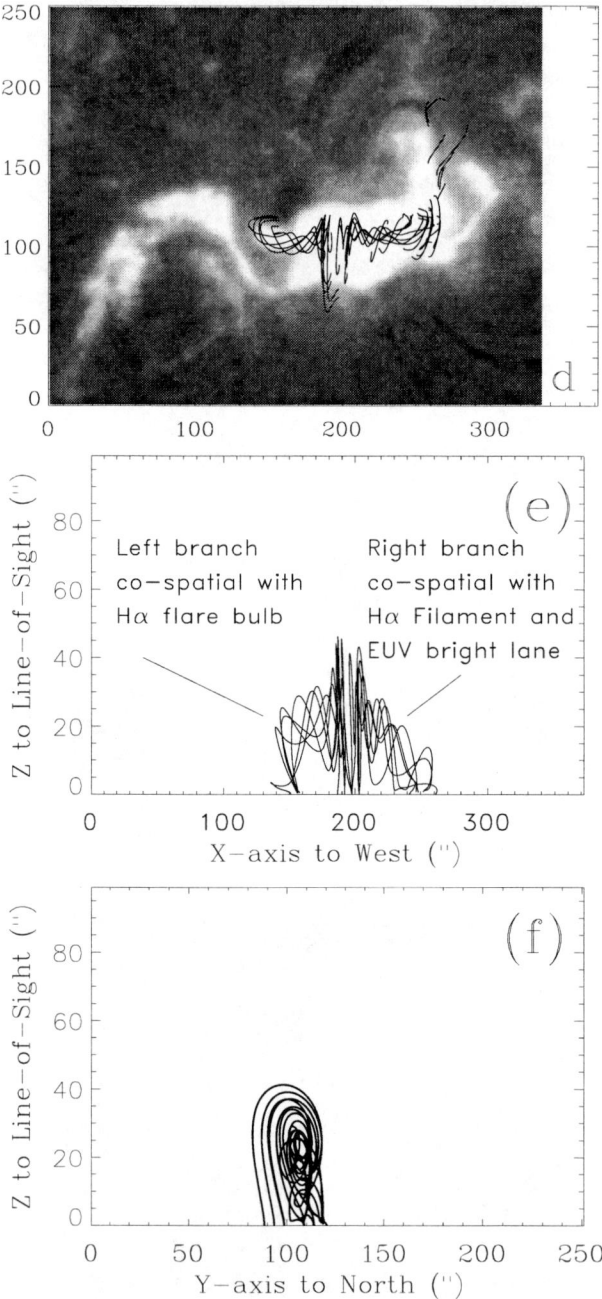

Fig. 1. (continued): (d) The alignment of the flux rope with Hα flare ribbons at 10:20:25 UT. The front (e) and side (f) views of the flux rope.

mass ejections, etc. ([21] and the references therein). Our method [34, 35] is able to recover the 3D coronal fields properly from boundary data under the general (non-constant-α) force-free field assumption. For the Bastille-day event, a pre-existing flux rope has been reconstructed that is sitting along the neutral line under the overlying field. This flux rope may lose its balance and trigger the energetic flare/CME event.

4 Conclusions

In summary, it is shown that for a finite energy force-free field, the field can be represented by the proposed boundary integral equation. The boundary integral equation can be solved by a standard numerical technique: the boundary element method (BEM). The applications by the BEM is promising and perspective. It is able to overcome the stability problem due to the errors in the boundary data and the method is numerically robust. It becomes a practical and convenient tool in solar magnetic field analysis.

It should be noted that many issues in the coronal field reconstruction are still open. The 180 degree ambiguity in the observed transverse field is intrinsic and can only be resolved under certain assumptions. Thus it adds further uncertainty to the reconstructed 3D coronal field with this resolved vector field as boundary conditions.

As an example to show the magnetic field configuration relevant to energy release, the non-constant-α force-free magnetic field above the NOAA 9077 active region was reconstructed for the famous Bastille-day event [33, 37]. The presence of a magnetic rope is, for the first time in [33], revealed from the extrapolation of the 3D magnetic field structure. This magnetic rope is located above the magnetic neutral lines at the place where the filament appeared. The calculated field lines of the rope rotate around its axis for more than 3 turns. Overlying the rope are multi-layer magnetic arcades with different orientations. These arcades are in agreement with TRACE UV/EUV observations. Such magnetic field structure provides a favorable model for the interpretation of the energetic flare processes as revealed by Hα, UV/EUV, SXR, HXR, and radio observations.

Acknowledgements

This work is supported by Ministry of Science and Technology of China (G20000-78403), NNSF of China (19973008, 19833050, 49990452) and Chinese Academy of Sciences. The author thanks the AvH Foundation for partial support. He acknowledges Dr. T. Neukirch for helpful comments on description of the FFF model and Dr. H. N. Wang for fruitful discussions on the treatment of boundary conditions. He also thanks Drs. Y. Liu and T.J. Wang for stimulative discussions. The author is grateful to Huairou and *TRACE* team for providing the observational data used in this research.

References

1. Aly, J. J.: Solar Phys. **120**, 19 (1989)
2. Amari, T., Aly, J. J., Luciani, J. F., Boulmezaoud, T., Mikic, Z.: Solar Phys., **174**, 129 (1997)
3. Amari, T., J. F., Boulmezaoud, T., Mikic, Z.: A&A, **350**, 1051 (1999)
4. Brebbia, C. A., Telles, J. C. F., Wrobel, L. C.: *Boundary Element Techniques* (Springer-Verlag, Berlin 1984)
5. Démoulin, P., Hénoux, J. C., Mandrini, C. H., Priest, E. R.: Solar Phys., **174**, 73 (1997)
6. Fletcher, L., Warren, H.P.: *this volume*
7. Gary G. A.: Mem. Soc. Ast. It., **61**, 457 (1990)
8. Gary G. A.: Solar Phys., **203**, 71 (2001)
9. Kliem, B., MacKinnon, A., Trottet, G., Bastian, T.: *this volume*
10. Liu, Y., Akioka, M., Yan, Y., Ai, G.: Solar Phys., **177**, 395 (1998)
11. Liu, Y., Akioka, M., Yan, Y., Sato J.: Solar Phys., **180**, 377 (1998)
12. Low, B. C., and Lou, Y.Q.: ApJ, **352**, 343 (1990)
13. Low, B. C.: ApJ, **399**, 300 (1992)
14. Manoharan, P.K., Tokumaru, M., Pick, M., Subramanian, P., Ipavich, F. M., Schenk, K., Kaiser, M. L., Lepping, R. P., Vourlidas, A.: ApJ, **559**, 1180 (2001)
15. McClymont, A. N., Jiao, L., Mikic, Z.: Solar Phys., **174**, 191 (1997)
16. Metcalf, T. R., Jiao, L., McClymont, A. N., Canfield, R. C., Uitenbruek, H.: ApJ, **439**, 474 (1995)
17. Mikic, Z., McClymont, A. N.: In: *Solar Active Region Evolution: Comparing Models with Observations*, ed. by K. S. Balasubramaniam, G. Simon (A.S.P., San Francisco 1994), p.225
18. Neukirch, T.: A&A, **301**, 628 (1995)
19. Priest, E. R.: *Solar Magnetohydrodynamics* (Reidel, Dordrecht 1982)
20. Priest, E.R.: 'Magnetohydrodynamics'. In: *Plasma Astrophysics* ed. by A.O. Benz, T.J.-L. Courvoisier (Springer, Berlin 1994), p.25
21. Priest, E. R., Forbes, T.: *Magnetic Reconnection: MHD Theory and Application* (Cambridge University Press, Cambridge 2000)
22. Roumeliotis, G.: ApJ, **473**, 1095 (1996)
23. Sakurai, T.: Publ. Astron. Soc. Japan, **31**, 209 (1979)
24. Sakurai, T.: Solar Phys., **69**, 343 (1981)
25. Sakurai, T.: Space Sci. Rev., **51**, 11 (1989)
26. Vršnak, B.: *this volume*
27. Wang, H. N., Yan, Y., Sakurai, T., Zhang M.:, Solar Phys., **197**, 263 (2000)
28. Wang, H. N., Yan, Y., Sakurai, T.: Solar Phys., **201**, 323 (2001)
29. Wang, J. X.: Fundamentals of Cosmic Physics, **20**, 251 (1999)
30. Wu, S. T., Sun, M. T., Chang, H. M., Hagyard, M. J., Gary, G. A.: ApJ, **362**, 698 (1990)
31. Yan, Y.: Solar Phys., **159**, 97 (1995)
32. Yan, Y.: Publ. Beijing Astron. Obs., No.4 (Special Issue), 65 (1998)
33. Yan, Y., Deng, Y., Karlický, M., Fu, Q., Wang, S., Liu, Y.: ApJ, **551**, L115 (2001)
34. Yan, Y., Sakurai, T.: Solar Phys., **174**, 65 (1997)
35. Yan, Y., Sakurai, T.: Solar Phys., **195**, 89 (2000)
36. Yan, Y., Liu, Y., Akioka, M., Wei, F.: Solar Phys., **201**, 337 (2001)
37. Yan, Y., Aschwanden, M. J., Wang, S., Deng, Y.: Solar Phys., **204** (2001), in press
38. Yan, Y., Yu, Q., Kang, F.: Solar Phys., **136**, 195 (1991)

39. Yan, Y., Yu, Q., Shi, H.: In: *Advances in Boundary Element Techniques*, ed. by J. H. Kane, G. Maier, N. Tosaka, S. N. Atluri (Springer-Verlag, Berlin 1993), pp. 447-469
40. Zhang, C., Wang, H. N., Wang, J., Yan, Y.: Solar Phys., **195**, 135 (2000)

The Energy Release Process in Solar Flares; Constraints from TRACE Observations

Lyndsay Fletcher[1] and Harry P. Warren[2]

[1] Department of Physics and Astronomy, University of Glasgow,
Glasgow G12 8QQ, U.K.
[2] Harvard-Smithsonian Center for Astrophysics, 60 Garden Street,
Cambridge, MA 02138, U.S.A.

Abstract. The Transition Region And Coronal Explorer Satellite, TRACE, launched in 1998, has proved a valuable tool in the study of solar flares. UV and EUV observations of the impulsive and gradual phases of many tens of flares have been made. TRACE's excellent spatial resolution and image cadence on the order of one second allow the rearrangement of the magnetic field to be tracked in some detail. The combination of these observations with data from other instruments, and with magnetic field reconstructions, have provided strong evidence for (a) UV emission as a beam proxy in the impulsive phase (b) long duration coronal heating in the gradual phase (c) very complex and varied magnetic geometries. We review the observational evidence for the above, discussing implications for energy release.

1 Introduction

The TRACE satellite (Handy et al, 1999), launched on April 2 1998, is the highest-resolution solar UV and EUV imager in operation, and at the time of writing has observed more than 200 GOES M-class solar flares and 20 GOES X-class flares, as well as numerous C-class events[3]. (This classification system refers to the soft X-ray flux provided by the flare in the 1-8 Å range, with X class flares providing a flux at Earth of more than 10^{-4} Wm^{-2}, M class more than 10^{-5} Wm^{-2} and C class more than 10^{-6} Wm^{-2}). The exceptional quality of these observations has led to the development of dedicated flare-observing programs, optimised to capture the rapid evolution during the initial phases of solar flares. An early summary of some flare phenomena observed with TRACE can be found in Sect. 15 of Schrijver et al. (1999).

TRACE observes in broadband white light (centred at 5000 Å), three UV wavelengths (1500 Å 1550 Å and 1600 Å) and three EUV wavelengths (171 Å 195 Å and 284 Å); the most often used wavelengths for flare observations are the 1600 Å channel (imaging plasma from $4 - 10 \times 10^3$K in the UV continuum and the lines of C I and Fe II and C IV), and the 171 Å and 195 Å channels, which nominally image 1 MK (Fe IX/X)and 1.5 MK (Fe XI/XII) plasma respectively (though see Sect. 2).

We review and summarise TRACE observations pertaining to flare energy release, but do not discuss the many TRACE observations related to other as-

---

[3] see http://hea-www.harvard.edu/SSXG/kathy/flares/flares.html

pects of solar flares, such as loop oscillations (e.g. Aschwanden et al 1999) or flare surges (e.g. Gallagher et al. 2000). After describing the instrument in Sect. 2, in Sect. 3 we discuss impulsive and pre-impulsive phase phenomena, including flare kernels and filament brightenings. Section 4 focuses on the relationship between TRACE and hard X-ray (HXR) data and Sect. 5 discusses the evidence from TRACE supporting hot coronal sources. A discussion in Sect. 6 of TRACE's contribution to studies of the reconnecting magnetic field is followed by our conclusions in the final Section. Note that, while no comprehensive survey of TRACE flare characteristics has been undertaken, it is the impression of the authors that the phenomena described here, and illustrated by single events, are nonetheless fairly widespread.

2 Suitability of TRACE as a Flare-Observing Instrument

TRACE's 0.″5 pixels, corresponding to a distance of 325 km at disk centre (a spatial resolution of 750 km) approach the observed size of elementary flux bundles within the solar photosphere. This has turned out to be a useful scale in examining flare evolution, imaging loops and their footpoints in groups with small transverse extent.

It is broadly accepted that magnetic reconnection facilitates the release of stored magnetic energy. Direct observation of a reconnecting current sheet would require resolving a region with thickness on the order of the ion gyroradius - beyond TRACE's capabilities. However, the intensely bright, narrow, elongated structures sometimes appearing before and during the impulsive phase of flares are suggestive of the presence of separators (the 3-D analogue of the current sheet); indeed in two instances their locations are consistent with coronal separator locations in magnetic field extrapolations (Sect. 6.1).

The chromosphere and corona can be imaged by TRACE within a few seconds of one another, with a cadence of less than a minute. The field-of-view is large enough to encompass the entire flaring active region and some of its surroundings. This resolution/FOV combination allows examination of large-scale coronal connections and response, while at the same time identifying the often very small sources which appear to be sites of particularly strong heating and activity. The time cadence delivered by TRACE can be as low as 1 s in the UV channels, while in the EUV channels 10 s is more typical.

The dynamic range of TRACE is nominally about 1.5 orders of magnitude (from a detector pedestal of 86 data numbers (DN) per pixel to a saturation value on the order of 4×10^3 DN/px), however during flares the dynamic range is increased *in the EUV channels* by diffraction patterns around bright features, caused by the fine mesh grids supporting the front entrance filter (Schrijver et al, 1999). The known characteristics of this diffraction pattern can be used to estimate the brightness in saturated features, thus adding a further 2 to 4 orders of magnitude in dynamic range (however, this requires deconvolution with the instrument characteristics).

Past flare programs have alternated images in 171 or 195 Å and 1600 Å at a cadence of ~ 30 s though, latterly, problems with the quadrant selection shutter have led to this kind of observation being avoided in favour of sequences in a single line. The EUV filters are relatively narrow-band filters, which pick out a subset of the coronal temperature structure. This has the advantage of allowing one to examine in detail the dynamics of constant temperature loops, but features are lost as they increase or decrease in temperature by more than about a factor 2. Of particular interest in EUV flare observations are (1) the presence of the Fe XXIV line in the 195 Å channel, imaging flare plasma at around 15 MK, and (2) free-free thermal bremsstrahlung emission in the EUV channels (Feldman et al, 1999), both of which strengthen during flares. This combination has been used by Warren & Reeves (2001) to determine temperatures for hot flare plasmas in a particularly strong flare.

The 1600 Å channel has a broad response, with a FWHM extending from ~ 1500 Å to 1650 Å. It contains a pair of C IV lines at 1548 Å and 1551 Å which are (a) enhanced during flares (Brekke et al, 1996) and (b) sensitive to pressure (Hawley & Fisher 1992). It has been pointed out by Warren (2000) that during flares the emission in the 1600 Å channel may be dominated by these lines, providing a further possible diagnostic use for TRACE imaging data, not exploited as yet.

3 Pre-impulsive and Impulsive Phase Emission

3.1 Flare-Associated Microflares

'Microflares' is the name given to the small-scale energy release events, observed at soft X-ray and EUV wavelengths, and occurring both in active regions and in the network of the quiet Sun (e.g. Krucker & Benz. 2000, Shimizu & Tsuneta 1997). With a typical energy of 10^{18} to 10^{20} J, they are often interpreted as the lower energy end of a spectrum of flare-like events. It is not at all clear at this stage that these transient releases are smaller-scale versions of larger flares, however this is the commonly adopted hypothesis. It has been proposed (e.g. Parker 1988) that the energy released in these events could contribute significantly to the heating of the solar corona, but this requires that the magnitude of the slope of the energy spectrum be greater than 2 (Parker 1988, Hudson 1991). TRACE measurements have extended the energy spectrum downwards by 0.5 orders of magnitude compared to the SoHO/EIT limit, into the 'nanoflares' regime of 10^{17} J, and below. The difficulty in correctly treating these tiny TRACE events, not much above the resolution and detectability threshold of the instrument, has added to the debate on the spectral index, and the significance for heating, with e.g. Aschwanden (1999) finding a slope with modulus less than the critical value of 2, and Parnell & Jupp (2000) finding a super-critical slope. Aschwanden & Charbonneau (2002) back up the sub-critical result with Monte-Carlo simulations of the effects of the TRACE and SoHO/EIT temperature biases.

TRACE has also been used to study the relationship of microflare sites to the local magnetic field, and to large flare events. Wang et al (1999) observed a cluster of 70 microflares in CIV 1550 Å during 3 hours of active region monitoring. They found that the majority of events occurred close to magnetic neutral lines, and that there was no clear distinction (in size, shape, time profile or peak intensity) between the microflares in this majority, and those which occurred in unipolar regions without observable neutral lines. 40% of the microflares occurred close to the site of a C5.2 flare. The authors suggest this may be evidence that the microflaring triggers a bigger flare, though no time or detailed spatial information is given to substantiate this: the C5.2 flare happens within the first 30 minutes of the 3 hour observation, so presumably a majority of microflares occur after the larger flare and are not therefore a part of the 'build-up'. Note, these authors found no corresponding detections in the lowest energy BATSE channels. Thus there is no clear evidence for particle acceleration in these events, in contrast to the work of Warren & Warshall (2001) (Sect. 4.1).

3.2 Pre-flare Observations

Because TRACE pointing tends to be fixed on one active region (unlike *Yohkoh* which repoints to the flare region when a flare occurs), good TRACE pre-flare observations are frequently available. The pre-flare context images prove invaluable in understanding the structures which were present (and occasionally bright) before the flare, but which become saturated and more difficult to interpret during the maximum. Flares occur in a wide variety of complex configurations; no two are alike at triggering, although the 'end state' of relaxed post-flare loops may look very similar from one event to another. The peak of the flare may be so overexposed as to lose all spatial detail.

For example, in Figs. 1 and 2 we show flare and pre-flare 171 Å images of two events in which peak EUV fluxes occur at very nearly the same location within an active region. These take place on 16-Jan-1999 19:44 UT (Fig. 1) and 18-Jan-1999 07:50 UT (Fig. 2). During their peaks, both events show long, narrow, saturated flare kernels, giving no clue to the flare configuration. However, the discrimination offered by the pre-flare observations is very valuable in interpreting the geometry of such events. In the former event, one or two loop-like structures brighten between 19:40 UT and 19:50 UT, suggesting that the interaction of coronal loops or the brightening of separators may be important. In the latter event no such structures are visible, and the flare appears to start in a narrow and possibly twisted filament or loop which brightens between 07:50 UT and 07:55 UT. There was no sign of a filament in the earlier event.

In general, then, pre-flare data may show locations of (sometimes persistent) pre-flare activity or heating, presumably significant in the overall flare geometry, which may be compared with model predictions of separator or current sheet locations, for example. Quite apart from qualitative morphological information, the UV channels may give evidence of an energetic build-up phase to the flare. Chromospheric emission in the form of discrete and persistent pre-flare 'kernels'

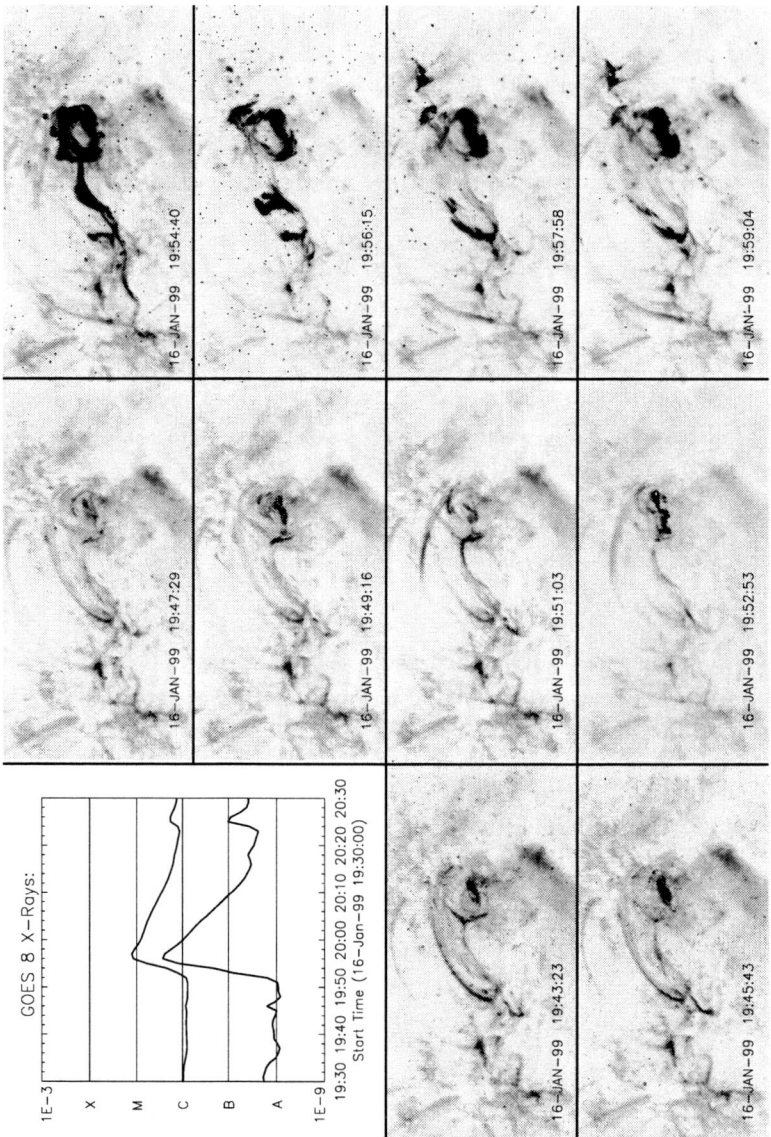

Fig. 1. Preflare and flare 171 Å images of a GOES M flare on 16-Jan-1999. The field of view is approximately 100,000 km × 50,000 km. The fieldlines illuminated in the pre-flare images suggest that the activation of, or interaction between, narrow, looplike structures is significant in triggering the flare.

implies the delivery of energy to particular locations in the chromosphere, possibly by conduction along specific sets of field lines, or possibly by particle beams linked to an acceleration site. We return to this topic in detail in Sect. 4. A final important role of pre-flare data is in providing a 'reality check' on magnetic field

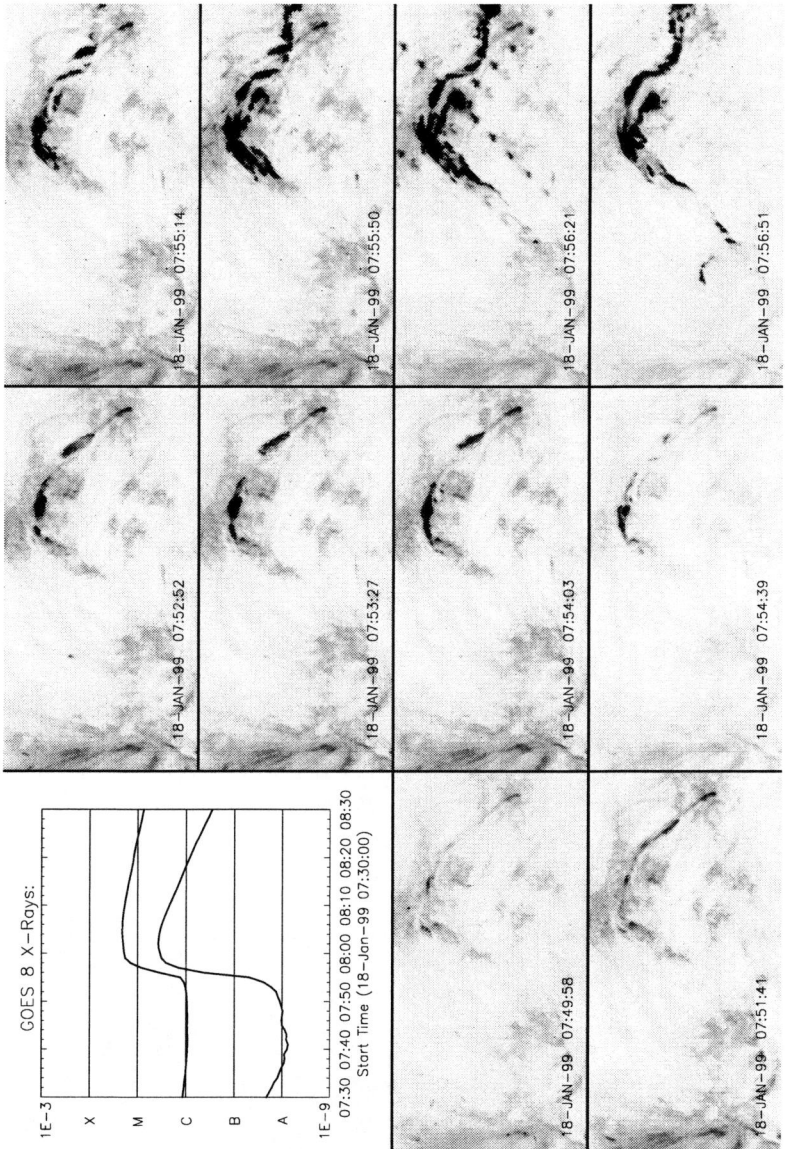

Fig. 2. Preflare and flare images of a GOES M flare on 18-Jan-1999, from the same region as seen in Fig. 1. The field of view is approximately the same. In this case, the flare involves the activation of a filament. No sign of a filament was seen in the flare on 16-Jan-1999.

extrapolations . Though it is not necessary to have a one-to-one correspondence between theoretical field lines and that subset which is made visible in the UV or EUV, the two should at least be consistent - i.e. the connections made by theoretical field lines should not be ruled out by what is observed.

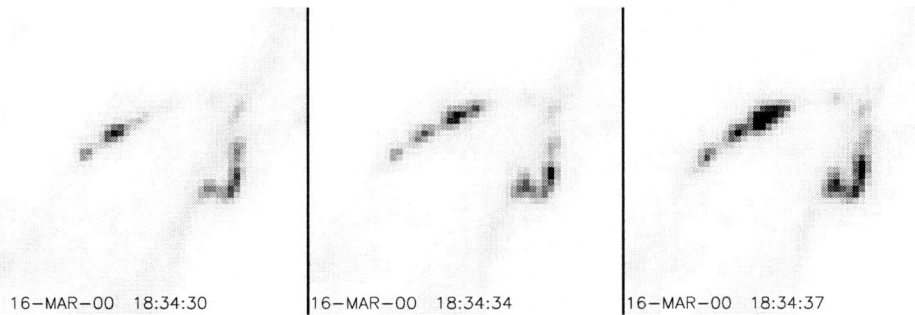

Fig. 3. Three 1600 Å images from the rise of a C9 flare on 16-Mar-2000. The TRACE pixels have a physical size of approximately 325 km × 325 km

3.3 Impulsive Phase Flare Kernels

The impulsive phase results in new UV kernels, apparently formed low in the atmosphere. At their appearance, the kernel size is comparable to the instrument point-spread function (PSF) of 2-3 pixels, indicating that the actual sources may be smaller still (e.g. Fig. 3). This suggests energy deposition fragmented on scales of hundreds of kilometres or less. If these kernels are the signature of electron beam precipitation at the chromosphere (and in Sect. 4 we present evidence for this) then the beam precipitation area per kernel is on the order of 10^{12} m^2. This is at least an order of magnitude smaller than that estimated by Canfield et al. (1991) from H$_\alpha$ and SMM Hard X-ray Burst Spectrometer data, using a model of thick-target emission by a power-law electron beam with low-energy cut-off of 20 keV.

The electrodynamic consequence of a propagating electron beam in a conducting plasma (see e.g. van den Oord (1990) and references therein) is an oppositely-directed return current. It is driven by the electrostatic field due to beam charge displacement, and the inductive field due to the beam current. The electron flux in the return current balances the beam flux, ensuring charge neutralisation and avoiding the presence of an embarrassingly large beam self-field. However, if the return current electrons are forced to flow too fast, the beam itself may be inhibited from propagating. It was pointed out by Canfield et al. (1991), and McClymont & Canfield (1986) that a small precipitation area (and high beam flux at precipitation) has severe consequences for beam propagation. Brown & Melrose (1977) show that the return current generated by a high flux electron beam will be unstable to the generation of ion acoustic turbulence if the electrons in the return current are forced to flow at higher than the ion sound speed. Ion acoustic turbulence will impede the propagation of the beam. According to Brown & Melrose (eq. 12) the beam can propagate freely only if

$$\frac{A_{14}\, n_{16}\, T_7^{1/2}}{n_{frag}} \geq 1,$$

for medium-sized events, where A_{14} is the beam area in units of 10^{14} m^2, n_{16} is the local ambient density in units of 10^{16} m^{-3}, and T_7 the ambient temperature in units of 10^7 K (assuming electron and ion temperatures equal in the ambient plasma) and the equation has been modified to include the effect of fragmentation of the total electron flux over n_{frag} locations. Assuming a coronal origin for the beam, the small precipitation areas from TRACE observations ($A_{14} \sim 0.01$) demand rather extreme coronal conditions. For example, if the pre-impulsive corona has an ambient temperature of 10^7 K, and there are 5-10 observed kernels, the ambient density must be on the order of $1 - 2 \times 10^{17}$ m^{-3} for the beam to propagate freely; lower ambient temperatures demand higher ambient densities. If the magnetic loop in which the beam is propagating expands in the corona, this reduces the requirements (by reducing the beam flux), but the proximity of the footpoint sources to one another suggests that the expansion of an 'elemental flux bundle' cannot be more than $\sim$ a factor 4 in area.

We can express this also as a condition on particle flux. Using the conditions of charge neutralisation and demanding that the return-current speed be less than the ion acoustic speed, we find that the upper limit to propagating particle flux F is given by (e.g. van den Oord 1990)

$$F < 10^{36} \, n_{16} \, T_7^{1/2} \, A_{14} \text{ electrons s}^{-1}$$

An observational estimate of F can be made from hard X-ray emission. The *Yohkoh* Hard X-ray Telescope (HXT) signature from the particular flare shown in Figs. 3 and 4 was analysed by Qiu et al. (2001) who give the total counts/s/HXT subcollimator during the 30 s between 18:35:00 and 18:35:30. In the M2 channel, this ranges between 15 and 30, corresponding to 1000 - 2000 counts/s in M2 over all 64 subcollimators. Approximately half of these counts come from HXR sources in the field of view shown in Fig. 3, though the HXT spatial resolution is too coarse to make identifications of HXT sources with individual UV kernels. However, supposing again 5-10 precipitation kernels in this region gives an average of 50-100 counts/s in M2 from each. Forward modelling by folding in the HXT spectral response and effective area (see Alexander & Metcalf 1997) allows us to determine the electron energy flux necessary to give this count level. Qiu et al. determine a spectral index in energy of ~ 4.4, and with this value we find that an electron energy flux of $\sim 5 \times 10^7$ J m^{-2} s^{-1} is necessary to explain the HXR signature. Assuming a 20 keV cutoff, this corresponds to a total electron precipitation rate of 9×10^{33} electrons/s, over the observed precipitation area of 10^{12} m^2. The theoretical limit on F from van den Oord thus demands a coronal temperature in the propagation channel of $> 10^7$ K and a coronal density of $> 10^{16}$ electrons m^{-3}, confirming the limits from considering the area alone.

Interpreting the impulsive phase UV/EUV sources as the locations of electron precipitation thus implies certain conditions in the pre-flare corona; if the beam propagates from a coronal location, both limits obtained above suggest that it can only do so along hot and dense magnetic structures.

Flare kernels brighten within a few seconds, and can also increase in size. The spreading is visible in Fig. 3, right-hand panel, where several saturated pixels

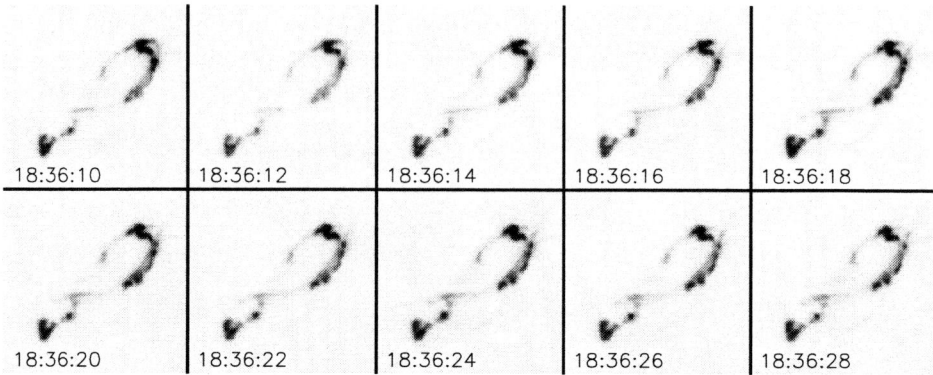

Fig. 4. Images from the peak of the flare show in Fig. 3, with the field-of-view extended to the south and east (total FOV size approximately 40,000 km × 40,000 km)

occur. This may happen due to CCD bleed, when the well-count (the number of photo-electrons per CCD pixel) exceeds some threshold, several times the 'saturation' value of $\sim 4 \times 10^3$ counts/pixel (T. Tarbell, private comm.) which implies that the counts per pixel may increase considerably beyond the saturation value. On the other hand, the spreading may be a result of an increased number of pixels brightening due to an increase in precipitation area. It is not possible to easily distinguish between the two, though the intensity profile at the edge of the source may help in distinguishing real sources (subject to the instrumental PSF) from CCD artefacts.

Figure 4 shows a series of 1600 Å observations made at a time resolution of approximately 2 s, during the impulsive phase of a C9 flare which occurred on 16-Mar-2000. The tiny scales of illumination are readily visible.

The 'dots' of emission along the ribbons are not stationary but move along the ribbon length, though not in an obviously ordered pattern. In the case shown, a dot moves its own diameter on a timescale of minutes, giving a projected speed of 30 km/s, comparable with the separation speeds of HXR footpoints measured by Sakao et al (1998). The association of the UV dots in Fig. 4 with faint loops which brighten in the last few frames, suggests that they are linked to the participation of individual loops in the flare. It is notable that, even in this example of a very small flare, the kernels are in the form of a ribbon (albeit a stationary one) rather than a 2-D grouping. This suggests that the energisation is occurring on a 'surface' of field lines which emanate from this quasi- 1-D locus of footpoints.

UV and EUV sources, which are widely spaced in distance, often brighten within seconds of each other, strongly suggestive of excitation by electron beams which have been simultaneously accelerated but propagate in different directions (Schrijver et al. 1999). However, there are also structures evolving in a way consistent with magnetic instabilities. This is exemplified by Fig. 5. This figure shows two frames from an EUV observation of an X-class flare on 07-Jun-2000. The flare was associated with the eruption of a filament towards the east (visible

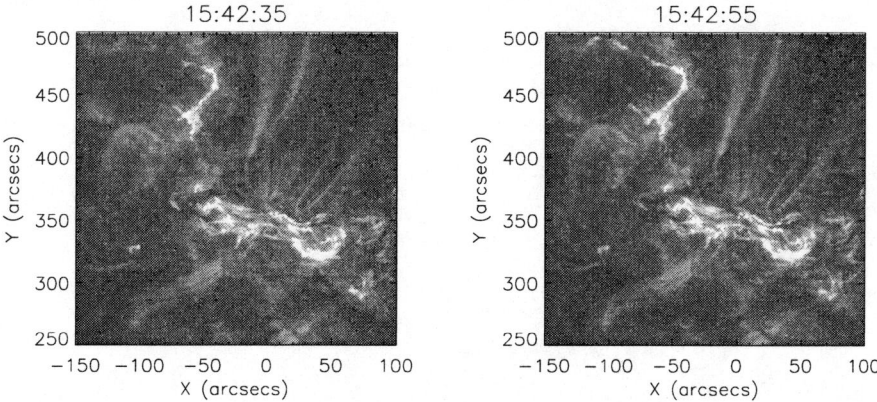

Fig. 5. An X-class flare which took place on 07-Jun-2000. The flare involved the rise and apparent ejection of a filament. During the second exposure shown here, a line of footpoints approximately 65,000 km long has brightened in the extreme north of the image, suggesting a very rapid signal propagation

as a dark structure centred around $\sim$ (-50″, 370″); note also a slight brightening at the filament's leading edge, around (-75″, 300″)). The flare resulted in two chromospheric ribbons and post-flare loops in the brightest central part. However, there was also a *single* EUV ribbon to the north, which brightened in the course of the eruption. It had no clear counterpart ribbon. In the image made at 15:42:55, which had an exposure time of 8 s this ribbon extends further to the north, over a de-projected distance on the order of 65,000 km. For this ribbon to brighten by means of a signal propagating along its length would require a propagation speed greater than 8,000 km/s. A magnetic disturbance propagating perpendicular to the magnetic field, along this ribbon, would travel at the fast magnetosonic speed, v_f where $v_f^2 = v_a^2 + c_s^2$, the sum of the squares of the Alfvén and ion sound speeds. In a coronal plasma with B = 0.01 T (100 G), $n_e = 10^{15}$ m^{-3} and $T_e = 10^6$ K, v_F is typically a few $\times 10^3$ km s^{-1}. Therefore it can be surmised that the brightening could also occur by means of a signal propagating perpendicular to the magnetic field.

At the peak of the impulsive phase, low-lying UV and EUV flare ribbons are often identifiable. These are typically a few TRACE pixels wide (comparable to the PSF) and show structure along their length. They are broken up into smaller elements, structured on a scale of about 1000-2000 km. This is a powerful indication that single 'monolithic' flare loops - such as would be inferred from SXT observations at low resolution - do not exist. TRACE suggests that the corona is further subdivided into groups of field lines, possibly rooted in the smallest photospheric magnetic flux bundles, which are activated as part of a common energisation process, but which respond according to their individual properties.

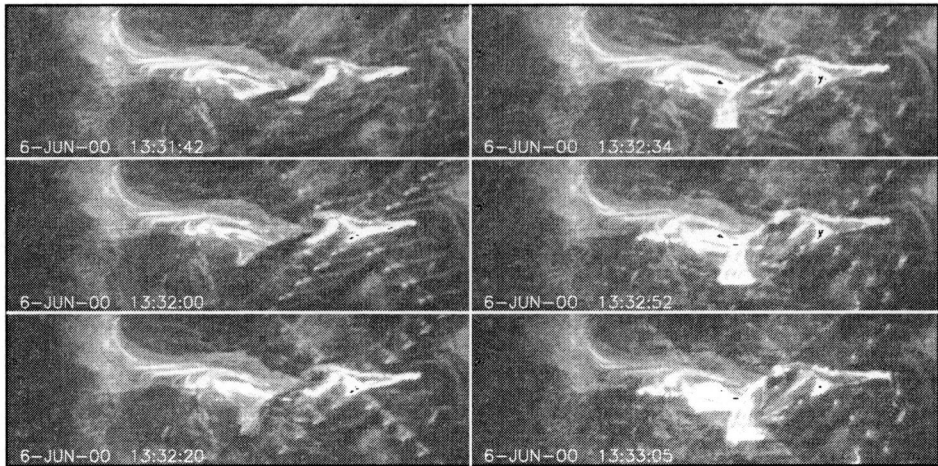

Fig. 6. The activation of a filament seen in TRACE 171 Å on 06-Jun-2000, at the beginning of an X1.1 flare. The field of view is approximately 125,000 × 62,500 km

3.4 TRACE Observations of Filaments in Flare Activation

The other obvious source of TRACE emission during the pre-flare and impulsive phase of many flares is an activated filament, seen in both the EUV and UV channels (often cool UV-emitting/EUV-absorbing material is apparently twisted together with hot EUV-emitting material). Studies of TRACE flare filaments are so far not reported in the literature; we give only a couple of recent examples. In the example of the 07-Jun-2000 flare mentioned above (Fig. 5), filament material starts to move rapidly (though not 'explosively') upwards by 15:41UT, before impulsive phase emission is observed in EUV or HXR. Although filament ejection or activation is not ubiquitous, this sequence is observed in several other X-class and many other M-class TRACE flares. The order of events suggests that the rapid rise of the filament may lead to, or provide conditions for, the instability which results in particle acceleration, rather than vice versa. Without a fuller survey of TRACE filaments in flares it would be premature to conclude any more on this important subject, however, it is worth noting that the same sequence of filament acceleration followed by impulsive phase onset was observed in four H_α and ISEE-3 flares by Kahler et al. (1988).

The 07-Jun-2000 flare was one of a series of three M and three X-flares from a single region, in which the motion of filament material was increasingly violent, and in which the larger scale field was increasingly affected. The earlier events show movement and heating along the filament, while in later events the material appears to 'bulge' out and twist, before flowing to a different location in the corona, accompanied by motion along the filament length. Figures 6 and 7 show two examples, both from 06-Jun-2000, in which twisted field appears to pop out from the confines of the filament. Again, this activity precedes the larger EUV bursts.

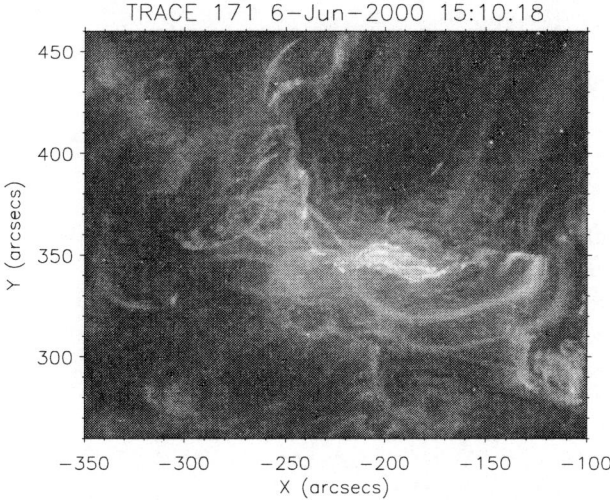

Fig. 7. The activation of a filament seen in TRACE 171 Å during an X2.3 event on 06-Jun-2000. These data have been logarithmically scaled to show the fainter twisted ejecta. The field of view of the panels in Fig. 6 extends between approximately $x = (-300, -120)$.

In the final flare of the series (not shown) filament material leaves the vicinity of the neutral line altogether and the large-scale field is apparently deflected by its passage, *or* has motion symptomatic of a larger-scale coronal rearrangement.

In all of these events it is noticeable how little the overall corona moves in response to even a major X-flare. There is no dectable shift prior to the event or as the filament starts to move, before the major EUV and HXR bursts. Small groups of loops sway if filament material approaches them, but a 'global restructuring' is not evident from the field lines illuminated in EUV, though field connectivities certainly do change in the filament's neighbourhood. This implies that the energy for the event does not come from a large coronal volume but is stored quite locally to the filament. To illustrate, in the event shown in Fig. 7, EUV loops which end 30,000 km from the very brighest part of the 06-Jun-2000 flare do not change position or intensity throughout the event. The magnetic energy density in a magnetic field B is $B^2/2\mu_\circ$ J m^{-3} (B in Tesla). So the total magnetic energy within a half-cylinder of radius 30,000 km (long axis aligned with the filament) surrounding the filament, with a length equal to the filament length of 150,000 km, is $8.4 \times 10^{28} B^2$ Joules. To power an X-flare (typical energy 10^{25} J) by reconnection would require 100% reconnection, throughout this whole volume, of a component of average strength 0.01 T or 100 G.

4 TRACE and HXR Impulsive Phase Emission

If it can be demonstrated that UV/EUV impulsive-phase emission is a proxy for electron beam deposition, this may provide a further useful nonthermal particle diagnostics. As EUV and UV are far easier to image than HXR, an enhanced understanding of flare magnetic configurations, and the paths taken by accelerated particles to the chromosphere, should follow.

A temporal relationship between UV and X-ray impulsive-phase time profiles has been known since the 1970s, from OSO-3 and OSO-5 data (e.g., Kane & Donnelly 1971; Kane, Frost, & Donnelly 1979) and was also found by SMM, when HXRBS and UVSP light curves were seen to be simultaneous to within 1 s (Woodgate et al. 1983). Spatial relationships could not be established at that time; following the early loss of SMM/HXIS there were no reported simultaneous UV/HXR imaged impulsive phase observations. However, in two flares, individual Si IV and O IV kernels observed with UVSP peaked within 1 s with individual bursts in the HXR time profile. The best time correlations were in HXRBS channels with energy greater than 50 keV (Cheng et al. 1981). Following the SMM repairs, some sources showed correlations to the 0.1 s level (Orwig & Woodgate 1986). A UV density diagnostic indicated that the UV emission came from high density regions, consistent with a low-atmosphere source. It was speculated at the time that the UV emission was produced in the chromosphere by the same beam which generates the HXR emission, through heating (e.g., Poland et al. 1984; Emslie, Brown, & Donnelly 1978) or direct collisional excitation by the non-thermal particles (e.g., Kane & Donnelly 1971). However, it was noted that (a) the near-simultaneity was a feature of 'selected' UV sources only, (b) the UV emission started to rise before the HXRs and (c) UV preflaring occurred which had no counterpart detected in HXRs (Cheng, Tandberg-Hanssen & Orwig 1984).

Reported observations to date with TRACE and *Yohkoh*-HXT show much the same picture. The time resolution provided by TRACE is poorer (at best 2 s) than that provided by UVSP, while that of HXT is 0.5 s, but both instruments of course have the advantage of imaging. As mentioned above, during flares TRACE shows compact and discrete flare kernels in both UV and EUV, and when overlaid with HXT images (e.g. Fletcher & Hudson 2001) it is found that some coincide well with HXT sources, but some do not. In the following we summarise some individual flare behaviours. The limited evidence so far points in the direction of at least the brighter UV or EUV flare kernels reflecting also the sites of HXR emission, provided that these sites are not also sites of pre-flare kernels. HXR imaging observations with a better dynamic range will be necessary to establish more detailed correspondence between EUV/UV kernels and HXR emission (though it may be possible with HXT data to make a statement about the minimum intensity of EUV/UV sources associated with HXR emission).

4.1 Preflare and Impulsive Phase Kernels

As reported by Warren & Warshall (2001), TRACE UV images show the excitation of emission during both the pre-flare and the flare impulsive phase. In the 17-Mar-2000 M1.1 flare, for example, two elongated ribbon brightenings were observed to develop approximately 360 s before the onset of hard X-ray emission. Because TRACE observations of this flare were taken at very high cadence (~ 2 s) it was possible to perform detailed correlations of the spatially unresolved BATSE hard X-ray light curves with the spatially resolved UV light curves determined from TRACE. This regression analysis showed that footpoints exhibiting *pre-flare* activity were not correlated with the initial burst of HXRs. In contrast, brightenings that were well correlated with the initial burst of HXRs, showed little or no pre-flare activity. The comparison of HXR images reconstructed from HXT with TRACE UV images for several flares has yielded similar results. The initial HXR burst in these events appears to be somewhat displaced from the pre-flare ribbon brightenings observed with TRACE. These results indicate that the energy release during the preflare and impulsive phases of the flare is occurring on different field lines. This is broadly consistent with what was found in a study of preflare brightenings using SXT by Fárník & Savy (1998) who observed that the majority (75%) of 32 observed preflare brightenings were not cospatial with the flare location (at the 5 arcsec SXT resolution limit), but were at best overlapping to some extent, and often separated by up to one degree. A still earlier study by Tappin (1991) using SMM Hard X-ray Imaging Spectrometer data identified 3.5 - 5.5 keV precursor events in more than 85% of a set of 86 HXIS flares. These had a broad range of spatial locations with respect to the main flare; from coincident to separated by several arcminutes. No statistical survey of the precursor-flare separations was carried out by Tappin, but a cursory examination of the data suggests that a similar majority of these preflare events were separated by more than 10 arcseconds from the main flare.

4.2 May 3 1999 Flare

In the M1.9 flare of 3-May-1999, analysed by Fletcher et al. (2001), there were of the order of five 171 Å kernels during the flare impulsive phase. Of these only one was located near the HXT source (to within the alignment accuracy, estimated to be 5″). HXT showed a single source in the LO (13-23 keV), M1 (23-33 keV) and HI (53-95 keV) energy bands, and in the M2 (33-53 keV) band two or three sources were present (Fig. 8, LH panel). The HXT source(s) were associated with the largest and brightest of the EUV sources, which had the form of a small loop. In the lower energy HXT channels (13-23 keV and 23-33 keV), the sources overlap this small EUV loop; in the higher energy channels they lie near its endpoints, suggesting that the lower energy channels may image hot plasma in the loop, while the higher energy channels image its footpoints. The time resolution of both TRACE 171 Å and HXT was poor; it was possible only to establish that the two peaked within 90 s (Fig. 8, RH panel).

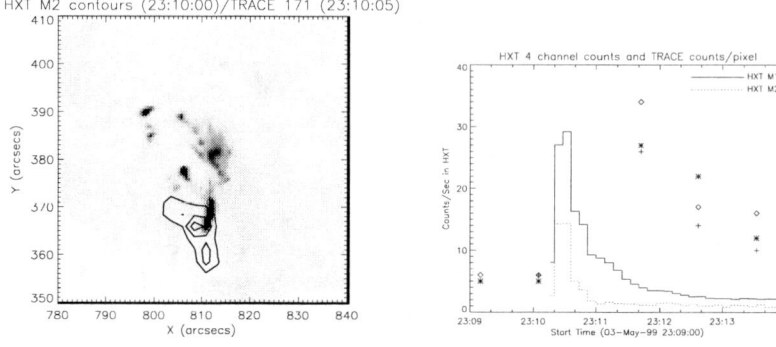

Fig. 8. Time profiles of TRACE EUV emission at three strong kernels (shown by symbols) and the HXT emission profiles in two channels; M1 (23-33 keV) and M2 (33-53 keV) during the 03-May-1999 M1.9 flare. HXT emission starts to rise while the EUV emission is still at its preflare level. The peak of the EUV emission is no more than 90 s after the HXR burst. Note, the right hand panel corrects Fig. 5 appearing in Fletcher *et al.* 2001.

Though shorter-lived and smaller, the other EUV kernels had comparable counts/pixel to the small loop source during the HXT integration time. However, due to their smaller size, their total overall EUV emission was smaller, and none of these other EUV kernels had an associated HXT source. However, it is dangerous to conclude that there was only one site of HXR emission. HXT has a small 'dynamic range'; the reconstruction algorithms (both Pixon and MEM) are poor at detecting more than two or three sources of unequal strength in HXT data if this inequality is more than a factor 4 or so. Multiple HXT sources, with fainter sources being a factor 4 or more less intense, would probably not be found by an HXT reconstruction. However, the intensity profiles of the discrete EUV sources are the same to within a factor 2, thus there may not be a linear relationship between EUV and HXR flux. The very similar time profiles suggest a related cause for the widely-spaced EUV kernels.

4.3 July 14 2000 Flare

The 14 July 2000 flare was unique in HXT observations so far, in that it exhibited two HXR ribbons, approximately 120,000 km long (Masuda et al, 2001). The flare occurred in two impulsive stages. First the western part of the region erupted, forming ribbons, followed by the eastern half. During the first 100 s of the flare observed by HXT - somewhere in the middle of the first impulsive burst - the HXR ribbons were at their most extended. For the remainder of the first impulsive phase the HXR ribbons broke up into a number of bright sources. This was repeated in the second impulsive phase. At the peak of the second impulsive phase, the HXT and EUV or UV ribbons were well aligned (see Fig. 9), and the combined time profiles of the brightest EUV sources tracked the total HXRs in M2 and HI, peaking within $\sim$ 20 s. The ratio between the two

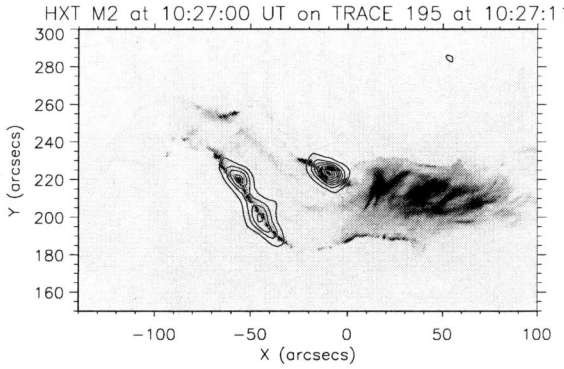

HXT M2 at 10:27:00 UT on TRACE 195 at 10:27:11

Fig. 9. The location of the HXR ribbons observed in the second phase of the Bastille Day 2000 flare, superimposed on TRACE 195 Å observations, showing that the HXRs come from the brightest EUV ribbon locations. The figure shows also the post-flare loops and EUV ribbons from the earlier eruption, of the western half.

count rates was approximately constant, with the HI channel tracking the EUV best overall, particularly in the first minute of the burst. The EUV counts fell off simultaneously with the HXR counts, rather than having the Neupert-like behaviour expected of a thermal signature (where the counts track the time-integral of the HXRs). This implies either that the EUV signature is thermally generated in an atmospheric layer where the radiative or conductive timescale is on the order of 10 s (density on the order of $10^{17} - 10^{18}$ m^{-3}) or that the EUV emission is in essence a 'non-thermal' signature, generated via direct collisional impact excitation of the atomic transition, by the same non-thermal beam that causes the HXR emission, or the secondary 'knock-on' electrons which this beam accelerates out of the background plasma in Coulomb collisions.

5 Hot Coronal Sources

As mentioned previously, TRACE is sensitive to high temperature flare plasma primarily because of the presence of the Fe XXIV 192 Å line in the 195 Å channel. Hot flare plasma can also be imaged in the 171 Å channel, where it produces thermal bremsstrahlung (though even in flares this is generally very weak).

TRACE 195 Å flare observations have shed new light on an outstanding problem in solar flares: is there a very hot component produced by energy release at the top of the flare arcade? Analysis of SXT data from the 13-Jan-1992 Masuda event by Tsuneta et al. (1997) revealed a localized region of hot plasma ($T_e > 1.5 \times 10^7$ K) sitting above the flare arcade. Since the emission measure of this hot component remained relatively constant while the emission measure of

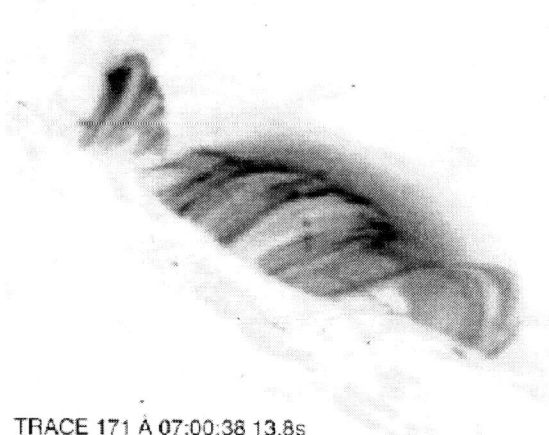

TRACE 171 Å 07:00:38 13.8s

Fig. 10. TRACE 171 Å image of an X2.5 flare showing a postflare arcade of loops and diffuse emission which is produced by plasma at temperatures in excess of 15 MK

the bulk flare plasma ($T_e \sim 1 \times 10^7$ K) rose significantly during the rise phase of the flare, they speculated that the hot region was the signature of a slow-mode shock. However, only about half of all flares appear to have hot components (Nitta et al. 2001, Doscheck 1999, McTiernan et al. 1999). It has also been suggested that the hot regions observed with SXT are strongly influenced by instrumental effects such as scattering (Nitta et al. 2001) or difficulties with accurate co-alignment of SXT images made through different filters (Siarkowski et al. 1996). Nitta et al., for example, have noted that the high temperature regions determined from the SXT filter ratios are often displaced from the brightest HXT L-channel emission.

TRACE observations have generally supported the existence of hot regions. Figure 10 shows a TRACE 171 Å observation of an X2.5 flare which occurred on 22-Nov-1998 showing a postflare arcade of loops, at about 1-2 MK surmounted by diffuse emission which is produced by plasma at temperatures in excess of 15 MK (Warren 2000). In another example, the 25-Jul-1999 M2.4 flare, the Fe XXIV 192 Å emission imaged with TRACE was shown to be consistent with SXT temperatures and emission measures in both the bulk of the flare plasma and in a hot region above a flare arcade (Warren et al. 1999). This suggests that the temperatures that have been derived from SXT filter ratios are at least qualitatively correct. The bulk of the flare plasma appears to lie at temperatures near 10 MK while there can be other regions where the temperature is systematically higher. TRACE observations of the 24-Mar-2000 X1.8 flare have provided direct evidence for a small region located above the brightest flare loops (Warren & Reeves 2001) with a temperature possibly as high as 20 MK. TRACE temperature measurements in flares are severely limited by the low sensitivity of the 171 Å channel to high temperature flare emission, but the emission measure in this compact event was high enough to generate significant counts in the TRACE 171 Å channel and allow temperatures to be inferred from the TRACE 195/171 filter ratios.

TRACE's broad temperature coverage provides a unique opportunity to study the cooling of flare plasma at high spatial and temporal resolution. Comparisons of TRACE images of hot flare loops and cooler post-flare emission have shown that the high temperature emission often appears to be very diffuse while the 1 MK post-flare emission appears as well defined loops. Warren (2000) conjectured that flare loops were composed of a large number of very fine "threads" instead of a few large scale loops. Reeves & Warren (2001) have developed a model of flare evolution based on this idea. In their model hot flare loops are formed at progressively larger heights and allowed to cool using the scaling laws for conductive and radiative cooling (Cargill, Mariska & Antiochos 1995). By incorporating the geometry of the flare loops derived from the high resolution TRACE images, they are able to reproduce SXT and TRACE light curves. This supports the idea of continuous coronal energisation lasting some minutes into the flare gradual phase.

6 Flare Magnetic Fields

The combination of TRACE images with magnetogram data, both scalar and vector, can be used to interpret the coronal magnetic field before and during flares. TRACE's narrowband coronal imaging allows a clear set of field lines to be identified and directly compared with the results of magnetic extrapolations. This can assist in the interpretation of the magnetic field in which the flare takes place and the identification of possible acceleration sites, as well as providing a reality check on the field extrapolations.

6.1 Coronal Nulls

In a study of the Bastille Day 1998 flare, Aulanier *et al.* (2000) used Mees vector magnetograms to perform a magnetic field extrapolation, and identified a coronal magnetic null in the 3-D field, with its associated 'spine' and 'fan' separator field lines (e.g. Lau & Finn 1990, Fig. 2; Priest & Titov 1996). These theoretical singular field lines could be identified with field lines and ribbons in TRACE EUV images, lending weight to the reconstruction, and to the interpretation of the flare being triggered by reconnection at a coronal null leading to magnetic breakout (Antiochos *et al.*, 1999). A similar study by Fletcher *et al.* (2001) compared not only a single magnetic extrapolation with TRACE observations, but also mimicked coronal field evolution under photospheric driving by a series of extrapolations of discrete source representations of the photospheric field. This work also supported the idea of reconnection at a 3-D null, and provided qualitative explanations for features such as the location of the observed flare ejecta (related to closed fields being opened by reconnection through the null), and of the HXR and EUV sources (related to the open fields which are being closed by reconnection).

6.2 Flare Ribbons

TRACE has also enabled detailed study of the evolution of flare ribbons, interpreted as the transition region/chromospheric footpoints of those coronal field lines which are at any instant involved (or recently involved) in flare energisation. The evolution of the flare ribbons thus somehow reflects the changes taking place in the coronal field. A quite elementary inspection of ribbon evolution reveals behaviour which is certainly mysterious, and may be inconsistent with the predominant theoretical ideas about the development of two-ribbon flares (e.g. Fletcher & Hudson 2001). In the usual Carmichael-Sturrock-Hirayama-Kopp-Pneumann model (cf. Kliem et al. 2002), a rising reconnection region, envisaged as an X-type neutral point (or, with translational symmetry, a line parallel to the chromosphere, built up of a series of X-type neutral points) produces a rising and expanding arcade of post-reconnection loops, at the outer edges of which are located two spreading ribbons of emission generated by accelerated electrons on just-reconnected field lines. If we interpret the impulsive flare ribbons observed by TRACE in this way, we must explain certain facts:

– There is often only one UV/EUV ribbon present, as in the flare shown in Fig. 5.
– In at least two cases where there are two ribbons in fields of opposite sign, overlays on SoHO/MDI line-of-sight magnetograms show that they do not sweep out equal magnetic fluxes in equal times (Fletcher & Hudson 2001, Saba et al. 2002) as would be predicted. The difference can be as much as a factor two, and is highest in the early phases of the ribbon evolution.
– Ribbons start off as fairly irregular structures, evolving as they spread towards a smooth curve or even straight line which is far more regular than the distribution of chromospheric network field.
– Ribbons are observed to form across the middle of supergranular cells (Fig. 11) where observations suggest only weak fields. This implies that weak field (presumably recently emerged in cell centres) can participate in reconnection high in the corona.

The first two points may simply be due to one of the ribbons being very faint along some or all of its length, which may be the result either of local chromospheric conditions, or of asymmetric beam precipitation. As well as this, the imbalance in fluxes traversed may be the result of 'patchy' coronal reconnection (Klimchuk 1997) or field structuring below the scale of the magnetogram or EUV images (both leading to a ribbon which is very fragmented along its length, but appears continuous) or highly non-vertical photospheric fields. Patchy coronal reconnection may also account for the initial irregularity of the flare ribbons, though this could equally be explained by smooth reconnection in a corona in

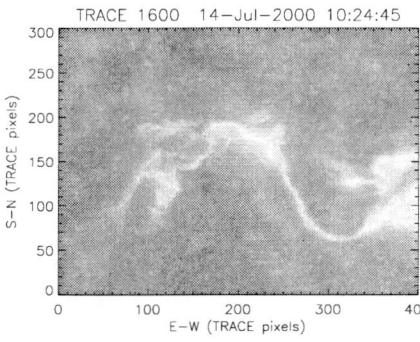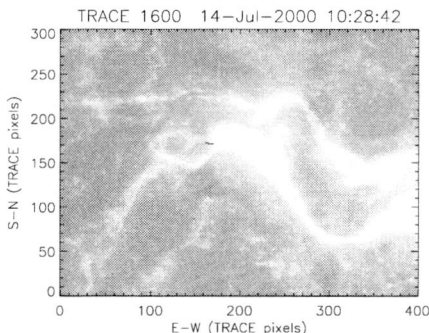

Fig. 11. Two 1600 Å TRACE images of the flare ribbons in the 'Bastille day 2000' flare. The cellular pattern of the chromospheric network can be seen, with the dark regions corresponding to supergranular cell interiors, where there is a relatively low field. In the left hand panel, the upper ribbon is not fully formed; four minutes later it has extended to the east. Note that the ribbon in the right hand panel extends in two places across network cells, apparently finding magnetic connections in the cell centre.

which adjacent reconnection sites do not map back to adjacent photospheric locations. The irregularity would then depend on the amount of braiding between corona and photosphere, and suggest that the mapping of chromospheric to coronal field becomes smoother at greater distances from the neutral line.

7 Conclusions

TRACE has opened up solar flares to detailed study in the UV and EUV parts of the spectrum, and in this paper we have reviewed some of the ways in which this data is being exploited to further our understanding of the energy release process, where the excellent resolution in time and space brings particular benefit. The TRACE flare data which have thus far been examined represent only a small fraction of the many sets available, and the many interesting features highlighted here have not been by any means fully explored. The overall impression from TRACE is that both the small-scale 'core' energy release region (filaments, kernels) as well as the form and topology of the large-scale magnetic field are important in determining the evolution of an individual reconnection event. We have also shown how UV and EUV observations of lower atmosphere flare signatures might serve as proxies for HXR emission and thus electron beam precipitation sites, giving additional clues to the location and nature of the primary electron accelerator.

References

1. D. Alexander & T. R. Metcalf: Astrophys. J **489**, 442 (1997).
2. S. K. Antiochos, C. R .Devore & J. A. Klimchuk: Astrophys. J. **510**, 485 (1999).
3. M. J. Aschwanden, L. Fletcher, C. J. Schrijver & D. Alexander: Astrophys. J. **520**, 880 (1999).
4. M. J. Aschwanden: Solar Physics **190**, 249 (1999).
5. M. J. Aschwanden & P. Charbonneau: Astrophys. J. **566**, L59 (2002).
6. G. Aulanier, E. E. DeLuca, S. K. Antiochos, R. A. McMullen & L. Golub: Astrophys. J. **540**, 1126 (2000).
7. P. Brekke, G. J. Rottman, J. Fontenla, P. G. Judge: Astrophys. J. **468**, 418 (1996).
8. J. C. Brown & D. B. Melrose: Solar Physics **52**, 117 (1977).
9. R. C. Canfield, D. M. Zarro, J.-P. Wülser & B. R. Dennis: Astrophys. J. **367**, 671 (1991).
10. P. J. Cargill, J. T. Mariska & S. K. Antiochos, Astrophys. J. **439** (1995).
11. C. -C. Cheng, E. Tandberg-Hanssen, E. C. Bruner, et al: Astrophys. J. **248**, L39 (1981).
12. C. -C. Cheng, E. Tandberg-Hanssen, & L. E. Orwig: Astrophys. J. **278**, 853 (1984).
13. G. A. Doschek: Astrophys. J. **527**, 426 (1999).
14. A. G. Emslie, J. C. Brown & R. F. Donnelly: Solar Physics **57**, 175 (1978).
15. F. Fárník & S. K. Savy: Solar Physics **183**, 339 (1998).
16. U. Feldman, J. M. Laming, G. A. Doschek, H. P. Warren & L. Golub.: Astrophys. J. **511**, L61 (1999).
17. L. Fletcher, T. R. Metcalf, D. Alexander, D. S. Brown & L. A. Ryder: Astrophys. J 554, 541 (2001).
18. L. Fletcher & H. S. Hudson, Solar Physics **204**, 71 (2001).
19. P. T. Gallagher, D. R. Williams, K. J. H. Phillips *et al*: Solar. Phys **195**, 367 (2000).
20. B. N. Handy, L. W. Acton, C. C. Kankelborg, C. J. Wolfson, D. J. Akin et al.: Solar Physics **184**, 229 (1999).
21. S. Hawley, G. H. Fisher: Astrophys. J. Supps. **78**, 565 (1992).
22. H. S. Hudson: Solar Physics **133**, 357 (1991).
23. S. W. Kahler, R. L. Moore, S. R. Kane & H. Zirin: Astrophys. J. **328**, 842 (1988).
24. S. R. Kane & R. F. Donnelly: Astrophys. J **164**, 151 (1971).
25. S. R. Kane, K. J. Frost & R. F. Donnelly: Astrophys. J. **234**, 699 (1979).
26. B. Kliem, A. MacKinnon, G. Trottet & T. Bastian: *this volume*
27. J. Klimchuk in: *Magnetic Reconnection in the Solar Atmosphere* ed. by R. D. Bentley and J. Mariska, ASP Conf. Proceedings **111**, 319 (1997).
28. S. Krucker & A. O. Benz: Solar Physics **191**, 341 (2000).
29. Y.-T. Lau & J. M. Finn: Astrophys. J. **350**, 672 (1990).
30. S. Masuda, T. Kosugi & H. S. Hudson: Solar Physics **204**, 55 (2001).
31. A. N. McClymont & R. C. Canfield: Astrophys. J. **305**, 936 (1986).
32. J. M. McTiernan, G. H. Fisher & P. Li: Astrophys. J, **514**, 472 (1999).
33. N. V.Nitta, J.Sato, & H. S. Hudson: Astrophys. J. **552**, 821 (2001).
34. G. H. J. van den Oord: Astron. Astrophys. **234**, 496 (1990).
35. L. E. Orwig & B. E. Woodgate, in *'Rapid Fluctuations in Solar Flares' (SEE N87-21785 14-92)*, p277, 1986.
36. E. N. Parker: Astrophys. J. **300**, 474 (1988).
37. C. E. Parnell & P. E. Jupp: Astrophys. J. **529** 554, (2000).

38. A. I. Poland, L. E. Orwig, J. T. Mariska, L. H. Auer, & R. Nakatsuka: Astrophys. J. **280**, 457 (1984).
39. E. R. Priest & V. S. Titov: Philos. Trans. R. Soc. London, A **354** 2951 (1996).
40. J. Qiu, M. D. Ding, H. Wang et al.: Astrophys. J., **554**, 445 (2001).
41. K. K. Reeves & H. P. Warren: in *'Multi-Wavelength Observations of Coronal Structure and Dynamics – Yohkoh 10th Anniversary Meeting'*, ed. by P. C. H. Martens and D. Cauffman, Elsevier, 2002 (in preparation).
42. J. L. R. Saba, T. Gaeng & T. D. Tarbell: in *'Multi-Wavelength Observations of Coronal Structure and Dynamics – Yohkoh 10th Anniversary Meeting'*, ed. by P. C. H. Martens and D. Cauffman, Elsevier, 2002 (in preparation).
43. T. Sakao, T. Kosugi & S. Masuda: in *'Observational Plasma Astrophysics: Five Years of Yohkoh and Beyond'*, ed. by T. Watanabe, T. Kosuga & A. C. Sterling, (Kluwer Acedemic Publishers, Dordrecht, 1998) p. 273
44. C. J. Schrijver, A. M. Title, T. E. Berger, L. Fletcher, N. E. Hurlburt et al.: Solar Physics **187**, 261 (1999).
45. T. Shimizu & S. Tsuneta: Astrophys. J **486**. 1045 (1997).
46. M. Siarkowski, J. Sylwester, J. Jakimiec & M. Tomczak: Acta Astronomica **46**, 15 (1996).
47. S. J. Tappin: Astron. Astrophys. Suppl. Ser. **87**, 277 (1991).
48. S. Tsuneta, S. Masuda, T. Kosugi & J. Sato: Astrophys. J. **478**, 787 (1997).
49. H. Wang, J. Chae, J. Qiu, C. K. Lee, & P. R. Goode: Solar Physics **188**, 365 (1999).
50. H. P. Warren, J. A. Bookbinder, T. G. Forbes, L. Golub, H. S. Hudson, et al.: Astrophys. J. **527**, L121 (1999).
51. H. P. Warren: Astrophys. J, **536**, L105 (2000).
52. H. P. Warren & K. K. Reeves, Astrophys. J. **554**, L103 (2001).
53. H. P. Warren & A. D.Warshall: Astrophys. J **560**, L87 (2001).
54. B. E. Woodgate, R. A. Shine, A. I. Poland & L. E. Orwig: Astrophys. J. **265**, 530 (1983).

Radio Diagnostics of Flare Energy Release

Arnold O. Benz

Institute of Astronomy, ETH, Zurich, Switzerland

Abstract. The radio emission of flares at wavelengths from millimeter to decameter waves includes a large variety of emission processes. They can be considered as different diagnostic tools particularly suited for the analysis of non-thermal electron distributions, enhanced levels of various kinds of plasma waves and plasma phenomena. Incoherent gyrosynchrotron emission at millimeter and centimeter waves provides higher sensitivity for observing MeV electrons than existing hard X-ray (HXR) and gamma-ray satellites. Very intense coherent emissions are observed at wavelengths longer than about 10 cm, weaker ones from about 4 cm. They are caused by plasma instabilities driving various wave modes that in turn may emit observable radio waves. Particularly important are type III bursts, caused by electron beams exciting Langmuir waves. Their trace in the corona points back to the acceleration region of the electrons. Less known are radio emissions from trapped electrons driving loss-cone unstable waves. This is the interpretation usually given to decimetric type IV emission. These types of coherent radio emission give clues on the geometry and plasma parameters near the acceleration region.

More speculative are emissions that are *directly* produced by the acceleration process. A possible group of such phenomena are narrowband, short peaks of emission. Narrowband spikes are seen sometimes at frequencies above the start of metric type III events. There is mounting evidence for the hypothesis that these spikes coincide with the energy release region. Much less clear and highly controversial is the situation for decimetric spikes, which are associated with HXR flares. More frequently than spikes, however, there is fluctuating broadband decimetric emission during the HXR phase of flares. The use of these coherent radio emissions as a diagnostic tool for the primary energy release requires a solid understanding of the emission process. At the moment we are still far away from an accepted theory. Only careful comparisons with complementary observations of energetic electrons and the thermal coronal background in EUV lines and soft X-rays can put coherent emissions into context and test the different scenarios. The comparison with HXR, millimeter and centimeter observations will be necessary to derive quantitative results on energy release. In combination with other wavelengths and their recent imaging capabilities, exciting new possibilities are now opening for radio diagnostics.

1 Introduction

A major issue of solar physics in the past three decades is the way magnetic energy in the corona is released. Convective processes and magnetic buoyancy transport ample free magnetic energy into the solar atmosphere. It corresponds to building up electric currents. Now, these currents seem to release their energy not by continuous ohmic heating like a constant resistor in an electric circuit. On the contrary, the release of electromagnetic energy occurs at least partially in extremely impulsive flares converting initially a considerable part of the energy into energetic electrons and ions. Indeed, synchrotron emission of electrons spiraling in the magnetic field, or its mildly relativistic form of gyrosynchrotron emission, is observed in the smallest flares (Fürst et al. 1982, Gary et al. 1997) and even in 10^{19} J micro-events of the quiet sun (Krucker et al. 1997a). Electrons at tens of MeV are observed in large flares by their bremsstrahlung (e.g. Rank et al. 1996) and by millimeter (Trottet et al. 2002) and sub-millimeter synchrotron emission (Kaufmann 2002, discussion by Luethi et al. in Kliem et al. 2002).

When a plasma is shaken by a major energy input such as a flare, oscillations in all eigenmodes may be expected. At the highest frequencies, eigenmodes include oscillations of electrons versus ions at the plasma frequency and electron oscillations at the electron cyclotron frequency. These highest eigenmodes can generally be excited by non-thermal electrons and emit coherent electromagnetic emissions in the radio range. Coherent means here that the emission is not radiated by individual particles, but in joint action (i.e. phase coherent) organized by a wave in the plasma.

Considering that flares generally accelerate non-thermal electrons that could emit coherent emissions, one may expect that all flares are accompanied by coherent radio emissions. The reality is different and more complex: *(i)* Simnett and Benz (1986) have pointed out that some 15% of large flares do not show radio emission in meter and decimeter wavelengths. This percentage needs to be revised using modern broadband spectra that cover also the range above 1 GHz. Nevertheless, the fact remains that the coherent radiation is sometimes much weaker than expected. *(ii)* Coherent radio emission appears in many different spectral forms varying enormously in bandwidth, polarization, fluctuation, duration and frequency drift. The origin of these widely different emissions does not appear to be identical and several emission mechanisms seem to be at work. It is thus not immediately clear which emission originates from the main acceleration region, refers to escaping or trapped electrons, or is produced by some secondary shocks.

2 Why Decimetric Radio Bursts?

The first radio observations of the sun and its flares have been made at meter wavelengths. This is the range where bursts have been classified from type I to V. All of them are now generally assumed to be coherent emissions. If coherent emission is emitted at the plasma frequency, the wavelength of one meter (300 MHz) corresponds to a density of 3×10^9 cm^{-3}, as the plasma frequency

$$\omega_p = \sqrt{\frac{e^2 n_e}{\epsilon_0 m_e}} = 2\pi\, 90 \sqrt{\frac{n_e}{10^8 \mathrm{cm}^{-3}}}\ [\mathrm{MHz}], \qquad (1)$$

where e is the elementary charge, m_e the electron mass and n_e the electron density. To reach the number of accelerated electrons observed in major flares (up to 10^{38}), particles in an enormous volume would have to be accelerated. Similarly, meterwave coherent emission at the electron gyrofrequency would suggest a magnetic field of less than about 100 Gauss, as the electron cyclotron frequency

$$\Omega_e = \frac{eB}{m_e} = 2\pi\, 280 \left(\frac{B}{100\ \mathrm{G}} \right) \quad [\mathrm{MHz}], \tag{2}$$

where B is the magnetic field strength. Thus for a flare energy of some 10^{25} J, an equally questionable flare volume would be necessary. If the coherent emissions were at harmonics of these eigenmode frequencies, the required volumes would be even bigger. For these reasons the meter wave radio bursts ($\nu < 300$ MHz) cannot be direct emissions of the energy release process of major flares.

Of course, the above arguments do not exclude energy release at lower densities, magnetic fields and thus higher altitudes. On the contrary, the better transparency of the high corona to radio waves emitted at the local plasma frequency and their direct interplanetary influence make high-altitude events particularly interesting. However, they release only little energy that is often not visible in HXR and Hα (Kane 1981), or is secondary to the main energy release (Hudson et al. 2001, Benz et al. 2002).

2.1 Observations of Coherent Emissions in the Decimeter Range

For a long time, decimetric emissions were the least studied radio phenomena of solar flares. They appear more diverse than their counterparts in the meter range, noted for its five relatively distinct types of bursts and the centimeter (or microwave) synchrotron emissions dominating at frequencies $\gtrsim 3$ GHz. Many decimetric types of bursts have been reported and classified in the literature. The first observers with broadband analog spectrographs already reported complex and unresolved features (Young et al. 1961). Differences between decimetric and metric bursts were remarked very early (Kundu et al. 1961). Nevertheless, the classification of the metric bursts has generally been applied to the decimeter waves whenever possible. For the other bursts, in particular broadband pulsating structures, narrowband spikes, patches of continuum emission as well as unresolved events, the abbreviation DCIM has been used by some observers.

As long as there is no definite way to relate burst types to emission processes, any classification into types and subtypes remains an artificial and accidental task. Nevertheless, ordering the decimetric bursts by similarity of their shape in the dynamic spectrum is necessary for theoretical work. It has become meaningful with sufficient resolution by digital spectrometers in the 1980s and 1990s. The first such extended surveys and classifications were made by Güdel and Benz (0.3 − 1.0 GHz, 1988), Isliker and Benz (1.0 − 3.0 GHz, 1994), Bruggmann et al. (6.5 − 8.5 GHz, 1990). The usual criteria taken into account were: *bandwidth, duration, drift rate, substructures, impulsiveness, order*. Well over 90% of the events can be assigned to 5 classes:

1. Decimetric type III bursts (fast drift bursts): Many decimetric emissions are shaped similar to metric type IIIs: Short duration (about 0.5 - 1.0 s), impulsive onset, high drift rates (usually >100 MHz/s), groups of some tens to hundreds. Reverse drift bursts are as common as normal drift bursts and dominate above about 1 GHz. Type III bursts are generally interpreted by electron beams interacting with the ambient coronal plasma to excite a bump-on-tail instability of Langmuir waves. The electromagnetic emission is generally assumed to be produced at the local plasma frequency or its harmonic. Type III bursts then are a diagnostic of the electron density of the plasma traversed by the beam.

2. Decimetric type IV events are continua of many minutes duration occurring in the 0.3–5 GHz range. The emission is usually modulated in time on scales of 10 s or less, and is often strongly polarized. Often, the emission carries fine structures in frequency and time (Bernold 1980). The phenomena are generally interpreted by electrons trapped in loop-shaped magnetic fields (Stepanov 1974, Kuijpers 1975).

3. Diffuse continua occur most frequently in the 1–3 GHz range, have various forms and sometimes drift in frequency. Their characteristic duration is between one and some tens of seconds, too long for a type III burst and too short for a type IV burst. They have also been noted in the 0.3–1 GHz range and have been called 'patches' in the literature. The circular polarization is usually weak. Continua have been interpreted by continuous injection of electrons (Bruggmann et al. 1990) or by proton beams destabilized by impacting the transition layer (Smith & Benz 1991).

4. Pulsations are broadband emissions (several 100 MHz) with periodic or irregular short fluctuations. They are sometimes quasi-periodic with pulses of 0.1 to 1 second separations, occurring in groups of some tens to hundreds and lasting some seconds to minutes. The drift rates exceed the type III bursts by at least a factor of 3. Also different from type III bursts are the well defined upper and lower boundaries in frequency. There are significant differences in modulation depths of pulsations, and some gradually drift to lower frequency in the course of the event. The emission has been interpreted by a loss-cone instability of trapped electrons (Aschwanden and Benz 1988), possibly in a reconnection geometry (Kliem et al. 2000).

5. Spikes of narrowband emissions lasting only a few tens of microseconds have been reviewed by Benz (1986). Individual spikes are very short (<0.1 sec), extremely narrowband (some MHz) intense emissions forming broadband clusters of some tens to ten thousands during some seconds to about a minute. Clusters are sometimes organized in small subgroups or chains. In some cases, harmonic structure is present. Metric spikes at 250 – 500 MHz form a different class of spike bursts, common near the starting frequency of metric type III bursts.

Figure 1 gives a simplified overview on the characteristic ranges of duration and bandwidth of the five classes. Note that there are more burst characteristics and that this scheme should not be used blindly. The burst duration usually decreases with frequency. There are probably more burst types in the unclassified

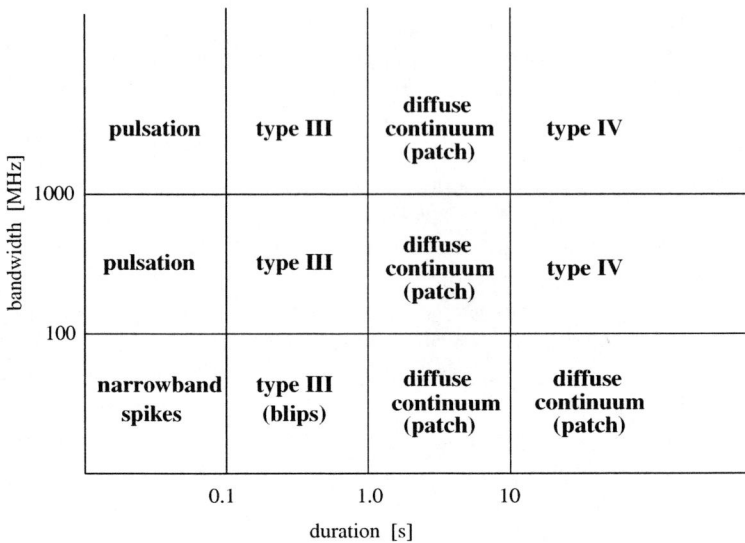

Fig. 1. Schematic overview on the five classes of 1–3 GHz bursts classified by duration and bandwidth of single bursts. The borders are not to be taken too literally since other characteristics must also be considered as described in the text (after Isliker & Benz 1994).

decimetric bursts, although some may be just borderline cases. A few peculiar events have been noted which did not fit at all into any of these classes, such as the drifting type II-like events (Benz 1976, Allaart et al. 1990) and "lace" bursts with rapid frequency variations (Karlický et al. 2001). Nevertheless, the diagram may be used to tentatively classify 0.3 – 8 GHz single frequency observations in the lack of spectrometric information.

In addition to the above emissions, the featureless, broadband gyrosynchrotron emission occurs in the decimetric band. It is well known from its microwave and even millimetric radiations (Tanaka 1961). Synchrotron emission has been shown to extend below 1 GHz in some flares (e.g. Batchelor et al. 1984). More information on decimeter wave observations can be found in Benz (2002) and in the review by Bastian et al. (1998).

2.2 Theories of Coherent Emission

Several processes have been proposed that can convert waves in a plasma into propagating electromagnetic emission. Scattering on ions, wave decay, sharp density gradients as well as emissions by non-linear waves such as solitons have been proposed (cf. summary in Benz 2002). In the solar atmosphere and where $\Omega_e \ll \omega_p$, the standard emission process seems to be wave-wave coupling, also termed *wave coalescence*,

$$\omega_1 + \omega_2 = \omega_3 \tag{3}$$

$$\mathbf{k_1} + \mathbf{k_2} = \mathbf{k_3} \tag{4}$$

The system of equations, often referred to as the parametric equations, expresses the conditions for effective transfer of wave energy between wave 1 (ω_1 and $\mathbf{k_1}$) plus wave 2 into a third wave that may be the electromagnetic wave observed as radio emission. Wave coalescence is the best confirmed process, corroborated by in situ observations in the interplanetary medium, where wave 1 was identified as a Langmuir wave and wave 2 either an ion acoustic wave (fundamental plasma radiation) or an oppositely directed second Langmuir wave (harmonic) (e.g. Lin et al. 1986). For the conversion rates the reader is referred to the monograph by Melrose (1980).

If $\Omega_e \gtrsim \omega_p$, an instability, called electron cyclotron maser emission, exists that converts free energy in a velocity distribution increasing in perpendicular velocity into electromagnetic emission at the electron gyrofrequency and its harmonics (Wu & Lee, 1979). As it can escape directly from the corona as radio emission, the maser process is extremely efficient. It is the generally accepted emission process for the terrestrial kilometric radiation and Jupiter's decametric radiation. For the sun, however, it has come out of fashion, as the above requirement is beyond all magnetic fields inferred up to now.

Of particular interest are waves that are predicted by existing acceleration models. For reconnection models requiring anomalous resistivity, some current instability is usually invoked. The low-frequency waves excited by currents are mostly of the ion acoustic or lower hybrid type. They may emit radio emission by the coalescence process described in Eqs. (3) and (4) if accelerated electrons provide the Langmuir waves. Such models have been proposed by Benz & Wentzel (1981) and Benz & Smith (1987), respectively, to interpret narrowband radio emissions.

A very popular acceleration model for solar flares is transit-time-damping (Miller et al. 1997, Schlickeiser 2002). The waves assumed in this model are even lower in frequency and are unlikely to couple into radio waves. Other models, in particular those based on electron beams producing waves to accelerate ions (e.g. Temerin & Roth 1992), are prone to predict intense radio emission. Whether coherent radio emission is emitted or not, is therefore an observable criterion for the validity of acceleration models.

3 Narrowband Bursts

Narrowband radio bursts must be coherent and can only be emitted from localized sources, as the relevant frequency, either the plasma frequency or the electron gyrofrequency, cannot have variations more than the observed bandwidth, $\Delta\nu$. One can express this in the form

$$\Delta\nu \lesssim \nu \frac{\Delta s}{H_f} \ , \tag{5}$$

where Δs is the size of the source and H_f is the scale length of the relevant frequency.

Narrowband bursts require very small source sizes. Therefore, they cannot be produced by propagating beams that tend to disperse in space and cause relatively large and thus broadband sources.

3.1 Metric Bursts

The narrowband type I bursts have been proposed to be signatures of acceleration events by Benz & Wentzel (1981). Indeed, type I burst activity has been observed to set in simultaneously with an enhancement of soft X-rays. An initial report by Lantos et al. (1981) has been confirmed by many more examples (Raulin & Klein 1994, Fig. 2). The observations can be interpreted as energy release in the high corona which accelerates particles and emits the type I bursts. Accelerated particles propagate along field lines to the chromosphere and heat it locally to coronal temperatures. The newly heated plasma emits enhanced soft X-rays. Krucker et al. (1995a) have found the type I continuum located on a loop with enhanced soft X-ray emission. In their example the bursts were separated by about 100" from the continuum and not located on a bright loop.

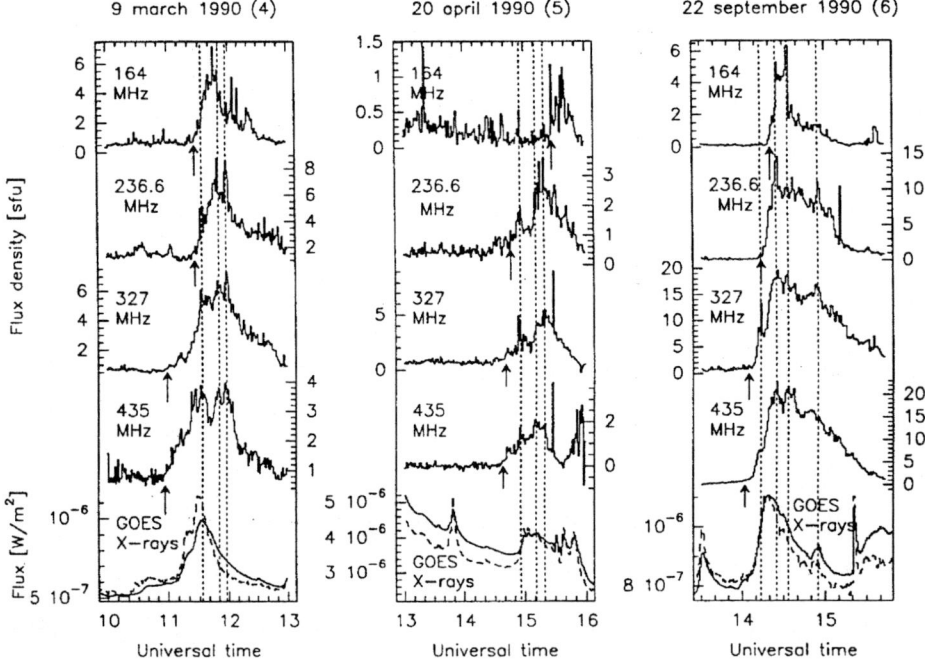

Fig. 2. Evolution of the strength of noise storms in solar flux units observed by the Nançay Radio Heliograph compared with the full sun soft X-ray flux recorded by GOES in the $1-8$ Å channel (full curve) and $0.1-4$ Å channel (dashed)(from Raulin & Klein 1994).

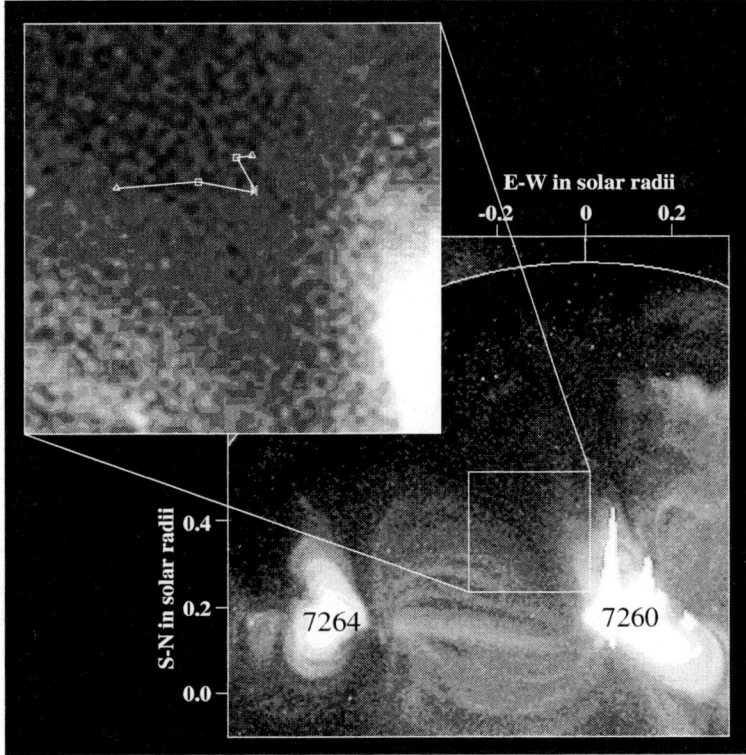

Fig. 3. Type III and metric spike sources overlaid on a *Yohkoh*/SXT image (insert). The centroid positions of the sources at different frequencies were observed in Nançay Radioheliograph and are connected by lines. The two lines originate from the same location at the highest frequency, where the ETH Zurich radio spectrometer has observed metric spikes (from Paesold et al. 2001).

More information on the acceleration process can be obtained from narrowband metric spikes. They are found in 10% of all meterwave type III groups near but slightly above the start frequency of the type III bursts (Benz et al. 1982). Small clusters of metric spikes correlate in time with individual type III bursts (Benz et al. 1996) thus with the acceleration of electron beams. The spikes are concentrated in the spectrogram on the extrapolation of the type III bursts to higher frequency. More precisely, Fig. 3 suggests that metric spikes are located on the extension of type III trajectories, supporting a model of energy release in or close to the spike sources. Thus for the first time, radio observations can approximately locate the energy release. Modeling coronal densities (Paesold et al. 2001) and spatially resolved observations of metric spike events (Krucker et al. 1995b; 1997b) put the sources at altitudes of 2×10^{10} cm and more. Not surprisingly, their associated soft X-ray emission, if any, is delayed by some tens of seconds (Krucker et al. 1997b).

Metric spikes have been found to be associated with impulsive electron events in the interplanetary medium (Benz et al. 2001a). The low energy cut off of the observed electron distribution defines an upper limit of the density in the acceleration region. The derived electron density is of the order of 3×10^9 cm^{-3}, consistent with the density in the source of metric spikes, assuming second harmonic plasma emission.

The similarity of spikes and type I bursts has been emphasized previously (Benz 1986, Klein 1995). Both noise storms and metric spike/type III bursts are not associated with regular, HXR and centimeter-wave emitting flares. The short duration and the narrow bandwidth suggest a small source size and therefore a high radio brightness temperature (order of 10^{15} K). Only a coherent mechanism can account for the emission, but none of the published mechanisms is generally accepted. Nevertheless, the evidence that short narrowband radio bursts in meter wavelengths are signatures of the acceleration process has steadily increased over the past decade.

3.2 Decimetric Spikes

The situation is much less clear for narrowband spikes at *decimeter* wavelength. The term 'narrowband, millisecond spikes' has been introduced for them referring to narrowband (few percent of the center frequency) and short (few tens of ms) peaks above 1 GHz (Droege 1977). Originally, they were reported to occur in the rise phase of centimeter radio bursts and thus to be associated with major flares (Slottje 1978). The association rate with HXR flares is high (Benz & Kane 1986; Güdel et al. 1991). However, a delay of spike groups relative to the HXR peaks of the order of a few seconds has been noted (Aschwanden & Güdel 1992). Moreover, spikes have been discovered also in decimeter type IV bursts occurring after the HXR emitting phase of flares (Isliker & Benz 1994). Contrary to their relatives at meter waves, decimetric spikes do not correlate with type III bursts and are extremely rare (2% of all HXR flares, Güdel et al. 1991).

The emission process of decimetric spikes is highly controversial. Originally, the loss-cone instability of trapped electrons has been proposed to produce electron cyclotron maser emission at the footpoints of flare loops (Holman et al. 1980; Melrose & Dulk 1982). To avoid the assumption of high magnetic field strength in the source, the model has been changed to emission of upper-hybrid and Bernstein modes (Willes & Robinson 1996). The scheme can interpret occasional harmonic emission in decimeter spikes (Benz & Güdel 1987). Alternatively, Tajima et al. (1990) and Güdel & Wentzel (1993) proposed the spike sources to be in the acceleration regions of flares and to result from waves produced by the acceleration process.

Benz et al. (2002) have recently located clusters of narrowband decimetric radio spikes that occur in the decay phase of solar flares at low decimeter frequencies (327-430 MHz). Contrary to previous observational claims and leading theoretical expectations, most of the observed spike clusters reported occured well outside the main energy release region of the flare and not at the feet of magnetic loops (Fig.4). Instead, the observations suggest that the electrons for

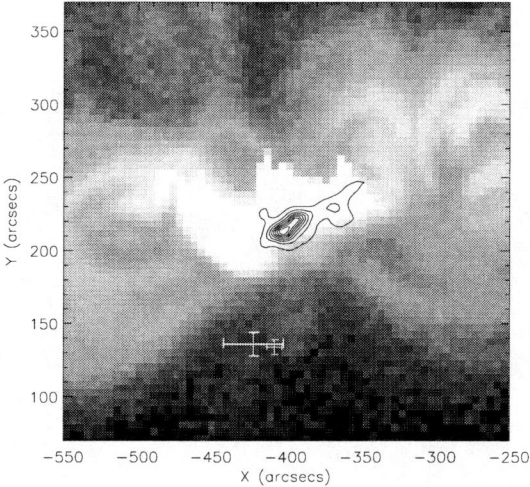

Fig. 4. Location of two clusters of decimeteric spikes at 432 MHz (error bars) in the course of the X1.9 event on 2000/07/12 superposed on an SXT/*Yohkoh* image. The hard X-ray intensity as observed by HXT/*Yohkoh* (M1 channel) is displayed by isophotes (from Benz et al. 2002).

these radio bursts are accelerated at secondary sites high in magnetic fields adjacent to the main flare site, as these fields adjust in response to the flare. In at least two cases the spikes are near loop tops. Similarly, reversed drifting type III bursts at frequencies above a spike cluster, suggesting electron beams propagating downward from the spike source, were found in high sensitivity observations (Benz et al. 2001b).

The role of spike emission in the decimeter range is not settled. It is still possible that they reveal a property of acceleration in general. However, if decimetric spikes do represent the main acceleration site as previously assumed, this could be only at high frequency.

4 Decimetric Emissions During Major X-Ray Events

We may invert the question now and ask what is observed during major HXR flares in the decimeter range. The two wavelengths must be compared exactly in time and the spatial location of the emissions should be close. This has been done only partially and very recently in a few cases.

4.1 Drifting Pulsating Structures

Much more often than narrowband spikes, decimetric type III bursts are associated with some decimetric continuum. They are usually pulsating at various

Phoenix-2, total flux

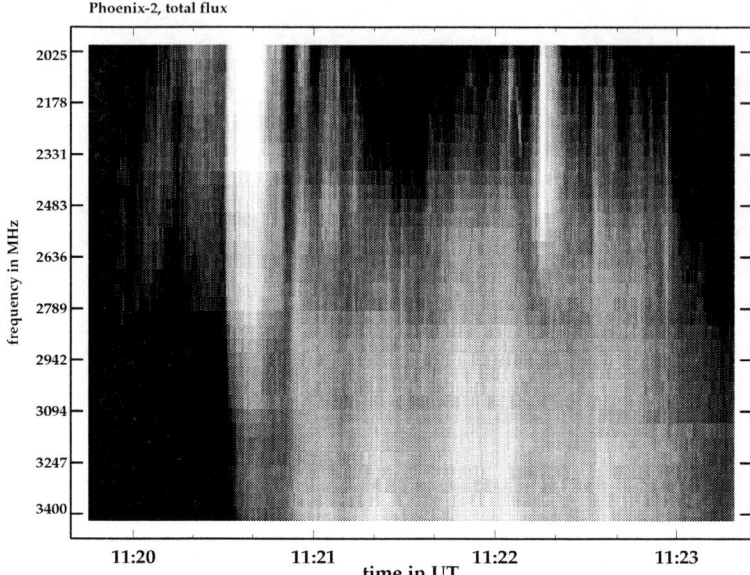

Fig. 5. Dynamic spectrogram of 2–4 GHz total flux observed the Phoenix-2 radio spectrometer on 2001/03/28. Enhanced emission is shown bright. In the upper part of the image, a "noisy" continuum is seen, extending to lower frequencies. In the lower right corner, the low-frequency part of gyrosynchrotron emission is visible.

degrees of modulation and sometimes drifting in frequency. Figure 5 shows an example together with gyrosynchrotron emission visible at high frequencies above about 2500 MHz and reported up to 15 GHz in Solar Geophysical Data. The gyrosynchrotron emission peaks at 11:21:02 UT, about 20 seconds after the first decimeter emission in the upper part of the image below 3000 MHz. The sense of circular polarization of the two emissions is opposite.

The decimeter continuum in Fig. 5 has a non-thermal spectrum that is too narrow for gyrosynchrotron emission. Thus a coherent process has probably emitted it. The emission is pulsating irregularly and does not have a smooth spectrum, thus it has a "noisy" appearance. Two minutes later, a burst with similar spectral shape, but more pronounced pulsations occurred.

Kliem et al. (2000) have studied a deeply modulated drifting pulsating structure in the 0.6–2 GHz range before the HXR peak. In the main HXR phase, the intensity increases and the pulsations become less regular. The decimeteric pulsations drift slowly to lower frequency during the event. The authors propose a model in which the pulsations of the radio flux are caused by quasi-periodic particle acceleration episodes that result from magnetic reconnection in a large-scale current sheet. Under these circumstances, reconnection is dominated by repeated formation and subsequent coalescence of magnetic islands. The process is known as secondary tearing or impulsive bursty regime of reconnection. Such a model, involving a current sheet and a growing plasmoid, is consistent with the

Yohkoh/SXT imaging observations of the same flare (Ohyama & Shibata 1998). The unified explanation of plasmoid formation and pulsating structure suggests that the current sheet in the considered flare was not formed by the ejection of the plasmoid (contrary to the often favored view of eruptive flare processes), but existed before and was the cause of the event.

The suggestion of a plasmoid does not mean that most of the electrons are contained. A large fraction of the accelerated particles may only temporarily be trapped in the plasmoid, escape and produce the HXR bremsstrahlung by precipitation. Nevertheless, the acceleration process itself may form an anisotropic velocity distribution unstable against electrostatic or electromagnetic waves that may cause the emission.

5 Type III Bursts

Type III bursts are drifting in frequency as time progresses. They are generally interpreted as the signature of electron beams propagating through the corona and interplanetary medium. As the beams excite plasma waves at the local plasma frequency, the frequency changes with density. Type III bursts thus trace the path of the beam from near the acceleration site toward the final destination of the electrons as long as the beam is capable to excite radio emission.

5.1 Evidence for Reconnection

Bidirectional type III bursts have been detected by Aschwanden et al. (1995) in radio spectra. The acceleration site was concluded to be at a plasma frequency of about 300 - 500 MHz. Thus a density in the acceleration site of a few times 3×10^8 to 3×10^9 cm^{-3} was inferred. Robinson & Benz (2000) interpreted the general weakness and slower drift of the downgoing branch by the combination of beam properties and magnetic geometry.

Imaging observations have shown that the type III sources often do not emerge single. Klein et al. (1997) have reported that down-propagating branches of type III bursts are sometimes double sources. Their simultaneous existence suggests a common origin. Paesold et al. (2001) have found double type III sources to diverge from the same spike source (Fig. 3). These findings support the hypothesis that narrowband metric spikes are closely related to the acceleration region. The multiplicity of the beam paths, on the other hand, is consistent with the predictions of the reconnection scenario for magnetic energy release where magnetic field lines from different directions meet closely.

5.2 Decimetric Type III

Decimeter type III bursts have durations of a few tens of milliseconds (Benz et al. 1983) and also occur in the HXR main phase. Thus they may be confused with narrowband spikes or pulsations if the spectrum is not known. Type III bursts at

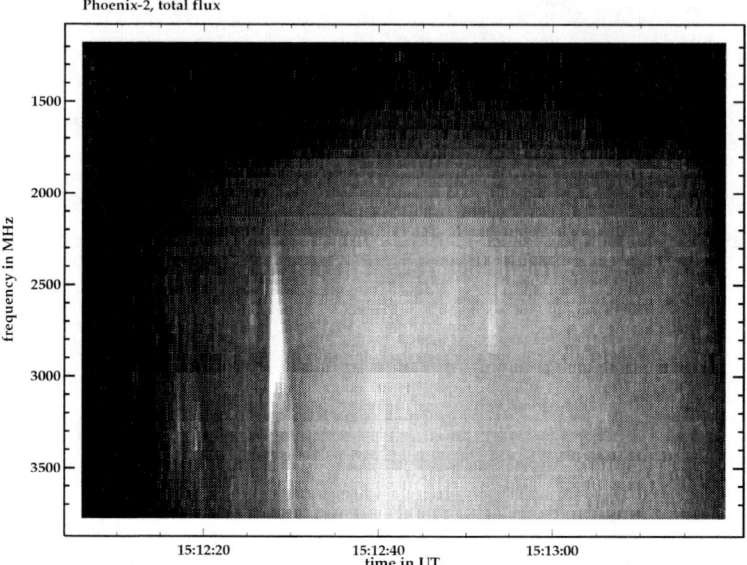

Phoenix-2, total flux

Fig. 6. Dynamic spectrogram of 2–4 GHz total flux observed by the Phoenix-2 radio spectrometer on 2001/09/09. Enhanced emission is shown bright. An intensive type III burst (U shape) ocurred at 15:12:28 UT. In the center of the image and below, the low-frequency part of the diffuse gyrosynchrotron emission is visible.

meter waves, 0.3–1 GHz (Benz et al. 1983), 3–5 GHz (Stähli & Benz 1987), and 6–8 GHz (Benz et al. 1992) have similarly low degrees of circular polarization, as well as frequency drifts and durations that scale with center frequency. Thus they are generally assumed to be produced by the same emission process. The total bandwidth at high frequency is often as low as a few 100 MHz. It seems to reflect the smaller scale of the magnetic field geometry at lower altitude.

It is interesting to compare gyrosynchrotron emission produced by high-energy electrons in the main phase of major flares with type III bursts in the decimeter range originating from electron beams. Figure 6 displays the association of a type III burst and gyrosynchrotron emission. Type III bursts are located in the spectrum below the peak of the gyrosynchrotron radiation. They occur often but not exclusively during the early phase of the gyrosynchrotron emission.

The appearance of high-frequency type III bursts is surprising for two reasons: *(i)* The escape of radio emission from high density plasma is much reduced by free-free absorption. Even the second harmonic cannot escape from the 7 GHz level in a spherically symmetric atmosphere at coronal temperatures. *(ii)* The densities for such plasma emission is at least 10^{11} cm^{-3}. If coronal, such densities are astonishing considering the early flare phase to which they refer. Whether the bursts occur in the transition region and what this would imply, warrants further studies.

It must be noted here that there is no quantitative theory yet of coronal type III bursts. Type III bursts appear under such different conditions than in interplanetary space that accepted theories may not simply be applied to coronal counterparts.

6 Conclusions

The evidence increases that some of the coherent radio emissions in the short wavelength meter radiation is emitted near or from the acceleration region. Although there are common emissions of decimeter waves during the main HXR phase of major flares, its origin is unclear. They are drifting, broadband pulsations rather than spikes. Very little is firmly known about the emission process. There are many proposals, but confirmation is difficult.

Even if this first step is not yet achieved, some conclusions on the acceleration physics can be inferred. In particular, proposed acceleration processes can be tested on their general predictions for radio emission. Mechanisms that necessarily produce excessive coherent radio emission can be excluded. An example is electron acceleration in the form of beams or electron beams as sources of low frequency waves. Such beams are unstable and would produce intense coherent radio emission in all cases, which is not observed. More direct inferences may become possible when the emission mechanisms of narrowband spikes and broadband drifting pulsations will be known.

Indirect emissions also can contribute to understanding acceleration. Type III bursts have been used to locate the energy release site of small flares in the high corona. Their role in major flares is not clear.

Imaging observations of coherent radio emissions will allow in the near future to put them into the context of the HXR emission and the thermal coronal plasma as outlined in EUV lines and soft X-rays. The location of the source relative to coronal loops allows testing the predictions of some emission models. Thus it will soon become possible to distinguish between emissions by trapped electrons from radiations originating near or at the acceleration site. The existing EIT and TRACE observations and the RHESSI mission are promising ingredients for a new era of research in coherent radio emission.

References

1. M.A.F. Allaart, J. van Nieuwkoop, C. Slottje, L.H. Sondaar: Solar Phys. **130**, 183 (1990)
2. M.J. Aschwanden, A.O. Benz: Astrophys. J. **111**, 113 (1988)
3. M.J. Aschwanden, M. Güdel: Astrophys. J. **401**, 736 (1992)
4. M.J. Aschwanden, A.O. Benz, R.A. Schwartz: Astrophys. J. **417**, 790 (1995)
5. T.S. Bastian, A.O. Benz, D.E. Gary: Ann. Rev. Astr. Astrophys. **36**, 131 (1998)
6. D.A. Batchelor, A.O. Benz, H.J. Wiehl: Astrophys. J. **280**, 879 (1984)
7. A.O. Benz: Kleinheubacher Berichte **20**, 376 (1976)

8. A.O. Benz: Solar Phys. **104**, 99 (1986)
9. A.O. Benz: *Plasma Astrophysics: Kinetic Processes in Solar and Stellar Coronae,* *2nd edition* (Kluwer, Dordrecht 2002)
10. A.O. Benz, D.G. Wentzel: Astron. Astrophys. **94**, 100 (1981)
11. A.O. Benz, M. Jaeggi, P. Zlobec: Astron. Astrophys. **109**, 305 (1982)
12. A.O. Benz, T.E.X. Bernold, B.R. Dennis: Astrophys. J. **271**, 355 (1983)
13. A.O. Benz, S.R. Kane: Solar Phys. **104**, 179 (1986)
14. A.O. Benz, M. Güdel: Solar Phys. **111**, 175 (1987)
15. A.O. Benz, D.F. Smith: Solar Phys. **107**, 299 (1987)
16. A.O. Benz, A. Magun, W. Stehling, H. Su: Solar Phys. **141**, 335 (1992)
17. A.O. Benz, A. Csillaghy, M.J. Aschwanden: Astron. Astrophys. **309**, 291 (1996)
18. A.O. Benz, R.P. Lin, O.A. Sheiner, S. Krucker, J. Fainberg: Solar Phys. **203**, 131 (2001a)
19. A.O. Benz, P. Messmer, C. Monstein: Astron. Astrophys. **366**, 326 (2001b)
20. A.O. Benz, P. Saint-Hilaire, N. Vilmer: Astron. Astrophys. **383**, 678 (2002)
21. T. Bernold: Astron. Astrophys. **42**, 43 (1980)
22. G. Bruggmann, A.O. Benz, A. Magun, W. Stehling: Astron. Astrophys. **240**, 506 (1990)
23. F. Droege: Astron. Astrophys. **57**, 285 (1977)
24. E. Fürst, A.O. Benz, W. Hirth: Astron. Astrophys. **107**, 178 (1982)
25. D.E. Gary, M.D. Hartl, T. Shimizu: Astrophys. J. **477**, 958 (1997)
26. M. Güdel, A.O. Benz: Astron. Astrophys. **75**, 243 (1988)
27. M. Güdel, M.J. Aschwanden, A.O. Benz: Astron. Astrophys. **251**, 285 (1991)
28. M. Güdel, D.G. Wentzel: Astrophys. J. **415**, 750 (1993)
29. G.D. Holman, D. Eichler, M.R. Kundu: In: *Radio Physics of the Sun*, Kundu M.R. and Gergeley T.E. (eds.), IAU Symp., Vol. **86**, 457 (1980)
30. H. Hudson, T. Kosugi, N. V. Nitta, M. Shimojo: Astrophys. J. **561**, L211 (2001)
31. H. Isliker, A.O. Benz: Astron. Astrophys. Sup. Ser. **104**, 145 (1994)
32. S.R. Kane: Astrophys. J. **247**, 1113 (1981)
33. M. Karlický et al.: Astron. Astrophys. **375**, 638 (2001)
34. P. Kaufmann: 2002, *this volume*
35. K.-L. Klein: In: *Coronal Magnetic Energy Releases*, Benz A.O. and Krüger A. (eds.), Lect. Notes in Phys., Vol. **444**, 55 (1995)
36. K.-L. Klein, H. Aurass, I. Soru-Escaut, B. Kalman: Astron. Astrophys. **320**, 612 (1997)
37. B. Kliem, M. Karlický, A.O. Benz: Astron. Astrophys. **360**, 715 (2000)
38. B. Kliem, A. MacKinnon, G. Trottet, T. Bastian: 2002, *this volume*
39. S. Krucker, A.O. Benz, M.J. Aschwanden, T.S. Bastian: Solar Phys. **160**, 151 (1995a)
40. S. Krucker, M.J. Aschwanden, T.S. Bastian, A.O. Benz: Astron. Astrophys. **302**, 551 (1995b)
41. S. Krucker, A.O. Benz, L.W. Acton, T.S. Bastian: Astrophys. J. **488**, 499 (1997a)
42. S. Krucker, A.O. Benz, M.J. Aschwanden: Astron. Astrophys. **317**, 569 (1997b)
43. J. Kuijpers: Solar Phys. **44**, 173 (1975)
44. M.R. Kundu, J.A. Roberts, C.L. Spencer, J.W. Kniper: Astrophys. J. **133**, 255 (1961)
45. P. Lantos, A. Kerdraon, G.G. Rapley, R.D. Bentley: Astron. Astrophys. **101**, 33 (1981)
46. R.P. Lin, W.K. Levedahl, W. Lotko, D.A. Gurnett, F.L. Scarf: Astrophys. J. **308**, 954 (1986)

47. D.B. Melrose: *Plasma Astrophysics: Non-thermal Processes in Diffuse Magnetized Plasmas*, v.2, (Gordon & Breach, New York 1980)
48. D.B. Melrose, G.A. Dulk: Astrophys. J. 447, **844** (1982)
49. J.A. Miller et al.: J. Geophys. Res. **491**, 14631 (1997)
50. M. Ohyama, K. Shibata: Astrophys. J. **499**, 934 (1998)
51. G. Paesold, A.O., Benz, K.-L. Klein, N. Vilmer: Astron. Astrophys. **371**, 333 (2001)
52. G. Rank et al.: In: *High Energy Solar Physics*, ed. by R. Ramaty et al. (AIP, Woodbury 1996) pp. 219–236
53. J.P. Raulin, K.-L. Klein: Astron. Astrophys. **281**, 536 (1994)
54. P.A. Robinson, A.O. Benz: Solar Phys. **194**, 345 (2000)
55. R. Schlickeiser: 2002, *this volume*
56. G.M. Simnett, A.O. Benz: Astron. Astrophys. **165**, 227 (1986)
57. C. Slottje: Nature **275**, 520 (1978)
58. D.F. Smith, A.O. Benz: Solar Phys. **131**, 352 (1991)
59. M. Staehli, A.O. Benz: Astron. Astrophys. **175**, 271 (1987)
60. A.V. Stepanov: Soviet Astronomy AJ **17**, 781 (1974)
61. T. Tajima, A.O. Benz, M. Thaker, J.N. Leboeuf: Astrophys. J. **353**, 666 (1990)
62. H. Tanaka: Proc. Res. Inst. Atmosph. Nagoya Univ. Japan **8**, 1 (1961)
63. M. Temerin, I. Roth: Astrophys. J. **391**, L105 (1992)
64. G. Trottet, J.-P. Raulin, P. Kaufmann, M. Siarkowski, K.-L. Klein, D. E. Gary: Astron. Astrophys. **381**, 694 (2002)
65. A.J. Willes, P.A. Robinson: Astrophys. J. **467**, 465 (1996)
66. C.S. Wu, L.C. Lee: Astrophys. J. **230**, 621 (1979)
67. C.W. Young, C.L. Spencer, G.E. Moreton, J.A. Roberts: Astrophys. J. **133**, 243 (1961)

Signature of Energy Release and Particle Acceleration Observed by the Nobeyama Radioheliograph

Kiyoto Shibasaki

Nobeyama Radio Observatory, Minamimaki, Minamisaku, Nagano 384-1305 Japan

Abstract. Microwave imaging observations of solar flares are presented and a new scenario for solar flares is proposed. Microwaves are effectively emitted by high-energy electrons gyrating in active region magnetic fields. Higher harmonics (10 - 100) of the gyro-frequency in active regions, excited by mildly relativistic electrons, correspond to microwaves. Imaging observations of strong microwave emission associated with solar flares make it possible to study where and how the high-energy electrons are crreated in solar flares, which is one of the long-standing questions of solar flares. Hot and dense plasma created by solar flares also emits microwaves by the free-free mechanism although usually weak compared to the non-thermal emission. It is shown that flares start in a small loop and also shown that hot plasmas and high-energy electrons are fed into a nearby larger loop from the small one. Based on these and other observations, it is proposed that "high-beta disruption" is the cause of solar flare phenomena.

1 Introduction

Microwaves and Hard X-rays are used for diagnostics of high-energy phenomena in solar flares. From the ground, microwaves are the only way to observe high-energy phenomena on the Sun directly. Most microwave observations of the Sun have been limited to total flux measurements by radiometers and spectrometers. Spatially resolved observations by large radio interferometers make it possible to study the generation and propagation processes of high-energy electrons if the spatial and temporal resolution are high enough. However, probability of detecting solar flares within the telescope field of view during the observing session is very low due to the limitations of the large radio interferometers such as small field of view, data acquisition cadence, inflexible observing time allocation etc. Solar dedicated large interferometers are needed to overcome such limitations. The construction of the Nobeyama Radioheliograph made it possible to overcome these limitations.

In the following section, we will explain, how the radio interferometer works to produce radio images, and the specific capabilities of the Nobeyama Radioheliograph. In Sect. 3, radiation mechanisms of microwaves in the short-cm range (frequency > 8 GHz) from the Sun are summarized. Flare observations by the Nobeyama Radioheliograph are presented in Sect. 4. Based on these observational results, a new solar flare scenario is presented in Sect. 5.

2 Radio Interferometer
and the Nobeyama Radioheliograph

Radio telescopes collect electromagnetic waves using antennas and measure their energy flux by receivers. In the microwave range, antennas consist of parabolic reflectors and horns to collect and introduce electromagnetic waves into receivers. With a single dish antenna, we can measure spatially integrated total power. If the antenna has a polarizer, we can measure polarized components in the received electromagnetic wave. For solar radio emission, polarization is mainly circular (right-handed and left-handed). To get spectral information of the intensity and/or polarization, sweeping-frequency or many fixed-frequency receivers are needed.

For imaging observations, we need good spatial resolution. To realize good spatial resolution, we need a large single dish antenna or a radio interferometer consisting of many antennas. In the case of the large single dish antenna, it is necessary to scan the target by physically moving the antenna. Hence, the temporal resolution is low. This method is not good for observations of quickly changing radio sources such as flare related radio emissions. In the case of the radio interferometer, there are two types: one is a beam-forming type and the other is a multi-correlator type. Due to the advancement of electronics and computer technologies, the multi-correlator type is mainly used for large-scale radio interferometers. The beam size of each antenna element covers a wide field of view, such as the full solar disk, and the combinations of antennas measure spatial Fourier components (or "visibilities") of the radio intensity distribution within the field of view [12]. We call the Fourier space as the uv-plane. The coordinates in the uv-plane are defined by the baseline vectors of pairs of antennas normalized by the observing wavelength. Radio images are synthesized by applying the inverse Fourier transformation to the visibility data in the uv-plane. Image restoration such as "CLEAN" is applied to minimize large side-lobes due to incomplete sampling of the uv-plane. The distribution of sampled visibilities in the uv-plane and also the weighting of these visibilities determine the size of the synthesized beam (or spatial resolution).

The Nobeyama Radioheliograph (NoRH) is a multi-correlator type radio interferometer dedicated for solar observations [4]. It has been operating at 17 GHz since 1992 and at 17 / 34 GHz since 1995. Imaging capability of the instrument is as follows: Spatial resolution is 10 / 5 arc seconds at 17 / 34 GHz. Observing hour is from 23 UT until 06 UT. Temporal resolution is 1 second for normal observations and 0.1 second for flare observations. Both right-handed and left-handed circularly polarized emission are measured at 17 GHz. At 34 GHz, only intensity is measured. The normal synthesis method, which we take, is without any weighting in the uv-plane. Due to dense sampling in the central part of the uv-plane determined by the array configuration, the synthesized beam size is about 50 percent larger than the maximum spatial resolution determined by the array length.

3 Microwave Emission from the Sun

Electromagnetic waves are emitted from charged particles with accelerated motion. Electrons are the major concern because of their small mass compared to ions. The accelerations are caused by collisions with ions or by cyclotron motions around the magnetic field. When the electron motion has a thermal origin, the emission is called thermal emission. When the electron population is different from thermal, it is called non-thermal emission. If the energy distribution of electrons is power law, the higher energy tail is enhanced compared to a thermal one. Higher harmonics of the gyration frequency are emitted efficiently in the microwave region from mildly relativistic electrons created by solar flares. Dulk [2] reviewed radio emission mechanisms on the Sun. In the microwave region (short-cm range), the main emission mechanisms are 1) thermal free-free emission, 2) thermal gyro-resonance emission, and 3) non-thermal gyro-synchrotron emission. In the following, we describe how the measured flux is related to the emissivity of charged particles on the Sun and then summarize the characteristics of the emissivity of each emission mechanism based on the review by Dulk [2]. A detailed treatment of the theory of gyro-synchrotoron emission and absorption by an arbitrary distribution of high-energy electrons can be found in the paper by Ramaty [8]. Gaussian cgs units are mainly used in this paper to be consistent with the majority of research papers in solar physics.

3.1 Relation Between Emissivity and Received Power

For each emission mechanism, the emissivity (η) and/or absorption coefficient (κ) of electromagnetic waves are calculated. The ratio between η and κ is called the "source function" (S). In thermal emissions, the source function is equal to the Planck function and we use the Rayleigh-Jeans limit in the microwave range:

$$S = \eta/\kappa = k_B T f^2/c^2, \tag{1}$$

where k_B is the Boltzmann constant, T is the temperature, f is the frequency and c is the light speed. As the emissivity and the absorption coefficients are related (Kirchhoff's law), either the emissivity or the absorption coefficient is enough to determine the emission characteristics. For non-thermal emissions, an effective temperature (T_{eff}) is defined in a manner similar to the thermal case:

$$S = k_B T_{eff} f^2/c^2. \tag{2}$$

In spatially resolved observations, we measure electromagnetic wave energy flux per unit area, unit time, unit frequency and unit solid angle. This wave energy flux is called the "specific intensity" (I) and is expressed by a temperature called the "brightness temperature" (T_b):

$$I = k_B T_b f^2/c^2. \tag{3}$$

The brightness temperature and the effective temperature are related by the "radiative transfer equation":

$$dT_b/d\tau = -T_b + T_{eff} \tag{4}$$

where $d\tau = \kappa d\ell$, τ is the optical depth, and ℓ is the line of sight length. In the special case of an isolated source with constant T_{eff}, such as a solar flare:

$$T_b = T_{eff}[1 - exp(-\tau)]. \tag{5}$$

For $\tau \gg 1$,

$$T_b = T_{eff}, \tag{6}$$

and for $\tau \ll 1$,

$$T_b = T_{eff}\tau = (c^2/(k_B f^2))\eta L, \tag{7}$$

where L is the source dimension along the line of sight. If an optically thin, constant T_{eff} isolated source is located between the background emitting body, such as the quiet sun, and the observer:

$$T_b = T_{eff}[1 - exp(-\tau)] + T_{b0}exp(-\tau) \tag{8}$$

where T_{b0} is the brightness temperature without the isolated source.

The specific intensity integrated over a source solid angle or a limited solid angle (Ω) within the source is called "flux" (F). The unit of the flux is called "solar flux unit" (SFU) (10^{-22} Wm^{-2}Hz^{-1}) and this unit is 10^4 times larger than Jansky (Jy) which is used for non-solar radio astronomy.

$$F = \int I d\Omega = k_B f^2/c^2 \int T_b d\Omega. \tag{9}$$

The circular polarization degree (r_c) is defined as follows:

$$r_c = (T_{b,x} - T_{b,o})/(T_{b,x} + T_{b,o}), \tag{10}$$

where subscripts x and o correspond to x-mode wave and o-mode wave respectively. In the presence of a magnetic field, electromagnetic waves are influenced by the electron gyration around the magnetic field; hence the emissivity and the absorption coefficients are different between x- and o- modes. The x-mode corresponds to right-handed circular polarization (RCP) and the o-mode to left-handed circular polarization (LCP) for N-polarity magnetic field. The sense of rotation of the x-mode corresponds to the sense of electron gyration. Polarity reversal can happen when the wave crosses the quasi-transverse (QT) magnetic field. This phenomenon is often observed when bi-polar sources are located near the solar limb (either East or West limb). The line of sight of the outer source crosses the QT-region of the dipole magnetic field; hence the polarization reversal is expected. For studies of polarization near the solar limb, we have to be careful.

3.2 Thermal Free-Free Emission

Emission from non-flaring plasma is mainly free-free emission except from the sunspot, which is explained in Sect. 3.3. The absorption coefficient of the coronal and the chromospheric plasma is as follows:

$$\kappa = \xi N^2 f^{-2} T^{-3/2} \text{ cm}^{-1}, \tag{11}$$

where ξ is 0.1 for the chromosphere and 0.2 for the corona in the microwave range ($\sim$10 GHz). An integral of the absorption coefficient along the line of sight of the emitting plasma is the optical depth. If the emitting plasma has uniform temperature:

$$\tau = \xi f^{-2} T^{-3/2} \int N^2 d\ell. \tag{12}$$

Line integral of density squared is called the (column) emission measure (EM, cm^{-5}). The optical depth of the hot coronal plasma at 17 GHz is:

$$\tau = 7 \times 10^{-4} EM_{27} T_6^{-3/2}, \tag{13}$$

where EM_{27} is the emission measure in units of 10^{27} cm^{-5} and T_6 is the temperature in units of 10^6 K. At chromospheric temperature, the constant 7 should be replaced by 3.5. Plasmas in active region corona or very hot plasmas (10^7 K) produced in solar flares are mostly optically thin ($\tau \ll 1$) unless the density is larger than 10^{11} cm^{-3} with 10^9 cm thickness. Low temperature and dense plasma such as prominences ($T < 10^4$ K, $N > 10^{11}$ cm^{-3}) are optically thick even at 34 GHz. The brightness temperature of the optically thick source is equal to the temperature of the plasma. The observed brightness temperature might be lower than the actual brightness temperature due to the beam size of the antenna or the array being larger than the source size. The brightness temperature of an optically thick source on the solar disk and that above the solar limb should be the same. Background emission is completely absorbed by the source.

In case of optically thin emission, the brightness temperature is:

$$T_b = T_{b0} + 7 \times 10^2 EM_{27} T_6^{-1/2}, \tag{14}$$

where T_{b0} is the brightness temperature of the background such as the quiet sun. Above the limb, this term can be ignored. At 17 GHz, T_{b0} of the quiet sun is around 10,000 K. This relation is very useful when we analyze thermal emission from hot plasma produced by solar flares especially above the limb. The temperature dependence of the brightness temperature is rather weak and in the opposite sense: $T_b \propto T^{-1/2}$. To estimate the emission measure from the observed brightness temperature, the assumed value of the temperature need not be precise. Also, the estimated emission measure is not temperature differentiated, but integrated over a wider temperature range. We can estimate total emission measure rather than temperature limited emission measure such as in EUV and Soft X-ray observations. Doppler shifts due to high line-of-sight velocities do not influence the estimated emission measure, but can be important for narrow line observations. The integrated flux (F) of an optically thin source has a flat frequency spectrum if the background emission is subtracted or is negligible.

3.3 Thermal Gyro-resonance Emission

In the presence of very strong magnetic field, such as in a sunspot umbra, the cyclotron frequency (f_B) and its lower harmonics (2 or 3) are in the range of microwave frequency:

$$f_B \sim 2.8 \times B \text{ MHz}, \tag{15}$$

where B is the magnetic field strength in Gauss. If $B = 2000$ Gauss, the third harmonic is equal to 17 GHz. Each thermal electron emits and absorbs harmonics of the local cyclotron frequency, which is shifted by Doppler effect due to its thermal motion. The absorption coefficient is a complicated function of harmonic number (s), density (N), temperature (T), and the angle (θ) between the wave propagation and the magnetic field. The absorption coefficient for the x-mode is about two orders-of-magnitude larger than that of o-mode. Almost 100 percent polarized emission is observed in the x-mode sense at 17 GHz. The effective thickness of the resonance layer (L) is:

$$L = L_B \times (v/c), \tag{16}$$

where L_B is the scale size of magnetic field strength, v is the mean thermal velocity of electrons and c is the light speed. The scale size of the magnetic field is roughly the size of the sunspot. Due to the small value of v/c, the resonance layer is very thin. The strong angular (θ) dependence of the absorption coefficient and the thin resonance layer cause a center-to-limb variation of the gyroresonance source flux. This is one of the reasons why the sunspot- (or active region-) associated source is called a "slowly-varying source" or "S-component". The dependence of the optical depth on density and temperature is:

$$\tau \propto NT^{s-1} \tag{17}$$

A typical value of an x-mode optical depth at 17 GHz with some combination of parameters ($N = 10^8$ cm^{-3}, $T = 2 \times 10^6$ K, $s = 3$, $\theta = 45$ deg, $L_B = 10^9$ cm) is about 0.1. With $s = 2$, the optical depth is about three orders of magnitude larger, and with $s = 4$, about three orders of magnitude less. So, if the 2nd harmonic layer is in a high temperature region (transition region or corona), it will be the major contribution. Normally the 3rd harmonic is the major contribution at 17 GHz in a non-flaring active region. During the flare, some part of magnetic field in the umbra might have a connection to the hot and dense region. This will create a flare-associated gyroresonance source. Due to the lower density in the umbral atmosphere, microwave observations are very powerful there compared to other wavelengths such as EUV and soft X-rays.

3.4 Non-thermal Gyro-synchrotron Emission

During solar flares, high temperature plasma and high-energy particles are created. So far, it is still unclear how and where they are created. This is one of the major concerns of solar physics and our present work is trying to answer this question. High velocity electrons, due to high temperature or high energy, emit higher harmonics of the gyration frequency around the magnetic field in active regions. Even in weak magnetic field compared to sunspots, harmonic numbers of 10 - 100 can be in the frequency range of microwave. Radio emission of the harmonic numbers in this range is called gyro-synchrotron emission. When the high-energy electrons have a power-law energy distribution in the mildly relativistic energy range, the number of electrons with higher energy is larger than that of a thermal distribution of very high temperature. The energy distribution of high-energy electrons is:

$$n(E)dE = KE^{-\delta}dE, \qquad (18)$$

where $n(E)dE$ is the number of electrons per volume with energies between E and $E+dE$, and K is a constant. Here, the major contribution to the microwave emission is from electrons with higher than 100 keV energy. The emissivity and the absorption coefficient of the gyro-synchrotron emission are complicated functions of δ, θ, B, and n even if we assume uniform pitch-angle distribution of gyrating electrons. Approximate formulae of the emissivity and the absorption coefficients are given by Dulk [2].

The frequency spectrum of the flux has a peak (f_{peak}) determined mainly by the magnetic field strength. We can estimate the magnetic field strength of the microwave-emitting region if fluxes at enough frequencies are measured. At frequencies lower than f_{peak}, emission is optically thick, and that higher than f_{peak} is optically thin. In most cases, both 17 and 34 GHz are in the optically thin side. In the optically thin region, the emissivity and the source size are the necessary parameters to calculate the brightness temperature and integrated flux. The frequency spectral slope of the integrated flux (α) is:

$$\alpha = 0.9\delta - 1.2. \qquad (19)$$

A larger α means a steeper negative spectrum (or soft spectrum). Thus, we can infer the electron energy spectrum from the measured microwave spectrum in the optically thin region. By combining the microwave frequency spectrum and the hard X-ray energy spectrum, we can discuss the high-energy electron production through two independent emission mechanisms from the same population of high-energy electrons. For the harder X-ray emission, a high density target plasma is needed, while for the microwave emission, magnetic field is needed. Spatially resolved observations at 17 and 34 GHz give us information about the inhomogeneity of the electron energy distribution due to the trapping effect and that of magnetic field strength along the loop. High cadence imaging observations with two frequencies in the optically thin region tell us where the high-energy electrons are created with certain energy spectrum and how they propagate into coronal loops with non-uniform magnetic field strength. The variation of the degree of circular polarization along the loop can be used to estimate the distribution of magnetic field strength. If the ambient plasma density exceeds 10^{11} cm^{-3}, Razin-Tsytovich suppression will influence the observed flux even at 17 GHz.

Two-dimensional model calculations of microwave emission from loops filled with high-energy electrons were done by Preka-Papadema and Alissandrakis [7] using various combinations of parameters. Comparison of the model calculations and the actual observation using VLA were done by Nindos et al. [5] (cf. also references in Kliem et al., *this volume*).

4 Microwave Observations of Solar Flares

Microwave observations of solar flares tell us the behavior of non-thermal high-energy electrons and hot plasmas. The main purpose of solar flare studies is to know the energy release processes such as production of non-thermal electrons and of hot plasma; where and how they are produced. For this purpose, we need to identify the emission mechanism of radio sources. We use characteristics of each emission mechanism such as brightness temperature, frequency spectrum, and polarization degree. Comparison of microwave observations with soft X-ray, hard X-ray, and EUV observations is very useful for the identification of emission mechanisms and to obtain physical parameters of flare plasmas and high-energy electrons (cf. Vilmer and MacKinnon, *this volume*).

The temporal development of solar flares can be divided into several phases: precursor, impulsive, main, and decay phases (Fig. 1). Some flares start without a precursor and some flares are lacking the impulsive phase or long decay phase. In the impulsive phase, non-thermal electrons are generated and strong microwave emission, hard X-ray, and gamma ray emission are observed. We will focus our attention to the precursor and impulsive phases using microwave observations.

4.1 Flare Geometry

Using the NoRH images of flares in the impulsive phase, Nishio et al. [6] and Hanaoka [3] found that the microwave sources involve more than two loops, often one small loop and one large loop. Identification of loops is based on the

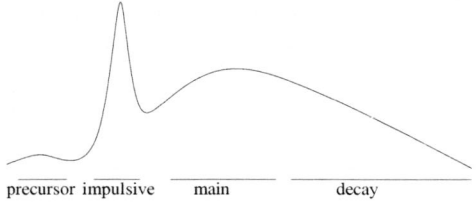

Fig. 1. Solar flare phases

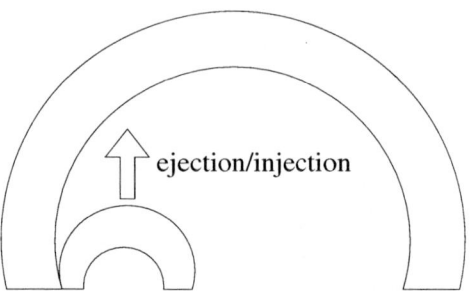

Fig. 2. Double loop configuration

microwave circular polarization and comparison with hard X-ray images and soft X-ray images from YOHKOH. Schematically it is shown in Fig. 2. Both authors interpreted this configuration as the evidence of magnetic reconnection of two loops, but the magnetic field directions of the small loop and of the large loop are parallel rather than anti-parallel. To interpret the observed configuration as the result of reconnection, complicated magnetic configurations have to be introduced. I interpret this configuration as the result of ejection and injection of thermal and non-thermal particles from the small loop to the large loop as the result of high-beta disruption without introducing the unseen magnetic structures. In the processes of ejection and injection, an effective cross-field plasma and particle transport is needed. The anomalous cross-field transport of plasma is well known in nuclear fusion experiments in tokamaks. High-beta disruption is explained in Sect. 5 in detail.

4.2 Flare Dynamics

Due to the high-cadence imaging capability of NoRH covering the whole Sun, we can observe flares from the precursor phase and follow their development. Shibasaki [10] studied a flare on the West limb from its very beginning. The emission is identified as optically thin thermal free-free mechanism from the frequency spectrum. Hence, the observed features are those of a dense plasma.

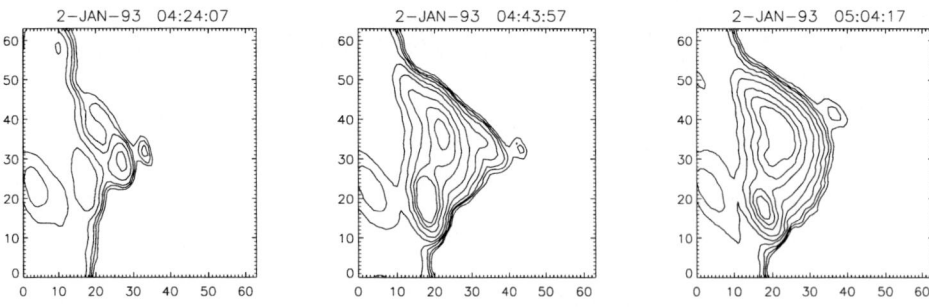

Fig. 3. Hot plasma ejection into large loop [10]

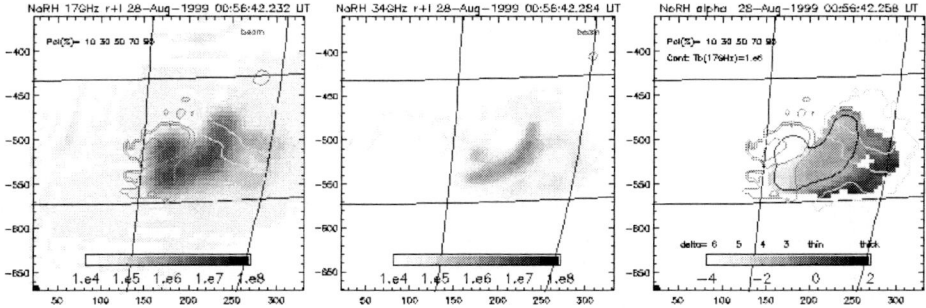

Fig. 4. The event of August 28, 1999. Left: 17 GHz brightness temperature. Brightness temperature unit is in Kelvin. A small countour near the upper right corner is the beam size. Center: 34 GHz brightness temperature. Right: The brightness temperature ratio of 17 and 34 GHz is expressed by the power-law index of the frequency spectrum [13]

In the beginning of the flare, a compact source appeared and another source was ejected from the compact source and moved upwards along a large loop. The whole loop became visible and expanded (Fig. 3). In this event, we could identify an ejected plasma cloud from the initial compact source. The configuration is very similar to that mentioned in the previous subsection for the non-thermal case. It is found that the plasma cloud is ejected from the top of the loop upwards and is trapped by the overlying large loop.

Yokoyama et al. [13] studied the best-resolved non-thermal event observed by NoRH. The event occurred on the disk on August 28, 1999 (Fig. 4). The event started from a compact source, and then a bright feature extended from one foot point to the other along a long loop located very close to the compact source. The speed of the extension along the loop was about 12,000 km/s. Another propagation of a brighter feature was found 10 seconds after the first one with the speed of 120,000 km/s. If we take into account the gyration around the magnetic lines of force and the magnetic field geometry, the microwave emitting electron speed is very close to the light speed. This speed is easily explained by the free propagation of high-energy electrons along the long loop. This observation was exactly what the NoRH was designed for. This event also started near one of the foot points.

Observations of high-energy phenomena by microwave and hard X-ray often show quasi-periodic intensity oscillations (Fig. 5). Spatially resolved observations by NoRH show that the oscillations are enhanced at one end of the long loop where magnetic field is strong while the hard X-ray oscillation is located in the compact source located near the other end of the long loop [1]. This result suggests that the electrons are accelerated in the compact source and the acceleration process is quasi-periodic.

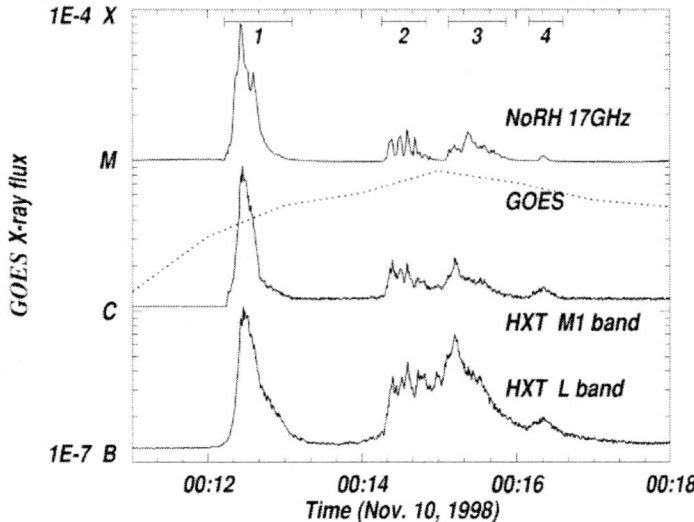

Fig. 5. Quasi-periodic oscillation of hard X-ray and microwave emission during the period marked "2" [1]

5 A New Solar Flare Scenario
Based on Microwave Observations

In this section, a new solar flare scenario is proposed based on microwave observations by NoRH and also by other observations such as TRACE [9]. The solar corona has been believed to be generally low beta (beta = gas pressure / magnetic pressure). In low-beta plasma, energy is stored mainly as magnetic free energy or current. To dissipate the stored magnetic energy impulsively in the highly conductive corona, a localized effective diffusion mechanism is needed. The activity in the outer layers of X-ray emitting magnetic loops in solar flares has been interpreted as the result of reconnection of magnetic field due to anomalous resistivity in the current sheets located above the flaring loops or located between the preexisting magnetic field and the emerged magnetic field. This is the standard solar flare scenario. However, little direct evidence for the reconnection has been presented so far.

5.1 High-beta Disruption

In the previous section, it is shown that flares originate in small loops. Thermal and non-thermal particles in larger loops are fed from the small loops located near the footpoints of the large loops. High cadence imaging observation by NoRH and by TRACE made it possible to witness this process. TRACE observations show that ejections and flows of plasma are ubiquitous on the solar surface. In small loops (in the lower atmosphere), the plasma beta is generally high (or finite) due to high plasma density, and the loop curvature radius is small. Bounded motions of charged particles (both thermal motion and the flow) along the magnetic lines of force create centrifugal force outwards. It can exceed the gravitational force under certain conditions, and can even exceed the magnetic tension force if the plasma beta is more than one. Under the balancing condition between the upward centrifugal force and the downward gravity and tension forces, a localized interchange instability can develop if the beta satisfies a certain condition (beta > loop diameter / curvature radius ~ 0.1). This is called the ballooning instability. This instability can develop into non-linear phase and explosive phenomena are expected. This is called "high-beta (or finite-beta) disruption". The high-beta disruption has many features in common with solar flares. Many flare-related phenomena can be interpreted in terms of high-beta disruption [11].

Studies of high-beta plasma are most developed in the field of nuclear fusion experiments in tokamaks. To realize economical nuclear fusion in magnetically confined conditions, a high plasma beta is necessary. Theoretical and experimental studies of the behavior of a magnetically confined high-beta plasma have been extensively done. Following is a list of the phenomena during high-beta disruptions known from tokamak experiments and computer simulations with application to the solar corona.

1. loop oscillation (ballooning oscillation)
2. plasma concentration at the loop top
3. plasma disruption at the loop top (high-beta disruption)
4. cross field plasma transport (anomalous cross field transport)
5. non-thermal electron production

Due to the limitations of the tokamak configuration, the highly developed ballooning instability or high-beta disruption has not yet been studied. Also, some of the items listed above are not yet fully understood theoretically, especially in the non-linearly developed phase. The flare phenomena that we can observe from the Earth are the result of non-linear development.

5.2 Flares and High-beta Disruption

Flares are often observed in emerging flux regions. New magnetic flux emerges from the lower atmosphere into the corona with dense plasma inside. The temperature is not high, but we can expect high beta due to high density. It is also expected that the magnetic field curvature is small in the early phase of the emergence. High-beta disruption is expected in this condition and this process corresponds to a small flare and can be a beta-loading process to preexisting larger loops above the newly emerged small magnetic loops. A loop type flare is expected after enough plasma (or beta) is supplied to the larger loop. High-beta disruption of the large loop results in larger energy release or flares due to the large plasma energy stored in the large loop.

Filament eruptions are known to be the cause of arcade type flares. When the lifting filament hits the overlying arcade of magnetic field, upward acceleration is generated. The filament plasma pushes the arcade of magnetic field upward.

This is also a suitable condition for the ballooning instability to develop all along the arcade because of high-density filament plasma. When the disruptions occur all along the arcade, ejected filament plasma has to cross most of the magnetic field lines that connect the preceding and the following parts of the active region, hence most part of the active region is involved in the arcade type flares.

Flare related phenomena can be interpreted as the result of the high-beta disruption phenomena listed in the previous subsection: turbulence (1), hard X-ray and microwave sources (5), hot plasma (5 + thermalization process such as electron beam induced turbulence), loop-top soft X-ray source (2), plasma ejection (3+4), quasi-periodic oscillation of hard X-ray and microwave emission (1+5), over-the-loop-top hard X-ray source (5+3), etc. These phenomena can be expected in loops with any size, if the beta value is large enough (beta >loop diameter / curvature radius) and the upward centrifugal force exceeds the gravitational force.

6 Summary

In this paper, spatially resolved microwave observations of solar flares by NoRH are presented. Then, emission mechanisms of microwaves from flares are summarized to interpret the observed features. Observations show that the flares involve more than two loops. Hot plasma and high-energy electrons are injected from one small loop into a nearby larger loop. Based on these observational results, a high-beta solar flare scenario is proposed instead of reconnection scenario. The behavior of high-beta plasma has been extensively studied in the fusion plasma physics, but not much in solar physics. The high-beta disruption has many features in common with solar flares. We need to develop studies of high-beta plasma in the solar atmosphere and to reanalyze observed phenomena from the high-beta point of view.

References

1. A. Asai, et al.: ApJ, **562**, L103 (2001)
2. G. A. Dulk: Ann. Rev. Astron. Astrophys. **23**, 169 (1985)
3. Y. Hanaoka: Solar Phys., **173**, 319 (1997)
4. H. Nakajima et al.: Proc. IEEE, **82**, 705 (1994)
5. A. Nindos et al.: ApJ, **533**, 1053 (2000)
6. M. Nishio et al.: ApJ, **489**, 976 (1997)
7. P. Preka-Papadema and C. E. Alissandrakis: A&A, **257**, 307 (1992)
8. R. Ramaty: ApJ, **158**, 753 (1969)
9. C. J. Schrijver et al.: Solar Phys., **187**, 261 (1999)
10. K. Shibasaki, In: *Magnetic Reconnection in the Solar Atmosphere*, ed. by R. D. Bentley and J. T. Mariska, (ASP Conference Series, **111**, 1996) p. 171
11. K. Shibasaki: ApJ, **557**, 326 (2001)
12. A. R. Thompson, J. M. Moran, and G. W. Swenson, Jr. *Interferometry and Synthesis in Radio Astronomy* (John Wiley & Sons, Inc., 1986)
13. T. Yokoyama et al.: ApJ. **576**, L87 (2002)

How Well Do We Understand Magnetic Reconnection?

Dieter Biskamp

Centre for Interdisciplinary Plasma Science,
Max-Planck-Institut für Plasmaphysik, 85748 Garching, Germany

Abstract. The present status of the theory of magnetic reconnection is reviewed. Quasi-Alfvénic processes arise, when the mechanism for magnetic diffusion is localized around the X-point. This is shown in the simple model of resistive MHD with artificially localized resistivity (it is well known that a uniform resistivity distribution leads to a macro-current sheet and slow reconnection dynamics). When collisionless effects dominate characterized by certain intrinsic plasma scale-lengths, localization comes about through the dispersion of hydromagnetic waves at wavelengths below these scales. As a consequence fast reconnection should be possible under most plasma conditions. Most of the energy released by reconnection is predicted to go into ion bulk motion, though the details of energy partition between ions and electrons and of the efficiency of suprathermal particle production are still under investigation. While energetic electrons may be generated by the runaway effect in the diffusion region, any super-Alfvénic ions seem to be due to some mechanism not related directly to the reconnection process.

1 Introduction

It is generally believed that most eruptive events observed in magnetized plasmas are driven by the fast release of magnetically stored energy by breaking the frozen-in constraint of the magnetic field, which is called magnetic reconnection, or reconnection, in short. Such processes occur under widely different plasma conditions ranging from collision-dominated plasmas such as in the solar convection zone, where reconnection is an essential mechanism in the generation of large-scale magnetic fields, to weakly collisional plasmas as in the solar corona, where reconnection is the basis of a host of eruptive processes, to truly collisionless plasmas as in the Earth's magnetosphere, where the substorm phenomenon appears to be closely related to reconnection of the stretched-out field in the tail.

The term magnetic reconnection is not precisely defined in the literature. Though in certain studies it is simply meant in the general sense of magnetic diffusion, there seems to be general consensus that reconnection is a localized process, which is, in fact, a necessary condition for the energy release to be fast. The basic picture, illustrated in Fig. 1, is that of two field lines carried along with the plasma, until they come close at some point where, due to weak nonideal effects in Ohm's law, they are cut and reconnected in different way. (The alternative term "field line merging", which was popular in the 1970s, has

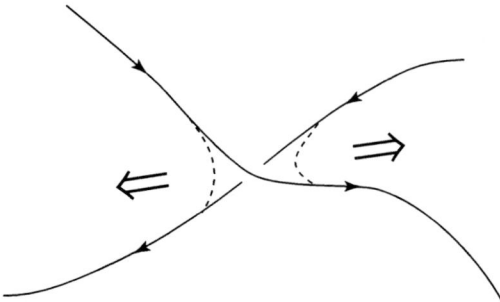

Fig. 1. Basic properties of field line reconnection: relaxation of magnetic tension and plasma acceleration by a local change of the field line connectivity.

gradually fallen out of use.) Simple as it may look this picture already comprises the characteristic features of the reconnection dynamics. Assume that in Fig. 1 the upper and lower endpoints of the field lines are moved around such as the footpoints of coronal field lines anchored in the photosphere. Physically the close encounter of field lines implies that magnetic field gradients become strong thus enhancing the formally weak nonideal process in Ohm's law. As we shall see in this article, the degree of localization essentially determines the time-scale of the reconnection process. After the local change of the field line connectivity in Fig. 1 the magnetic tension relaxes accelerating the plasma away from the reconnection site (the "sling shot" effect), which is the basic conversion mechanism of magnetic into kinetic energy.

The physics of the reconnection process itself is contained in the nonideal part $\mathbf{R}$ in Ohm's law,

$$\mathbf{E} + \frac{\mathbf{v}}{c} \times \mathbf{B} = \mathbf{R}, \tag{1}$$

where $\mathbf{v}$ is the plasma (bulk) velocity. The nonideal term comprises both dissipative effects, primarily resistive diffusion, and nondissipative effects connected with certain collisionless small-scale processes in the plasma. The essential point is that in most cases of interest these effects are formally very weak being proportional to small plasma parameters. Mathematically, the problem of reconnection arises from the ideal conservation of magnetic flux. In an electrically conducting fluid magnetic field lines have a concrete physical meaning as thin flux tubes, where the change of magnetic flux $\psi = \int_{\delta S} \mathbf{B} \cdot d\mathbf{S}$ across a surface element δS moving with the fluid is determined by R,

$$\frac{d\psi}{dt} = - \oint \mathbf{R} \cdot d\mathbf{l},$$

the line integral being taken along the boundary of the surface element δS.

There are mainly two different aspects to be considered in a theory of eruptive magnetic events such as flares. The first one concerns the global properties of the magnetic configuration, which determines the amount of energy to be released, i.e., the magnitude of the event. This is not only determined by the strength of the magnetic field but, more importantly, by the complexity of the field, i.e., the twist and shear of field lines, which reconnection allows to relax into a structurally simpler state. The configuration also determines the location of the reconnection process, which is not simple to predict in a complicated three-dimensional field. In the solar corona we are dealing essentially with force-free fields (cf. Vršnak, *this volume*) $\mathbf{j} = \nabla \times \mathbf{B} = \alpha\mathbf{B}$ with $\mathbf{B} \cdot \nabla\alpha = 0$ (because of $\nabla \cdot \mathbf{B} = 0$). In a theoretical treatment one either considers relatively simple model fields and their stability properties or one tries to reconstruct, under certain assumptions, the magnetic configuration from the field measured in the photosphere in regions where subsequently a major flare event was observed. The amount of free energy can be estimated by comparing the magnetic energy of the preflare configuration with that of the potential field with the same photospheric boundary conditions. In view of the broad variety of flare structures observed (see the article by M. Scholer in this volume) it is, however, very difficult to define a "typical" flare configuration. The problem of the global configuration will not be further discussed in this article.

The second aspect concerns the local properties in the reconnection region, the physics of the actual reconnection mechanism, where one considers

(a) the time-scale τ. It depends on the free energy W of the global field configuration being the shorter the larger the W, simply because higher velocities are generated. But τ is also expected to depend, more or less strongly, on the small parameters in the nonideal term R. More specifically we have $\tau \sim L/v$, where L is the spatial scale of the configuration and v a typical velocity, which should be proportional to $v_{\mathrm{A}} = \delta B/\sqrt{4\pi\rho}$, the Alfvén speed corresponding to the change of the magnetic field δB, $B\delta B \sim W$. Thus one may write

$$\tau \sim (L/v_{\mathrm{A}})/f(R) = \tau_{\mathrm{A}}/f(R). \tag{2}$$

The function $f(R)$ represents the influence of the particular reconnection mechanism which dominates under the given plasma conditions. For a fast process the R-dependence must be sufficiently weak. We can also write (2) in terms of the reconnection rate $E = d\psi/dt$,

$$E \sim \delta Bl/\tau = v_{\mathrm{A}}\delta B f(R), \tag{3}$$

where in the usual 2D X-point model E is, as we shall see below, the electric field E_z at the neutral point;

(b) energy partition. The magnetic energy released is transformed into (i) bulk plasma motion, (ii) electron and ion thermal energy, and (iii) acceleration of a certain number of particles to high suprathermal energies. The energy partition depends on the character of the reconnection process. A fundamental question concerns the efficiency of particle acceleration, in particular whether the high particle energies observed in an eruptive event are generated by the reconnection mechanism itself or result from a secondary process, which occurs outside the reconnection site proper;

(c) threshold conditions for reconnection onset. In general a certain amount of free energy will be accumulated before rapid relaxation occurs. Here the problem is the trigger mechanism responsible for the sudden relaxation, the explosive character typical for many magnetic events.

Most studies on reconnection theory have dealt with the time-scale problem, where considerable progress was achieved in recent years (for a review see [1]) and which also is a major topic in this article. I will, in particular, point out the intimate relation of a fast time scale and the localization of the actual reconnection process by revisiting the paradigm of resistive reconnection (Sect. 2). While there is no obvious physical mechanism of localizing the resistivity, collisionless processes naturally lead to a localized reconnection process and hence fast reconnection rates (Sect. 3). Section 4 addresses the problem of particle acceleration.

2 Localization, the Clue to Fast Reconnection

As indicated in Eq. (3) the reconnection rate is characterized by $f(R)$, the functional dependence on the dominating reconnection mechanism. Extensive numerical studies in the past have shown that one may distinguish between two basically different situations. Either the reconnection mechanism is "efficient" so that the acceleration of the plasma in the outflow region away from the reconnection site by the release of the magnetic stress in turn pulls plasma from the upstream region toward the reconnection site. In this case the inflow velocity is as fast as can be provided by the magnetic configuration, i.e., scales with the Alfvén velocity, and is essentially independent of the reconnection mechanism, typically $f = f(\ln R)$. Here we speak of Alfvénic reconnection. Or reconnection is "inefficient", which leads to piling up of the magnetic field in front of the diffusion region slowing down the inflow velocity and stretching the diffusion region into a macro-current sheet. In this case reconnection rate depends sensitively on the smallness parameter, hence $f \ll 1$ for $R \ll 1$. The dominant nonideal effect in R determines the efficiency of the process, and it is a major theoretical problem to relate this behavior to the algebraic structure of corresponding term(s) in R. In this section I want to illustrate a condition necessary for reconnection to be Alfvénic.

We treat the classical case of resistive reconnection, $\mathbf{R} = \mathbf{j}/\sigma$, $\sigma=$ electrical conductivity, which should, however, only be considered as a mathematical model since, as will be discussed in Sect. 3, in most plasmas of interest, resistivity is not the dominant reconnection effect. It is well known [2] that for uniform resistivity a macroscopic current sheet, called a Sweet–Parker sheet, is formed resulting in a relatively slow process $E \sim S^{-1/2}$, where the inverse Lundquist number $S^{-1} = \eta/Lv_A$, the usual nondimensional measure of the resistivity or magnetic diffusivity, to be precise, $\eta = c^2/(4\pi\sigma)$, is the small parameter in this case. (Only if S exceeds some, rather high, critical value, tearing instability breaks up the macro-sheet, which may accelerate the reconnection dynamics.) To account for the seemingly much faster processes observed in space, it is often assumed that the resistivity is locally enhanced in the diffusion region, for instance by a current-driven instability, which is modeled either by a phenomenological expression $\eta_{\mathrm{anom}} = \eta_0 (j - j_{\mathrm{crit}})^\mu$ (see, e.g., [3, 4]), or just by a resistivity $\eta(x, y)$ localized around the reconnection position (e.g., [5]), which both lead to fast reconnection. Here the question arises, whether this behavior is due to a relatively high value of the local resistivity in the diffusion region, i.e., effectively $S^{-1} \sim 1$, or simply to the localization of η.

This question has recently been studied in [6], where the diffusion region computed numerically is matched to Petschek's ideal external solution [7] leading to a simple algebraic expression for the reconnection rate in terms of the localization scale length d of the resistivity, in particular to Petschek's result in the case of a localized resistivity $d \sim \eta$. The matching procedure is, however, only approximate, leaving open the question of how fast reconnection actually becomes when it is not enforced by stationary boundary conditions but by a finite reservoir of free energy. It is therefore useful to investigate the effect of a localized resistivity in a self-consistent system, which can only be done numerically. As an example we study the coalescence of two magnetic flux bundles in a box of edge size 2π by solving the incompressible 2D magnetohydrodynamic (MHD) equations

$$\partial_t \psi + \mathbf{v} \cdot \nabla \psi = \eta j, \tag{4}$$

$$\partial_t \omega + \mathbf{v} \cdot \nabla \omega - \mathbf{B} \cdot \nabla j = \nu \nabla^2 \omega, \tag{5}$$

where $\mathbf{v} \cdot \nabla \psi = (\mathbf{v} \times \mathbf{B})_z$, $j = j_z$ is the current density and $\omega = \omega_z$ the vorticity,

$$j = \nabla^2 \psi, \quad \mathbf{B} = \mathbf{e}_z \times \nabla \psi,$$

$$\omega = \nabla^2 \phi, \quad \mathbf{v} = \mathbf{e}_z \times \nabla \phi.$$

Here ψ is essentially the z-component of the vector potential, $\psi = -A_z$. ψ equals the poloidal magnetic flux within a strip of unit length along z. Hence $\partial_t \psi = d\psi/dt|_X = E$ is the reconnection rate, the change of the flux per unit time at the X-point. The equations are written in conventional Alfvén time units $\tau_A = L/v_A$ with $v_A = B_0/\sqrt{4\pi\rho_0}$, the Alfvén speed corresponding to a typical magnetic field B_0. The initial state is [8]

$$\psi(x, y) = \sum_{j=1,2} \exp\{-r_j^4/4\} \tag{6}$$

with $r_j^2 = (x - x_j)^2 + (y - y_j)^2$, $x_1 = y_1 = (\pi/2) + 0.6$, $x_2 = y_2 = (3\pi/2) - 0.6$. The resistivity is localized at the X-point

$$\eta = \eta_0 \exp\{-r_X^2/2d^2\}, \tag{7}$$

where r_X is the distance from the X-point, $r_X^2 = (x - x_X)^2 + (y - y_X)^2$, $x_X = y_X = \pi$. We compare four runs with $\eta_0 = 3 \times 10^{-3}, 1.5 \times 10^{-3}, 7.5 \times 10^{-4}, 3.75 \times 10^{-4}$, the size of the resistive region is $d = 6.66\eta_0$, where the value of the coefficient is not essential, the important point being the proportionality $d \sim \eta_0$. The viscosity is uniform and small, $\nu/\eta_0 = 0.25$, such that viscous effects play no role in the reconnection process. The restriction to 2D geometry

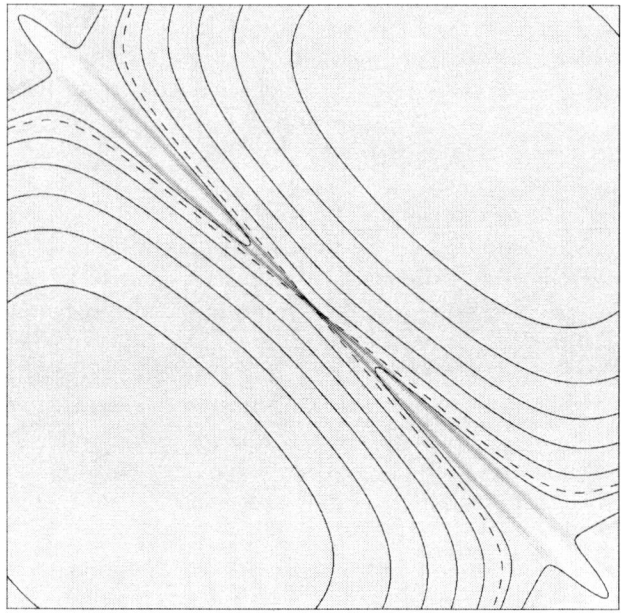

Fig. 2. Coalescence of two flux bundles. Grey-scale plot of the current density of the central region together with contour lines of ψ, where the separatrix line is dashed.

does not severely limit the generality of the result, since, except for the case of strictly antiparallel fields, the magnetic field in the third direction B_z, usually called the guide field, suppresses rapid variations along z, so that in the vicinity of the X-point the configuration can indeed be regarded as two-dimensional.

In the four cases considered the system evolves in a very similar way. Between the coalescing flux bundles one finds a finite-angle X-point configuration which, as shown in Fig. 2, agrees with Petschek's configuration exhibiting two pairs of slow shocks located inside the separatrix connected to the central micro-current sheet of length $l \simeq d$ and thickness $\delta \simeq 0.1d$. (Note that slow shocks survive in the incompressible limit, in fact, Petschek derived his solution in this limit.) Once set up, this configuration remains rather invariant during the coalescence process, in particular the angle formed by the separatrix, the downstream wedge, does hardly change in time, which is hence not directly related to the global geometry, i.e., the momentary size of the flux bundles.

The self-similarity of the process becomes clear, when comparing the time development of the reconnection rate in the different runs, Fig. 3.

While at the beginning, when the current distribution is still smooth, the reconnection rate is proportional to η_0, the fast phase of the process is almost independent of the resistivity. The decrease of the maximum value is only logarithmic, as shown in Fig. 4, where the simulation results are rather accurately fitted by the expression

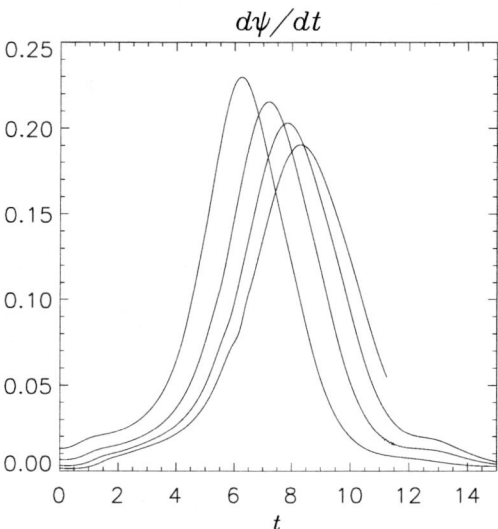

Fig. 3. Reconnection rate $E(t) = d\psi/dt|_X$ from four simulation runs with localized resistivity, η_0 decreasing from left to right by a factor of 2 for each curve.

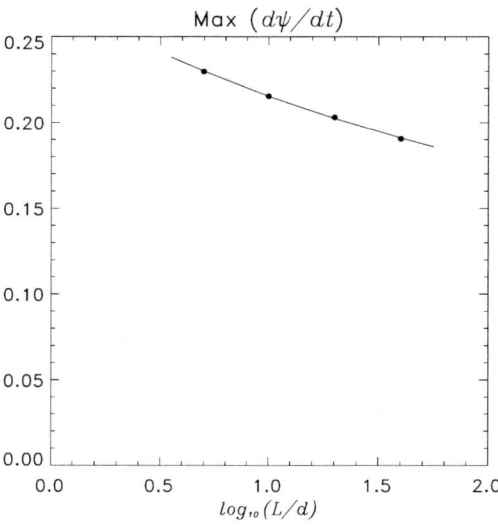

Fig. 4. Maximum values of the reconnection rate vs. $\log_{10} L/d$. The continuous line is the function (8).

$$E_{\max} \simeq \frac{1}{3.66 + \log_{10}(L/d)} = \frac{1}{|\log_{10}(a\eta_0)|}. \tag{8}$$

Here L is the macro-scale of the system and also the coefficient a depends on the global configuration, in our case $L = 1$ and $a = 1.4 \times 10^{-3}$. Since the Alfvén velocity of the upstream field is $B_0 = v_A \simeq 2$, we find $E/v_A B_0 = u/v_A \simeq 0.05$, which is consistent with the relatively small angle $\alpha = u/v_A$ of the outflow wedge seen in Fig. 2. Since the reconnection rate $E = \eta_0 j_0$ is nearly independent of η_0, the ohmic heating rate decreases with η_0, $\int \eta j^2 d^2x \sim \eta_0 j_0^2 d^2 \propto \eta_0$.

The results show that localization of the resistivity suffices to make the reconnection dynamics quasi-Alfvénic, i.e., almost independent of the actual reconnection mechanism. This confirms the feeling often expressed by numerical analysts that it does apparently not matter much which specific anomalous resistivity model was adopted. Here we find that for a localized resistivity, $d \sim \eta_0$, reconnection occurs as predicted by Petschek's model reproducing even the logarithmic η-dependence of the reconnection rate. Does this imply that Petschek's model is correct in spite of the criticisms raised during the past fifteen years? The essential difference is that Petschek assumed a uniform resistivity distribution arguing that η should only be important in the diffusion region taken to be microscopic. We now know that this argument is not correct. In fact numerical simulations have shown beyond reasonable doubt that for uniform resistivity a macro-current sheet is formed giving rise to the slow Sweet–Parker reconnection rate, see [2], which led to rejecting Petschek's model as a fundamental resistive MHD process. Though the ideal external configuration in Petschek's model is correct (and even stable), it does not match to the diffusion region for uniform resistivity. The origin of the difference compared to a localized resistivity is the magnitude of the current I induced in the diffusion region. In the localized case, I is small, since j is restricted to a region of size $O(d^2)$, $I = \int j d^2x \sim \eta$, and hence does not affect the magnetic configuration, which thus preserves the X-point character. In the uniform resistivity case, however, the resistivity invites the system to drive the current in a wider region, such that I is finite modifying the magnetic configuration which finally stretches into a macro-current sheet as shown by Syrovatsky [9].

However, collisional, or classical, resistivity does not lend itself to localization since, because of $\eta \propto T_e^{-3/2}$, the resistivity *decreases* in the diffusion region due to ohmic heating instead of increasing. Assuming a localized resistivity can therefore only be regarded as a mathematical model to illuminate the conditions which must be imposed on a physical mechanism to allow an Alfvénic reconnection rate.

3 Localization by Wave Dispersion in Collisionless Reconnection

The fact that classical resistivity does not allow fast reconnection, does not severely hamper our understanding of eruptive magnetic processes observed in space, since in most fully ionized plasmas of interest the resistive term is not the dominant nonideal effect in Ohm's law, even in cases, where collisions are still important. Let us write down the generalized Ohm's law, more properly speaking, its z-component, assuming again 2D geometry in the vicinity of the X-point,

$$\partial_t \psi + \mathbf{v} \cdot \nabla \psi = d_i (\mathbf{j} \times \mathbf{B} - \nabla \cdot \mathcal{P}_e)_z + d_e^2 \frac{dj}{dt} + \eta j, \qquad (9)$$

which generalizes the resistive equation (4). The reconnection rate is again given by $E = \partial_t \psi|_X$. As before the equation is written in Alfvén time units, where $\eta = S^{-1}$, d_i and d_e are the (normalized) ion and electron inertial lengths,

$$d_i = \frac{c}{\omega_{pi} L}, \quad d_e = \frac{c}{\omega_{pe} L},$$

and $\mathcal{P}_e$ is the electron pressure tensor, which contains the electron kinetic effects (remember that Ohm's law is essentially the electron equation of motion, hence there are no ion kinetic effects in this equation). The collisionless terms on the r.h.s. of (9) are the Hall term, the electron pressure term, and the electron inertia term, which are characterized by the intrinsic length scales, d_i and d_e. The resistive term becomes unimportant, when the collisionless scales exceed the resistive scale, where we define the latter by $\delta_\eta = \sqrt{\eta \tau_A}$ [1]. For instance, electron inertia dominates over resistive diffusion, if

$$\frac{\delta_\eta}{d_e} = \left(\frac{\tau_A}{\tau_e} \right)^{1/2} < 1,$$

where $\tau_e = \nu_{ei}^{-1}$ is the electron collision time. Inserting typical numbers one finds that in the solar corona collisionless effects dominate by a large margin. Nonetheless, resistivity, or some collisionless dissipative effect, are not completely negligible, since the actual reconnection process requires a genuine irreversible mechanism. The nondissipative terms on the r.h.s. in Ohm's law provide the necessary localization contrary to resistive MHD, as will now be discussed.

The theory of collisionless magnetic reconnection has recently received considerable attention, which has led to numerous articles in the literature, for instance [10]–[16]. These are mainly numerical studies using either a two-fluid approximation, or a fully kinetic description by particle simulation, or a hybrid plasma model, where only the ions are treated on a kinetic level while for the electrons a simple fluid description is adopted. Though differing in details most of these diverse attempts give rather similar results. A comparative study has been performed in the Geomagnetic Environmental Modeling (GEM) Magnetic Reconnection Challenge published jointly in a series of papers [14]. Here the initial equilibrium was chosen to be a Harris sheet, $B_{0y}(x) = B_0 \tanh(x/a)$ with plasma density $n(x) = n_0 \mathrm{sech}^2(x/a) + n_\infty$ and uniform temperatures, partly because of

---

[1] It should be emphasized that, contrary to the collisionless scales, the resistive scale is not a proper inherent plasma scale length, since it depends on the macroscale L, $\delta_\eta \propto L^{1/2}$, which is the reason why resistive diffusion is not automatically localized.

the relevance as a local model of the magnetotail, where important reconnection processes are supposed to occur related with the substorm phenomenon, partly because it is the simplest exact kinetic plasma equilibrium.

The GEM study shows that the different collisionless approaches lead to essentially the same fast reconnection rate, while resistive MHD yields a much lower rate. The results indicate that the most important collisionless effect in Ohm's law is the Hall term, which arises by the difference of ion and electron motions in the reconnection region (outside this region this difference is negligible and the MHD approximation is valid).

The configuration studied in the GEM initiative has antiparallel magnetic fields, i.e., there is no (mean) guide field, $B_{0z} = 0$. This does not mean that during the reconnection process B_z remains strictly zero. In the region where the magnetic field is weak the ion excursion is of the order of the ion inertial length, such that field structures smaller than d_i can only be followed by the electrons, while the ion velocity becomes negligible, i.e., for $l < d_i$ the current density is essentially carried by the electrons, $\mathbf{j} = \nabla \times \mathbf{B} \simeq -en\mathbf{v}_e$. Hence $\mathbf{v} \times \mathbf{B}$, the $\mathbf{v} \cdot \nabla \psi$ term on the left of (9), can be neglected compared to the Hall term $\mathbf{j} \times \mathbf{B}$ on the right. The in-plane electron flow gives rise to an out-of-plane field B_z, $\mathbf{j}_\perp = \nabla \times \mathbf{e}_z B_z$. In the two-fluid approximation the electron dynamics in the region $x, y < d_i$ is described by the electron MHD (EMHD) approximation, for a review see [1]. One can show that in this region the electrons support an X-point configuration with v_e accelerating toward the X-point in the inflow, or upstream, region and decelerating when moving away from the X-point in the outflow, or downstream, region. Reconnection occurs in a tiny region around the X-point, caused either by electron inertia, which gives rise to a microcurrent sheet of size d_e, or by kinetic effects due to nonadiabatic electrons (the $\mathcal{P}_e$-term in (9), see [17]), which takes place in a slightly larger region. In this micro-layer the electrons reach velocities largely exceeding the Alfvén speed, the plasma velocity along a resistive current sheet, $v_{e\,\mathrm{max}} \sim v_{Ae} = v_A \sqrt{m_i/m_e}$. For the ions restricted to scales $l > d_i$ the electron dynamics in the region $x, y < d_i$ acts much in the same way as a localized resistivity, such that the ions do not form a macrocurrent sheet and the reconnection rate is Alfvénic. We may say that the Hall term enforces a localization of the true reconnection mechanism, whatever it is.

As discussed in [16] the dynamics in the electron-dominated region $x, y < d_i$ can be related to the dispersion properties of the corresponding linear mode, which in the case of no (or, more generally, sufficiently weak) guide field is the whistler. In two-fluid theory the dispersion relation in a homogeneous plasma embedded in a uniform magnetic field is

$$\left[\omega^2(1 + k^2 d_e^2) - k_\parallel^2 v_A^2\right]^2 - \omega^2 k_\parallel^2 v_A^2 k^2 d_i^2 = 0, \quad k_\parallel = \mathbf{k} \cdot \mathbf{B}/B. \tag{10}$$

While for long wavelength $kd_i < 1$ one recovers the shear Alfvén wave $\omega^2 = k_\parallel^2 v_A^2$, in the short-wavelength regime $kd_i > 1$ the mode becomes dispersive,

$$\omega^2 = k_\parallel^2 v_A^2 \frac{k^2 d_i^2}{(1 + k^2 d_e^2)^2}, \tag{11}$$

in particular

$$\omega \simeq v_A d_i k_\parallel k, \quad d_i^{-1} < k < d_e^{-1}, \tag{12}$$

the frequency of the whistler, a circularly polarized wave rotating in the sense of electron gyration. Note that the dispersion relation (12) is independent of both the ion and the electron mass,

$$v_A d_i = \Omega_i d_i^2 = \Omega_e d_e^2 = cB/(4\pi n e),$$

where $\Omega_{i,e} = eB/m_{i,e}c$ are the Larmor frequencies. The important property of the dispersion relation (12) is that the phase velocity increases with k, $v_{\mathrm{ph}} = \omega/k \sim k$. Hence if one loosely connects the state in the outflow region with a (nonlinear) whistler [16], the widening of the region between the separatrix, i.e., a decrease of k, is accompanied by a slowing of the hyper-Alfvénic electron velocity down to the Alfvén velocity, where the electron flow reaches the ion velocity at a distance $l \sim d_i$ from the X-point.

When the guide field is strong, $B_{0z} \gg B_\perp$, which corresponds to a low-β plasma, there is no weak-field region in the system such that the ions (and of course also the electrons) are magnetized everywhere, in which case the Larmor radius $\rho_s = (\beta)^{1/2} d_i$ takes the role of an intrinsic ion scale length. In the drift approximation the particle velocities consist of the sum of ExB drift and the respective diamagnetic drift, hence the cross-field current in the Hall term in (9) is given by the diamagnetic current, since the ExB drift cancels. One thus obtains a set of equations valid for a plasma of low, but still finite, β, the three-field model [18], which has been solved in various numerical studies, for instance in [10] in cylindrical geometry or in [12] in plane geometry. In its simplest form this model reduces to the two equations generalizing the resistive equations (4), (5),

$$\partial_t \psi + \mathbf{v} \cdot \nabla \psi = -\rho_s^2 \mathbf{B} \cdot \nabla \omega + d_e^2 \frac{dj}{dt} + \eta j, \tag{13}$$

$$\partial_t \omega + \mathbf{v} \cdot \nabla \omega - \mathbf{B} \cdot \nabla j = \nu \nabla^2 \omega. \tag{14}$$

The principal result is that for not too low plasma pressure $\beta = 8\pi p/B^2$, $\beta > m_e/m_i$ such that $\rho_s > d_e$, where the first term on the right in (13) dominates, an X-point configuration develops as shown in Fig. 5 and the reconnection rate is essentially Alfvénic. If, on the other hand, $\rho_s \ll d_e$, i.e., if β is truely small and electron inertia is the dominant nonideal term, a macro-current is formed, as seen in Fig. 6, which implies that reconnection is relatively slow, $E \sim d_e$. (The fast explosive process claimed in [19] can only persist for a short transitory period.)

As in the high-β case, this behavior can be related to the dispersion properties of the linear modes. From (13) and (14) one derives

$$\omega^2 = k_\parallel^2 v_A^2 \frac{1 + k_\perp^2 \rho_s^2}{1 + k_\perp^2 d_e^2}. \tag{15}$$

For $\rho_s > d_e$ we have

Fig. 5. Greyscale plot of the current density in the nonlinear evolution of the kink mode in a collisionless plasma column with $\rho_s > d_e$ (from [1]).

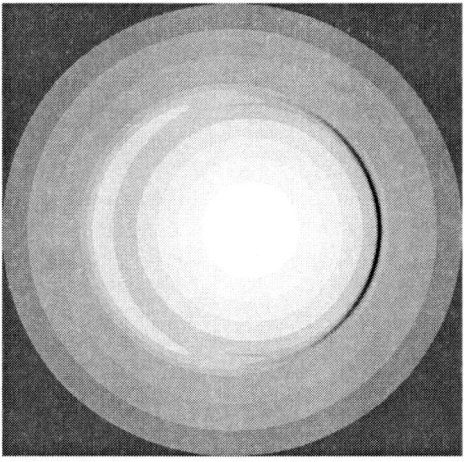

Fig. 6. Greyscale plot of the current density showing a state of the kink mode evolution similar as in Fig. 5 but with $\rho_s \ll d_e$ (from [1]).

$$\omega \simeq v_A \rho_s k_{\parallel} k_{\perp}, \quad \rho_s^{-1} < k < d_e^{-1}, \tag{16}$$

hence the phase velocity $v_{\mathrm{ph}} \sim k$ increases with k. In the opposite case $\rho_s \ll d_e$ the dispersion relation (15) gives

$$\omega \simeq \frac{v_A}{d_e} \frac{k_{\parallel}}{k_{\perp}} = \sqrt{\Omega_i \Omega_e} \frac{k_{\parallel}}{k_{\perp}}, \quad d_e^{-1} < k < \rho_s^{-1}, \tag{17}$$

hence $v_{\mathrm{ph}} \sim k^{-1}$ is decreasing with k. This dispersion behavior does not support the flow in a finite-angle outflow region, such that the configuration should collaps into a macro-current sheet, as is in fact observed, an example being shown in Fig. 6.

4 Particle Acceleration Processes by Reconnection

Until recently reconnection theory has focused on the time-scale problem, which now appears to be essentially solved, as briefly reviewed in the previous sections. Seen from the observational point of view it is, however, equally important to understand the distribution of the free magnetic energy into the different energy channels, ions or electrons, bulk motion, heating or single particle acceleration. The basic energy conversion by magnetic reconnection consists of accelerating the plasma, i.e., the ions, to the Alfvén speed in the outflow region, while dissipation processes, mainly Ohmic heating, are restricted to the diffusion region. In the paradigm of fast resistive reconnection by a localized resistivity the energy fraction which is dissipated Ohmically and thus goes into the electron channel becomes negligible for small η. To explain the various heating and acceleration phenomena observed to occur during eruptive events on the sun a fully kinetic theory is needed.

Let us first estimate the ion energy correponding to the Alfvénic outflow for typical plasma parameters in the corona. $W_i \sim m_i v_A^2 \sim (\delta B)^2/n$, where δB is the reconnected field component, which is largest for antiparallel fields. Note that the outflow velocity does not depend on the reconnection rate, for a relatively slow event the outflow wedge may be narrow, i.e., the number of ions accelerated may be small, but their velocity is always v_A. Since temperatures do not vary strongly across the corona, pressure balance indicates that roughly $(\delta B)^2 \sim n$, hence $v_A \sim const$. With $\delta B \sim 10^{-2}$T and $n \sim 10^8 \text{cm}^{-3}$ we have $W_i \sim 10$ MeV, which is indeed roughly the main energy range observed for the ions. But it is clear that bulk acceleration cannot explain the origin of the GeV ions occasionally observed. The most likely process leading to ion energies in this range is diffusive acceleration at the blast shock wave generated by the flare.

The main difficulty is to understand the energy in the electron channel. Estimates of the relative energy fractions received by the ions and the electrons vary considerably, since it is difficult from individual diagnostics such as particle energies in a certain energy range or radiation in a certain frequency range to estimate the integral amounts of energy. Though most reviewers seem to claim that the ions are the primary recipients, it is undisputed that a sizable fraction of the energy set free does go into the electrons, showing up mainly as soft X-radiation from the region close to the reconnection site and hard X-radiation from the dense footpoint regions (cf. Fletcher & Warren and Vilmer & MacKinnon, *this volume*).

Electron acceleration occurs mainly by the parallel electric field along the neutral line. The energy gained by an individual electron depends on the time it spends in the diffusion region. Intuitively energization should be most efficient in the case of nearly antiparallel fields, where the field in the diffusion region is weak such that particles are nonadiabatic. The basic acceleration mechanism has been studied by Speiser [20], who considers a magnetotail-like configuration with a small normal field component $B_n = B_x \ll B_y$ and an electric field E_z in the current direction. Moving along field lines particles enter the current sheet, where they are accelerated by E_z until they are deflected by the weak normal field B_n and leave the sheet region. The energy gained is $eE_z\rho_{nj}$, where $\rho_{nj} \propto B_n^{-1}$ is the Larmor radius in the normal field. During a single sweep ions gain much more energy than electrons, but the latter may return more frequently after being reflected by the strong magnetic field far away from the reconnection region to be accelerated further, such that the final energy gains of electrons and ions may not strongly differ. It is, however, difficult to estimate the number of particles accelerated in this way, and this mechanism seems to be restricted to the simple 2D geometry assumed by Speiser.

Moreover, this single-particle picture neglects collective processes which are excited by the accelerated particles or give rise to acceleration of a small number of particles by resonance effects. Such processes must be studied in a self-consistent nonlinear kinetic theory, for which three-dimensional electron-ion particle simulations are the only viable tool.

Three-dimensional electron-ion particle simulations using up to 10^9 macroparticles have recently been performed by several groups. In one study the authors consider the magnetotail situation with no guide field [21]. Electrons are heated, especially if they are rather cold in the upstream region, such that their broadened velocity distribution suppresses electron instabilities along the current direction, the Buneman instability and the electron shear flow instability, contrary to two-fluid modeling where both instabilities are excited making the system turbulent at the X-point and along the separatix. In the kinetic simulations the reconnection dynamics remains surprisingly laminar. Ion and electron temperatures are comparable, but most of the energy resides in the bulk outflow, which is not readily thermalized.

In a similar study [22] a relatively large guide field is included $B_z = 5B_y$, such that all particles are adiabatic. Electrons can only be heated parallel to B. Indeed, efficient heating occurs due to the Buneman instability, which makes the plasma turbulent in the reconnection region and along the separatrix. But again the electron energy is small compared with the energy of the Alfvénic bulk outflow.

5 Conclusions

It appears that a major problem in the theory of magnetic reconnection, the problem of fast time-scale, has now been solved. Quasi-Alfvénic reconnection with rates almost independent of the special mechanism occurs under rather general conditions. The essence of a fast process is the localization of the reconnection physics, as is demonstrated in the theory of resistive reconnection. While a uniform resistivity distribution leads to formation of a macro-current sheet and a correspondingly slow reconnection rate $E \sim \eta^{1/2}$, localizing the resistivity around the X-point makes the process essentially independent of the actual value of η. Though resistivity does not lend itself naturally to a local increase in the reconnection region, as Ohmic heating leads to decrease, not increase, of η, it serves as a model underlining the importance of a localization. In real plasmas noncollisional effects usually dominate over resistivity. These are characterized by certain intrinsic scale-lengths, notably the ion and electron inertial lengths, which give rise to dispersion of hydromagnetic waves at small scales. In fact, the presence of a mode with positive dispersion, i.e., phase velocity increasing with wavenumber, may be related to a fast reconnection process, the most important modes being the whistler in a high-β plasma and the kinetic Alfvén wave in a plasma embedded in a strong magnetic field. No, or negative, dispersion leads to macro-sheet formation and slow reconnection dynamics.

Direct observations of the change of the magnetic field by a reconnection process is, in general, very difficult, and most measurements refer to its result, the sudden increase of ion and electron energies. In the basic MHD model reconnection converts magnetic energy into ion flow energy, while electron heating is inefficient. This picture is expected to essentially persist also in a collisionless plasma, even if kinetic processes quantitatively modify the energy partition between electrons and ions. Strongly superthermal electrons may be generated by the runaway effect in the region, where strong electrostatic turbulence gives rise to a large anomalous resistivity and hence large parallel electric field. Such behavior is found in numerical simulations in the presence of a strong guide field, the case more typical for the solar corona than for the Earth's magnetotail. Super-Alfvénic ions can hardly be produced during the reconnection process itself and would need a separate acceleration mechanism, most probably diffusive shock acceleration at the blast wave produced by the magnetic eruption.

References

1. D. Biskamp, *Magnetic Reconnection in Plasmas* (Cambridge University Press, Cambridge 2000).
2. D. Biskamp, *Phys. Fluids* **29**, 1520 (1986).
3. T. Sato and T. Hayashi, *Phys. Fluids* **22**, 1189 (1979).
4. M. Ugai, *Phys. Plasmas* **2**, 388 (1995).
5. R. Hautz and M. Scholer, *Geophys. Res. Lett.* **14**, 969 (1987).
6. N.V. Erkaev, V.S. Semenov, and F. Jamitzky, *Phys. Rev. Lett.* **84**, 1455 (2000).
7. H.E. Petschek, in *AAS-NASA Symposium of the Physics of Solar Flares, NASA-SP 50*, edited by W. N. Ness (Washington, DC, 1964), p. 425.
8. D. Biskamp, E. Schwarz, and J.F. Drake, *Phys. Rev. Lett.* **75**, 3850 (1995).
9. S.I. Syrovatsky, *Sov. Phys. JETP* **33**, 933 (1971).
10. A.Y. Aydemir, *Phys. Fluids B* **4**, 3469 (1992).
11. M.E. Mandt, R.E. Denton, and J.F. Drake, *Geophys. Res. Lett.* **21**, 73 (1994).
12. R.G. Kleva, J.F. Drake, and F.L. Waelbroeck, *Phys. Plasmas* **2**, 23 (1995).
13. M.A. Shay, J.F. Drake, B.N. Rogers, and R.E. Denton, *Geophys. Res. Lett.* **26**, 2163 (1999).
14. J. Birn *et al.*, *J. Geophys. Res.* **106**, 3715 (2001).
15. M. Hesse and D. Winske, *J. Geophys. Res.* **103**, 26479 (1998).

16. M.A. Shay *et al.*, *J. Geophys. Res.* **106**, 3759 (2001).
17. B.N. Rogers, R.E. Denton, J.F. Drake, and M.A. Shay, *Phys. Rev. Lett.* **87**, 195004 (2001).
18. C.T. Hsu, R.D. Hazeltine, and P.J. Morrison, *Phys. Fluids* **29**, 1480 (1986).
19. M. Ottaviani and F. Porcelli, *Phys. Rev. Lett.* **71**, 3802 (1993).
20. T.W. Speiser, *J. Geophys. Res.* **70**, 4219 (1965).
21. A. Zeiler *et al.*, to be published in *J. Geophys. Res.* (2002).
22. J.F. Drake and M. Swisdak, private communication (2002).

Part II

Energetic Particles at and from the Sun

What Can Be Learned About Competing Acceleration Models from Multiwavelength Observations?

Nicole Vilmer[1] and Alexander L. MacKinnon[2]

[1] LESIA, Observatoire de Paris, Section d'Astrophysique de Meudon, 92195 Meudon-Cedex, France

[2] University of Glasgow, Department of Adult & Continuing Education, Glasgow G 3 1LP United Kingdom

Abstract. We review the available evidence from various wavelength ranges, alone and in combination, bearing on solar particle acceleration. Radio, X-ray and γ-ray observations yield direct information on ion and electron acceleration at the Sun. We describe the main spectral features in the X/γ domain, outline the means by which they yield information on accelerated particles, and summarise results obtained using them on numbers and energies of flare fast ions and electrons. Relative numbers and energy content of electrons and ions may vary from flare to flare, and in the course of a single event. In general, both electronic and ionic species appear to embody significant fractions of the total flare energy and either can be dominant, although there is great uncertainty over accelerated particle minimum energies. Rapid fluctuations in X/γ-rays point to a fragmented accelerator, acting on timescales of 100 ms or less, even after particle transport effects have been considered. Millimeter wave observations also reveal spatial fragmentation. Together with distributions of overall event size, such fragmentation suggests a scale-invariant energy release process, such as would occur in a state of Self-Organised Criticality. There is good evidence from X/γ and cm/mm observations for hardening of the electron distribution towards the MeV energy range. Intercomparisons of X/γ rays and cm/mm wave observations emphasise the importance of MeV energy range electrons in the latter. 'Electron-rich' events, characterised by a hard electron population extending to relativistic energies, may occur during individual flares. Existing instrumental capabilities mean that the absence of γ-ray lines does not rule out significant, simultaneous ion acceleration. Radio observations indicate these spectral changes are associated with changes in spatial structure. Spatially resolved radio observations indicate that primary particle acceleration takes place moderately high in the corona (10^7 to 10^8 m), and have recently been made to yield information on accelerated electron pitch angle distribution. Throughout, we emphasise questions on which the unprecedented capabilities of the RHESSI mission will shed new light.

1 Introduction

Like many astrophysical systems, the behaviour of the solar corona is governed by the interplay between magnetic fields and plasmas. This results in explosive phenomena of magnetic energy conversion leading to the production of energetic particles at all energies. Results obtained during recent years have clearly shown that supra-thermal particles are an essential key for understanding energy release in the solar magnetised plasmas. They certainly play a major role in the active

Sun since they contain a large amount of the energy released during flares and since fast particles with energies well into the relativistic regime are a universal feature of energetic solar flares. Radioastronomers have known for decades that the production of suprathermal electrons is also a phenomenon associated with the existence of active regions on the solar disk. This is true even in the absence of flares, although we need to note that the definition of a solar flare may be dependent on observing wavelengths as well as on the sensitivity of the instruments. Indeed, observations accumulated over the last two solar cycles from e.g. the SMM, YOHKOH, SOHO and TRACE missions show a whole distribution in magnitude of solar energy release events recognisable at UV and X-ray wavelengths but unaccompanied by the clear signature at optical wavelengths commonly used to classify flares. Also, radio observations have shown that noise storms, which are commonly observed as a signature of non thermal electron production outside flares, are associated at their onsets with the appearance of new sources at centimeter wavelengths [1] in the active region as well as flare-like signatures such as soft X-ray (SXR) or $\simeq$ 10 keV brightenings [2] and even weak magnetic field annihilation [3].

In this review, we shall focus on the most quantitative observations of energetic particles interacting at the Sun which may bring precise constraints to the question of acceleration processes, and to the flare associated phenomena for which a large number of quantitative analyses have been performed in the literature. We shall not address here the relationship between interacting and escaping particles in the interplanetary medium (cf. Dröge, *this volume*) but we focus on the following key questions that we shall address with respect to competing acceleration models:

- How many particles are accelerated during solar flares?
- What are the acceleration time scales?
- What are the characteristics of the flare accelerated particles: energy spectrum, electron to proton ratio, ion abundances?
- Where are the particle acceleration regions and where are the interaction sites? What are the effects of particle transport in the solar atmosphere on the determination of the characteristics of the accelerated particles?
- To what extent is the acceleration process temporally and spatially fragmented?

In the following, we shall review the observational constraints on accelerated particles using the most complete set of remote sensing diagnostics. Unfortunately, radiative signatures tell us about the state of particles in the regions where they radiate, which are evidently often not identical with the acceleration region. Inevitably then, attempts to constrain the primary accelerator involve us in lengthy digressions into details of transport and radiation mechanisms.

2 Hard X-ray and γ-ray Diagnostics of Accelerated Electrons and Ions

The only direct and quantitative diagnostics of both interacting electrons and ions lie in the hard X-ray/γ-ray domain (HXR/GR). Indeed, energetic electrons (i.e. with energies above $\simeq 10$ keV) produce bremsstrahlung continuum emission in the solar atmosphere by their braking in the Coulomb field of ambient ions and, above $\simeq 500$-700 keV, of ambient electrons (e.g. [4]). This continuum is dominant in the X-ray/γ-ray spectrum below 1 MeV and again from $\simeq 10$ MeV to 50 MeV. For a fraction of flares, this component can extend to a few hundreds of MeV (even to GeV) (e.g. [5, 6, 7]) allowing enquiry into the production of ultra-relativistic electrons.

Energetic ions interacting with the solar atmosphere produce a wealth of γ-ray emissions. A complete γ-ray line (GRL) spectrum is produced through interactions of ions in the $\simeq 1$ MeV/nuc to 100 MeV/nuc range and consists of several nuclear deexcitation lines, neutron capture and positron annihilation lines (see e.g. [8]). If the spectrum of accelerated ions extends above a few hundred MeV/nuc, their interaction with the ambient medium leads to nuclear reactions in which secondary products such as pions and neutrons are produced. Pion production then leads to a broad-band continuum decay radiation at photon energies above 10 MeV (with a broad peak around 70 MeV from neutral pion radiation) (e.g. [9]). The neutrons, if energetic enough, may escape from the Sun and be directly detected in the interplanetary space (≥ 10 MeV neutrons) or at ground levels (≥ 200 MeV neutrons) (see e.g. [10, 11, 12]). The temporal and spectral characteristics of all these radiations provide strong constraints on acceleration timescales, electron and proton energy spectra and numbers as well as energetic ion abundances. In the following, we focus on the GRL emission from ions which has been observed so far in many more flares than the high-energy continuum, and which gives quantitative constraints on the bulk of the flare fast ions. It must be recalled that the solar electromagnetic radiation above 100 keV is one of the last spectral domains where no spatially resolved observations have yet been obtained. Furthermore, the quantitative constraints from HXR/GRL spectroscopy have been deduced so far from observations with limited spectral resolution, thus leading to still large uncertainties in the derived parameters. Finally, no information has been obtained so far on the flare ions below 2 MeV from GRL spectroscopy. This may lead to a strong bias in the estimation of flare energy budgets since a large fraction of the energy contained in ions may well lie in this low energy part.

3 Radio Diagnostics of Accelerated Electrons

As has been known for more than 40 years, observations at radio wavelengths constitute useful and sensitive tools to detect the production of suprathermal electrons, even in low numbers, in the solar atmosphere. Quantitative diagnostics on the spectra and numbers of energetic electrons from a few hundred keV to a few MeV as well as some indication on the magnetic field and density in the emitting sites can be obtained through the observations and modelling of gyrosynchrotron emission in the centimeter/millimeter wavelengths range. A lot of results on energetic electrons in flares have been obtained using spectral observations of this component from different instruments (see e.g. [13] for a review and references therein). Together with a wide frequency coverage provided by e.g. the spectral observations between 3 and 50 GHz in Bern, spatially and temporally resolved observations are now available from different instruments to study electron acceleration, transport and interaction in the solar atmosphere. The Owens Valley Solar Array provides observations at up to 45 frequencies in the 1-18 GHz range with imaging possibilities for 10 to 12 frequencies (see e.g. [14]). Images are obtained with the Nobeyama Radioheliograph at 17 GHz with a time resolution of 100 ms and a spatial resolution of $10''$ and since 1996 at 34 GHz with a spatial resolution of $5''$ ([15, 16]; Shibasaki, *this volume*). At even higher frequencies (86 GHz), the Berkeley-Illinois-Maryland Array (BIMA) now provides observations of the Sun using a 10 element array interferometer. BIMA reveals properties of the radiation from MeV energetic electrons with a spatial resolution which can reach $6''$ (see e.g. [17]). As shown in many works, the multiple beam technique which has been applied for solar observations at 48 GHz at Itapetinga (see e.g. [18, 19, 20]) provides another valuable tool to determine the positions and fast temporal evolution of solar radio bursts, provided that their angular extent is small with respect to the beam size (position determination of less than $5''$ for $1'$ sources). This technique has been recently extended to much higher frequencies (212 GHz and 405 GHz) at the solar submillimeter telescope (SST) at El Leoncito observatory ([21] and also Kaufmann, *this volume*) and will lead to new tools to analyse with high spatial localisation (a few arc seconds) the production of relativistic electrons in flares. Coherent plasma radiations in the decimetric/metric domains provide additional sensitive diagnostics of supra-thermal electrons (around a few tens of keV) accelerated in the low and middle corona in connection with solar flares. In addition to spectral observations obtained in a wide frequency domain (from a few GHz to 40 MHz) with many spectrographs (especially in Europe: Zürich, Potsdam, Ondrejov, Artemis, Porto,...), the Nançay Radioheliograph [22] provides 2D images of the radio sources observed in the middle corona between 450 and 150 MHz with a spatial resolution of 0.3 to $0.6'$. The combined spectrally and spatially resolved observations provide crucial information on the location of the electron acceleration sites associated with flares of all sizes (see e.g. [23, 24]). However, the coherent and non-linear nature of the emission mechanisms poses a severe, even insurmountable obstacle to quantitative deduction of energetic electron numbers and spectra.

4 Optical Diagnostics of Accelerated Particles

The impact of non-thermal electron and proton beams on deeper layers of the atmosphere contributes significantly to particle excitation and ionization, leading to enhanced line and continuum emission. As reviewed by e.g. [25], the line profiles are related to the depth dependence of non-thermal particle energy deposition rate, allowing possible discrimination between the two particle species. Optical (e.g. Hα) observations performed with high time resolution can also be used as tracers of the particle energy transport and deposition in the chromosphere (see also Heinzel and Karlický, *this volume*). The observations of correlated fast time structures (on time scales of the order of one second) have also enabled the demonstration in one event that energetic particles, most probably electrons, are the dominant form of energy transport to the chromosphere [26]. Even if particle energy deposition plays an important, even major role in some flares, however, other observations suggest this is not always the case. In a few flares, it has indeed been shown that the non-thermal electron beam energy is not sufficient to power the whole flare development, or even the soft X-ray emitting plasma (e.g. [27]). The additional observations of impact linear line polarization in solar flares (e.g. at Hα wavelengths) provide information on the low energy particles, most probably ions (below 1 MeV) produced in flares (see [25] for a review, [28]). Given some assumptions on the atmospheric column density between acceleration sites and Hα emitting layers, they could provide constraints on the flux of low energy (≥ 200 keV) protons which cannot be detected through usual γ-ray spectroscopy. Another possible diagnostic for these lower energy protons would be the Doppler shifted spectral lines that result as they slow down to energies of tens of keV, recombine and emit while still in motion [29]. We await a convincing use of this diagnostic in the solar context, although redshifted line emission possibly produced in this way has been detected for one flare star [30].

5 Fast Time Structures and Implications for Particle Acceleration and Propagation

5.1 Elementary Timescales of Flare Energy Release

Since accelerated, non-thermal particles appear to be a major product of the flare energy release process, the temporal development of their most direct radiation signatures (i.e. bremsstrahlung X-rays, gyrosynchrotron radio radiation and GRL's) should reflect the temporal development of this process. Impulsive phase time profiles in these wavelength ranges are spiky, often showing several maxima and giving the general impression of being composed of many smaller, impulsive events. We may hope to learn from them about degree of fragmentation, any fundamental time/length scales involved in the energy release process...

Attempts to do this probably begin with [31]. From TD-1A deka-keV X-ray data they formed the impression that X-ray flares of tens of seconds duration could be decomposed into 'Elementary Flare Bursts' (EFBs), each of 1 - 3 seconds duration. Of course such a finding could guide theory. Instrumental capabilities have continued to improve, however. Substructures on shorter timescales have been clearly detected, and it now appears that the energy release process is active across a wide range of timescales, the shortest of these apparently masked by the finite source lifetimes of the radiating particles. We return to this point below.

Various measures of overall flare 'size' show a power-law distribution (e.g. [32]). This is a strong indication that the flare process takes place in a scale-invariant way. [33] suggested that the magnetised solar atmosphere might be in an overall state of Self-Organised Criticality (SOC - [34, 35]). The flux braiding picture modelled by [36], with a power-law distribution of energy release events, might represent a concrete physical implementation of these ideas. Scale-invariance would result naturally in SOC, although it must always be emphasised that observed power-laws do not necessarily imply SOC (e.g. [37]); in particular, a power law event size distribution might reflect properties of the driver, rather than the response of the driven system [38]. In any case, scale-invariance would have some relevant corollaries:

– substructures of large flares may be regarded as separate events in their own right, with the flare as a whole representing the overlapping occurrence of all its substructures;
– the distribution of substructures of individual events would probe essentially the same phenomena as the distribution of flare sizes - complicated by the simultaneous occurrence of several substructures at any particular time in a single event;
– interest centres on the statistical distribution of occurrence rates across timescales, rather than the presence in data of any particular timescale;
– on the other hand, departures from scale-invariance at the shortest timescales are important: definitive diagnosis of a smallest event scale would be a valuable clue to the workings of the energy release process, and would point to a total flare coronal heating rate in the same way as a smallest flare size (e.g. [39]).

The distribution of discernible flare substructures in BATSE/CGRO hard X-ray data has in fact been studied [40]. An automatic analysis routine was used to identify local intensity maxima and minima, and thus to pick out identifiable substructures in large flare time profiles. Substructure size distributions could then be constructed for individual flares. These vary in form, sometimes being best fit by power-laws, with a range of indices, sometimes by functions falling off exponentially. This valuable study does not yet directly test the ideas here: it uses only identifiable substructures, not any sort of decomposition of the totality of the flare into sub-events, the majority of which will not be separately resolved. The relationship of the 'tip of the iceberg' identifiable substructures to the totality of sub-events is as yet an un-answered question.

The question of a smallest event size, here identified with a shortest energy release timescale, is of course crucial for the role of micro- and nano-flares in coronal heating. Only in the mm wavelength range has such a smallest event size been diagnosed. [41] presented one month's patrol data at 22 GHz from Itapetinga. The smallest events seen during this period had peak fluxes of 1 sfu, well above the instrumental threshold. A recent study of Nobeyama 17 GHz data appears to confirm this finding [42]. Moreover, elementary structures at this flux level are consistent with 'ripple' seen in time profiles of larger events [43]. However, the apparent absence of such findings in most other wavelength ranges suggests that this indicates some sort of event size threshold for electron acceleration to the MeV energies necessary to produce mm wave gyrosynchroton radiation, rather than a phenomenon intrinsic to the primary energy release.

In practice, attempts to confirm or refute the scale-invariant character of flare energy release, and ideas that would follow from it, may be frustrated by finite acceleration and particle transport timescales, discussed in the following sections.

5.2 Transport Timescales

Fast particles do not simply radiate in the moment of their acceleration. Once produced they may move throughout the wider magnetic structure(s) surrounding the acceleration region, their velocities continuously changing under the influence of single-particle and collective interactions with ambient, thermal plasma. Thus they may continue to radiate during an extended period, and we must deal with the consequences for radiation signature temporal development if we wish to learn about the primary accelerator. Here we restrict attention to the consequences of electron transport for bremsstrahlung X-rays. The more involved issues for gyrosynchrotron radiation are dealt with below.

There is general agreement that particle acceleration takes place in the low-density corona. Subsequently, particles may be trapped by magnetic field inhomogeneity in low density regions or they may be able to proceed directly to the higher-density chromosphere. The binary collision energy loss timescale (τ in s) for an electron of energy E (keV) in a medium of density n (m^{-3}) is

$$\tau = 1.4 \times 10^{14}\, E^{3/2}/n \tag{1}$$

(e.g. [44] - this holds for E $\leq$ 160 keV). Typical low corona ambient densities n in the range $10^{15} - 10^{17}$ m^{-3} then imply deka-keV electron lifetimes in the 0.1 - 100 s range. Spectral hardening will result from the increase of electron lifetime with energy, and indeed some long-duration, extended hard X-ray bursts appear reconcilable with this picture (e.g. [44, 45, 46]).

Electrons able to precipitate directly to the chromosphere, on the other hand, rapidly encounter densities great enough to bring them to a halt effectively instantaneously. If we then assume a suitably localised acceleration region and pulsed acceleration, spectral softening will result from the later arrival of lower energy electrons, on the characteristic timescale of the transit time from acceleration region to chromosphere (e.g. [47]).

In practice both of these effects may occur simultaneously, with large pitch-angle electrons remaining trapped for significant times and small pitch-angle electrons precipitating directly. [48] first analysed CGRO/BATSE data in these terms, finding that data filtered on a 2-5 s timescale showed delays to higher photon energies, consistent with coronal trapping, while data filtered on shorter timescales ($\leq$ 1 s) displayed delays to lower photon energies, consistent with the loop transit time effects expected from directly precipitating electrons. The implication is a pulsed acceleration process producing electrons of all pitch angles, both inside and outside the loss-cone appropriate to the particular magnetic structure.

The presence of delays is established by cross-correlating time profiles in different photon energy ranges. Individual spikes cannot be identified in the short-timescale filtered data and examined separately; only a general delay to lower photon energies is identified across the whole of an event. One might worry, for instance, that the greater noise levels at higher photon energies could produce artefacts in the cross-correlation process. [49] point out that variations in the energy distribution of the primary accelerated electrons, on timescales comparable to loop transit times, will invalidate the cross-correlation procedure. The more general question of disentangling accelerator intrinsic behaviour from time-of-flight effects has been discussed by [50]. The linear correlation of loop lengths obtained from short-timescale spectral softening and Yohkoh/HXT images [51] does lend independent support to the above interpretation, however.

Evidently these investigations demonstrate consistency of existing data with a simple picture of trapping and precipitation in simple magnetic structures, possibly loops. Additional arguments or assumptions are necessary to make further statements about the accelerator, however. The short-timescale filtered behaviour might be taken as evidence for a highly fragmented acceleration in several short bursts, each attended by its own time-of-flight delays. But it is also possible that these bursts of precipitation correspond to brief episodes of enhanced pitch-angle scattering and filling of the loss cone. [52], for instance, highlighted the possible role of the electron cyclotron maser in this respect. Caveats of this sort need to accompany statements of likely accelerator timescales, to which we now turn.

5.3 Acceleration Timescales

The temporal characteristics of HXR/GR bursts at high time resolution provide strong constraints on the acceleration/transport process(es) at work in solar flares. Reviews on the constraints derived from the different timescales in HXR and GR observations can be found in e.g. [53, 54, 55]. We shall briefly summarise here the main results on HXR timescales derived from two systematic studies performed on HXR bursts recorded with fast time resolution respectively with BATSE/CGRO and PHEBUS/GRANAT. [48] analyzed more than 600 solar bursts recorded with BATSE/CGRO with a time resolution of 64 ms in the 25-100 keV range and found that, in more than 70 % of the bursts, time structures ("HXR pulses") with durations comprised in the 300 ms-3 s range were present. They further showed using a multiresolution analysis (wavelet transform) that for strong flares, the shortest detected timescales are in the 100 ms -700 ms range. For smoothly varying flares, the shortest timescales are in the 500 ms -5 s range, which is the likely result of the convolution between acceleration and trapping timescales. At higher photon energies (above 100 keV),

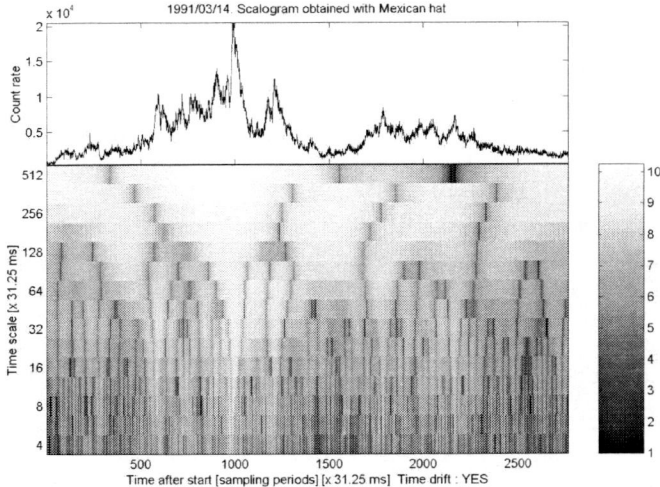

Fig. 1. Count rate of the 14 March 1991 event observed by PHEBUS/GRANAT in the 0.1-1.6 MeV range and scalogram showing the different time structures discussed in the text. The white features in the scalogram correspond to time scales and time periods with strong power deduced from the wavelet analysis (the log scale is shown on the right side of the figure).

[56] showed that for most of the bursts ($\simeq 100$) observed with a time accumulation of 31.25 ms by PHEBUS/GRANAT, time structures with rise time in the 100 ms -1 s range are systematically detected. As an example, the scalogram shown in Fig. 1 shows that many of these time structures can be found in a single burst with time scales ranging from 100 ms to 500 ms. The results deduced from both systematic analyses thus indicate upper limits for acceleration timescales of electrons in the 25-100 keV range of a few 100 ms. The information is more scarce regarding electron acceleration timescales in the MeV to 10 MeV range and ion acceleration time scales due to limited sensitivity of GR detectors to investigate short time structures. GR peaks attributed to bremsstrahlung radiation of ultrarelativistic electrons and lasting a few seconds to a few tens of seconds have been observed in several events by GRS/SMM (e.g. [57]) and one example of a subsecond time structure has been reported by [58] showing that ultra-relativistic electrons must interact in dense regions and be also accelerated on time scales less than one second. Regarding the ion acceleration timescales, no detailed analysis of the temporal evolution of prompt γ-ray line emissions has been performed systematically, but simultaneous peaking within $\pm$ 1 s of emission in the GRL domain and of HXR emission has been observed for a few events (e.g. [59]). This indicates that ions must also interact in most cases in a dense region and that ion acceleration timescales to a few MeV must be less than 1 s. Additional information on ultrarelativistic electron acceleration timescales may be provided by the more sensitive diagnostics of relativistic

electrons provided by their optically thin gyrosynchrotron emission. [60] thus reported observations both at 48 GHz and at HXR energies ranging from 25 to 325 keV of fast time structures of 200-300 ms half power duration in excess of a component slowly varying on timescales of 10 s. The correlation within 64 ms of the fast time structures observed in both wavelength domains indicates in the context of time-of-flight flare models, acceleration time scales less than 100 ms for electrons up to MeV. More information on these acceleration timescales are expected in the future using RHESSI (Reuven Ramaty High Energy Solar Spectroscopic Imager- [61]) observations combined with millimeter wave observations at high temporal resolution.

6 Spectra, e/p Ratio and Abundances of Accelerated Particles Deduced from X/γ-ray Spectroscopy

6.1 Energy Spectra of Flare Accelerated Electrons at Subrelativistic Energies

The exact shape of the interacting energetic electron spectra is in fact poorly known. In principle the measured photon spectrum may be directly inverted to determine the energy distribution of the emitting electrons [63], but most HXR observations to date have been performed with scintillation detectors whose limited energy resolution prohibits such direct inversion. All these observations have however provided good estimates of the energy contained in non-thermal electrons above 25 keV (see next section).

At the time of this writing, still rare are the published observations like [64], using a germanium detector with much finer energy resolution ($\simeq$ 1 keV). A key finding of this work was the first detection of a 'superhot' X-ray component corresponding to a thermal source with a temperature of 3.5 10^7 K, evident at hν < 35 keV in the decay phase of the flare. This component shows a transition between thermal and non-thermal bremsstrahlung radiation. One interpretation of the gradually varying component of the HXR emission is that the whole shape of the HXR spectrum, both 'thermal' and 'non-thermal' components, results from the same energy release mechanism, namely acceleration and heating in multiple current sheets (e.g. [65, 66]). At higher energies, the HXR spectra of the non-thermal component have been numerically inverted to determine the radiating electron spectrum [62]. Figure 2, from [62], shows both the photon spectrum and the derived electron spectrum for the spiky component of the HXR emission (i.e. once the gradually varying component, attributable to thermal flare plasma, has been removed - this use of temporal behaviour is one way of attempting to separate 'thermal' and 'non-thermal' contributions to an observed spectrum). This electron spectrum clearly shows a peak at $\simeq$ 50 keV and has a shape reminiscent of electron distribution functions measured in auroral arcs (e.g. [67]). It may suggest acceleration by a DC electric field, with the peak energy corresponding to the total potential drop. Clearly such spectra, obtained with high spectral resolution and for more flares, will be crucial for investigating

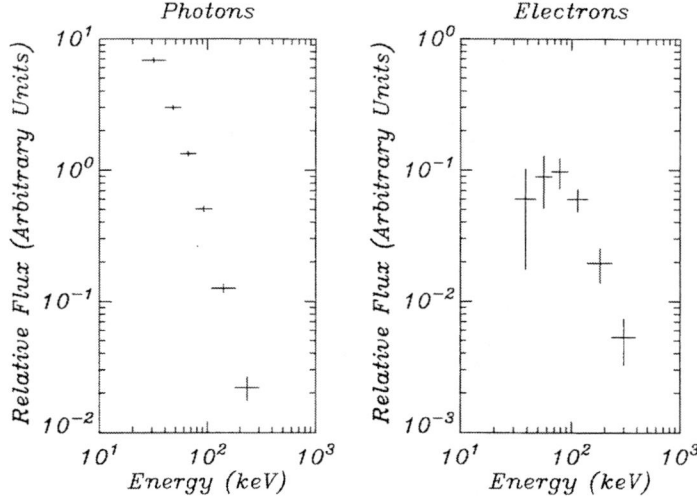

Fig. 2. Left: Hard X-ray spectrum of the main spike of the 27 June 1980 flare with the slowly varying component subtracted. Right: deconvolved electron spectrum (see text for more information) (from [62]).

the conditions of flare electron acceleration. The first spectral measurements made by the newly launched RHESSI mission let us expect that such a goal could be achieved in the near future.

6.2 Energy Spectra of Flare Accelerated Electrons at Relativistic Energies: Evidence of Spectral Hardening at Relativistic Energies

Hardening of the photon spectra above a few hundred keV (and thus most probably of the emitting electron spectra) has been reported in many HXR/GR events observed by SMM, Hinotori (e.g. [69, 70, 71]). A few events observed by PHEBUS/GRANAT with a photon spectrum extending up to 10 or 100 MeV exhibit a clear hardening of the spectrum at high energies, to a greater degree than that expected from cross-section and electron energy loss rate energy dependences. This is the case both for events with strong γ-ray line emission [72], and for electron-dominated events [68, 73] i.e. those without detectable γ-ray line emission (Fig. 3). As shown on Fig. 3, the break energy can also vary in the course of the event from peak to peak.

6.3 Energy Spectra and Abundances of Flare Accelerated Ions

As recalled in Sect. 2, a complete γ ray line spectrum is produced through interactions of ions in the $\simeq 1$ MeV/nuc to 100 MeV/nuc range. In this paper, we shall focus on the observations of nuclear deexcitation lines which may be either narrow or broad depending on whether they result from the bombardment

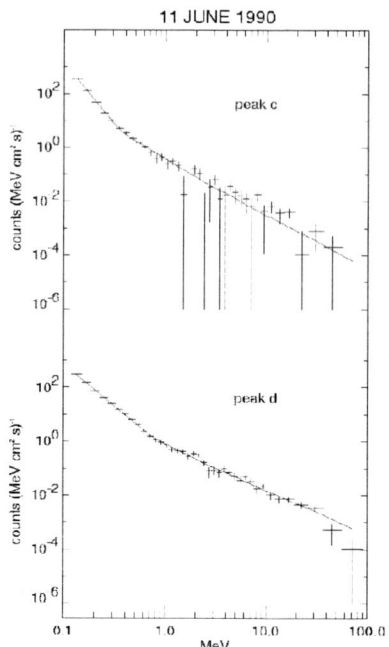

Fig. 3. Background subtracted HXR/GR spectra observed by PHEBUS/GRANAT during two succesive peaks (20 s apart) of the 11 June 1990 flare. For each spectrum, the solid curve represents the best fit model (from [68]).

of ambient nuclei by accelerated protons and α particles or from inverse reactions in which accelerated C or even heavier nuclei collide with ambient H or He. As shown in Fig. 4 from [74], strong deexcitation lines are found at 6.129 MeV from ^{16}O, 4.438 MeV from ^{12}C, 1.779 MeV from ^{28}Si, 1.634 MeV from ^{20}Ne, 1.369 MeV from ^{24}Mg and 0.847 MeV from ^{56}Fe. The broad lines merge into a quasi continuum dominating the bremsstrahlung emission in the $\simeq$ 1-8 MeV range. The broadening of these inverse reaction lines results from kinematic effects because the excited C, N, O, Si, Ne, Mg, Fe nuclei continue to move rapidly after their collisional excitation.

Deexcitation GRLs tell us about flare ions of energies above 2 MeV/nucl. The form of the flare site ion distribution below this energy is essentially unknown, but of particular interest for the total ion content. Additionally to the two possible low energy proton diagnostics mentioned in Sect. 4, weak, radiative capture γ-ray lines potentially constrain the proton distribution down to about 0.4 MeV [75]. Although not yet detected, the next generation of gamma-ray detectors may yield useful constraints on the strengths of these lines (Share, Murphy and Newton, 2002, Solar Phys, in press).

Since the first detection of solar γ-ray lines in 1972, many GRL flares have been observed with detectors aboard SMM, Hinotori, GRANAT, CGRO and YOHKOH. The analyses of these observations have led to quantitative results

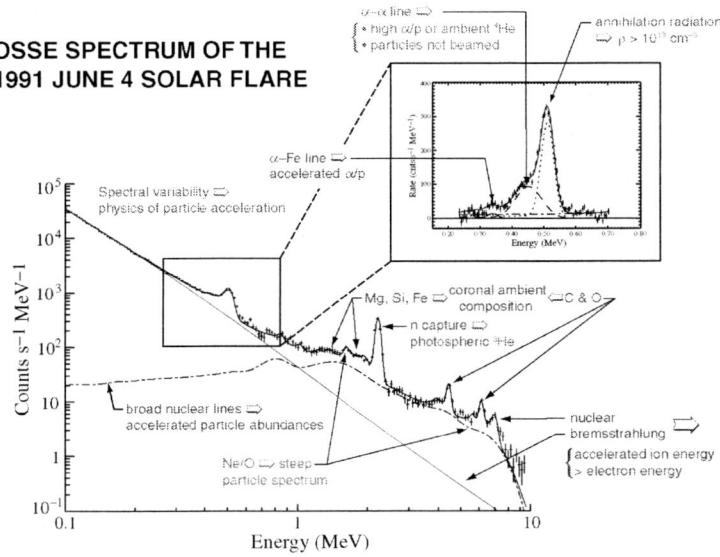

Fig. 4. OSSE spectrum of the 1991 June 4 flare summarising the different components seen in the 0.1 to 10 MeV range (from [74]).

for more than 20 GRL events, yielding information on both atmospheric elemental abundances, and on fast ion energy spectra. Secondary neutron production results from ions of higher energies than those which produce the various deexcitation lines. Thus the ratio of the fluence in the 2.223 MeV neutron capture line to the deexcitation line fluence in e.g. the 4 - 7 MeV photon energy range can in principle be used to deduce the form of the ion energy distribution in a parametric way (e.g. energy power-law spectral index). This diagnostic was applied to SMM data from various events during the 1980s (e.g. [8]). Exciting particle threshold energies vary from one deexcitation line to the next, however, so ratios of pairs of appropriately chosen deexcitation lines may also be employed in this way. In particular, [76] used the ratio of the fluxes in Neon and Oxygen lines at respectively 1.63 and 6.13 MeV to deduce fast proton energy spectra for 19 flares observed with GRS/SMM. The threshold energies for excitation of these lines by fast protons are significantly different, $\simeq 2$ and $\simeq 8$ MeV respectively, so the ratio of their measured fluences can be used as a measure of the proton energy distribution. The lines are not distinguished from one another in this way when αs do the exciting, however, so the conclusions of this process depend strongly on the α/p ratio. With this caveat, [76] found that the accelerated proton spectra should extend as unbroken power laws down to at least 2 MeV/nuc if a reasonable ambient Ne/O abundance ratio is used (i.e. in agreement with measurements of the Ne/O abundance ratio in the corona - [77]). This analysis

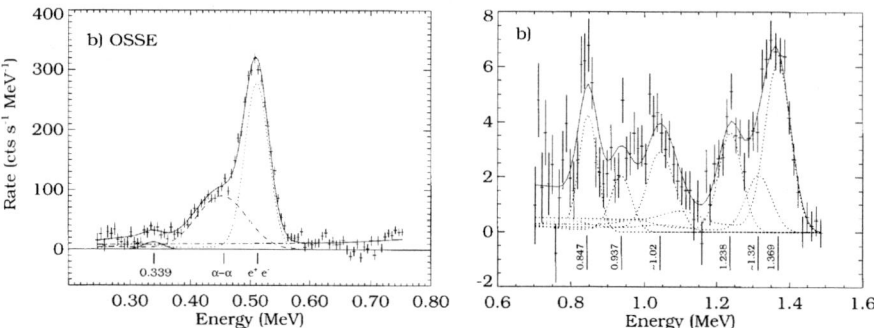

Fig. 5. Left: Summed spectrum from four CGRO/OSSE detectors during the last part of the intense 4 June 1991 flare. The intense bremsstrahlung component has been removed. Curves show fits to the data. Solid line: overall model with line at 0.339 MeV. Dotted line: 511 keV line and positronium continuum. Dashed line: α- α line. Right: Summed spectrum from 19 flares observed by SMM/GRS between 1980 and 1989 covering the energy band containing lines relevant to He studies. Best fit bremsstrahlung and broad lines have been subtracted. Solid line: overall fit. Dotted lines: fit to individual lines (from [79]).

was repeated for data from one CGRO/OSSE flare [78]. The spectrum deduced is thereafter used to estimate the energy contained in accelerated protons above 1 MeV (see [77]).

Information on α/p has been deduced for five flares [79, 80] from the fluence ratio of two lines: the prompt Fe line at 0.847 MeV which is produced by the interaction of accelerated protons and αs on ambient iron and a pure α line at 0.339 MeV (see Fig. 5 left) which only results from the interaction of energetic α particles on iron producing an excited state of nickel. It was found by [80] that for these five flares α/p exceeds the standard value of 0.1 and can even reach 0.5. Nuclear interactions of accelerated ^{3}He with ambient ^{16}O result in three γ-ray lines at 0.937, 1.04 and 1.08 MeV. While the fluence of the line at 0.937 MeV could be determined for a few flares observed by GRS/SMM and CGRO/OSSE, the other two lines cannot be separated from α lines at 1.05 and 1.00 MeV [79] (Fig. 5 right). All these lines yield to an unresolved feature centered at 1.02 MeV. Using combined information from this unresolved feature and from the line at 0.937 MeV, information could however be obtained on the ratio of accelerated ^{3}He with respect to accelerated ^{4}He, showing in seven flares enhancement of this ratio (0.1 to 1) with respect to coronal values [80]. Of course, more observations are needed to make stronger conclusions on acceleration processes. Finally, the limited energy resolution of the present observations renders such an analysis difficult. Improving spectral resolution will also allow the profile of $\alpha/$He (i.e. α-α) fusion lines around 0.452 MeV to be investigated in more detail than with GRS/SMM or CGRO observations (Fig. 5). This provides a potential diagnostic of the angular distributions of energetic particles and thus of the acceleration models. So far, it was shown that the line shape of $\alpha/$He fusion lines in two flares was inconsistent with the production by a downward beam and was rather consistent with an isotropic or a fan beam distribution (i.e. as obtained at a magnetic mirror point) in the interacting site [76, 79]. Even though most of the energy contained in ions resides in protons and α particles, crucial constraints for the acceleration processes also arise from estimations of the abundances of heavier accelerated ions. From the analysis of several events observed with GRS/SMM and CGRO (see e.g. Fig. 4), it was shown that accelerated heavy ions such as Ne, Mg, Fe, Si are overabundant with respect to their coronal composition [81, 82, 78] as is also observed for the composition of impulsive solar energetic particle events [83]. Observations of the behind-the-limb flare of 1 June 1991 with PHEBUS/GRANAT [72] have moreover shown that the enhancements in heavy ions may increase with time in the course of the flare, reaching towards its end the highest values observed for solar energetic particles in space [84, 85]. As a thin target production of the emission is required to account for the very high ratio observed of the 1.1-1.8 MeV flux to the 4.1-7.6 MeV flux, the temporal evolution of the abundances of accelerated ions is to be related to the evolution of the accelerated particles themselves. It is noteworthy that this flare, although behind the limb, was associated with one of the largest GRL fluences observed so far. It was also surprisingly associated with the observation of a strong flux of neutrons by the OSSE experiment [86]. As the neutrons are also expected to be produced as a thin target, a hard ion spectral index (power law index around -2) is deduced by [86] up to at least 50 MeV.

6.4 Electron and Ion Energy Contents in GRL Flares

Studies based on the quantitative analysis of several thousands of hard X-ray bursts above 20 or 25 keV (e.g. [32, 87]) have shown using thick target computations of non-thermal X-ray emission [63] that a large fraction of the flare energy goes to accelerated electrons with energy contents ranging from $\simeq 10^{19}$ J for microflares [88] to 10^{27} J for the giant flares [89]. As a result of the work of [76] and [77], described above, it is now widely accepted that the energy contained in > 1 MeV ions may be comparable to the energy contained in subrelativistic electrons. The derived ion energy content (above 1 MeV) lies in the $4 \; 10^{22}$ to $3 \; 10^{25}$ J range [77, 90]. We emphasise again that Ne line strength, central to this discussion, may depend on other factors, uncertain to various degrees: the source region Ne abundance (Ne behaves distinctively in many other contexts e.g. [91]), the appropriate value of α/p, and the length of time fast protons spend in a region where 'warm target' energy losses may be appropriate [92], so some uncertainty still surrounds this conclusion. For the giant flares of 1 and 4 June 1991 observed either by PHEBUS/GRANAT or OSSE/CGRO, the energy contained in > 1 MeV/nuc has also been deduced and lies around 10^{26} J [85, 78]. Comparison with deductions of the >20 keV electron energy content for the same flares indicates that, although there is a large dispersion of the relative electron and ion energy contents from one flare to the other, the energy contained in the ions sometimes exceeds the energy in the electrons. Crudely speaking, the large fraction of the flare energy released in accelerated particles is thus partitioned between electrons and ions, with however a variation from flare to flare of the relative importance of both components. While such deductions have a clear meaning, as total energy manifested as electrons/ions above some lower cutoff, the appropriate values of these cutoffs for estimating total fast particle energy are not well known. For the electron distribution, determination from observations of the low energy cut-off of the non-thermal distribution is almost impossible in the absence of measurements with high spectral resolution covering the energy range between a few keV and a few tens of keV such as the ones obtained for the flare of 27 June 1980. The only other determination of a low-energy cut-off is through the modelling of the thermal and non-thermal components of the X-ray spectrum observed with a limited spectral resolution, but in a wide energy band. This has been done for a few cases showing that the low energy cut-off of the non thermal electron distribution may be as low as 12 keV [93] or even 5 keV [94]. Such values of the low energy cut-off for electrons will considerably modify the estimate of the electron energy content. On the other hand, the low energy cut-off of the non-thermal ions is quite unknown: as recalled in Sect. 2, no information can be obtained below 2 MeV from GRL spectroscopy and other diagnostics of flare accelerated ions must be found (see Sect. 4).

6.5 Electron and Ion Energy Contents in Electron-Dominated Events

Extreme cases of the variability of the electron bremsstrahlung component with respect to the nuclear line component observed from one GRL flare to the other occur during the short duration (a few seconds to a few tens of seconds) bremsstrahlung transient bursts observed above 10 MeV any time during a flare and referred to as electron-dominated events (e.g. [57]). They are characterised by weak or no detectable GRL emission and by hard ≥ 1 MeV electron bremsstrahlung spectra. They were first reported from SMM/GRS observations (e.g. [95]), were afterwards observed by GAMMA1 [5, 6], PHEBUS and SIGMA experiments aboard GRANAT [58, 96], CGRO (e.g. [97]) and YOHKOH [98]. The spectral analysis of 12 electron-dominated events observed by GRS/SMM [99] confirmed the hardness of the bremsstrahlung spectra above 1 MeV (mean value of the power slope around -1.84) (see also Fig. 3 for an event observed by PHEBUS/GRANAT). The mean value of the spectral index between 0.3 and 1 MeV ($\simeq -2.7$) does not differ significantly from that of other flares. The apparent lack of GRL emission does not rule out a simultaneous production of electrons and ions, however [68, 73]. Indeed, if one assumes that the energy content in ions above 1 MeV is similar to that contained in electrons (i.e. around a few 10^{22} J for cases studied), as appears to be the case for GRL flares, one finds that no detectable GRL fluence could be observed, given the hardness of the electron bremsstrahlung component and the limited spectral resolution (and thus line sensitivity) of the experiment. The only remaining question would then be to understand why the spectrum of accelerated electrons is so much harder in these events than at most other times. Observations performed with a better spectral resolution should be able to clarify this point by increasing the contrast of the line with respect to the continuum even in the case of a hard continuum spectrum.

7 Variability of Particle Spectra and e/p Ratio in the Course of a Flare

Variability of particle spectra and of e/p ratio not only occurs from flare to flare as discussed above, but also on time scales of tens of seconds within individual events (see e.g. [100] for GRL events and [68] for high-energy electron-dominated events). A question of obvious interest, one impossible to answer directly in the absence of imaging observations at energies above 100 keV, is whether these changes correspond to spatial source evolution. However, several multiwavelength studies have combined HXR/GR spectral measurements with optical or meter/decimeter radio imaging observations (e.g. [100, 101, 68]) to show that the variability from one HXR/GR elementary peak to the next is usually linked to spatial variability observed at these other wavelengths, so that there is a link between the characteristics of the accelerated particles and the magnetic configuration in which the particles are produced. In addition, the variation of electron spectra on time scales of 10 s (see e.g. Fig. 3) corresponds to stepwise changes of metric/decimetric spectra clearly showing the involvement of different magnetic structures. In particular, the most intense electron-dominated high energy peak (peak d), interpreted in the present case as a signature of an upward moving population of relativistic electrons strongly beamed along the magnetic field, occurs when the radio emission previously observed at shorter (cm) wavelengths suddenly extends to the decimetric range. This suggests that electrons suddenly have access to large scale magnetic structures [68]. As suggested by [102], such an impulsive electron acceleration to high energies with a hard spectrum could result from the direct electric field associated with magnetic reconnection which would occur in the present case when different magnetic features are involved and interact.

Similar results have been obtained for a less energetic flare combining the spatially and spectrally resolved HXR observations from YOHKOH/HXT with spatially resolved observations from the NoRH [103, 104]. The HXR emission above 20 keV (YOHKOH/HXT M1 band, 22.7-32.7 keV) arises as in many flares from a double X-ray source with components separated by $20''$. The relative strengths of these components vary with time as shown on Fig. 6. In addition to this double HXR component, a remote HXR source is seen for $\simeq$ 10 s corresponding to the edge of a weak microwave source at 17 GHz remote from the bright main microwave source which is cospatial with the double HXR main source. Figure 6 shows the variation of the HXR count-rates in three energy bands (indicative of the HXR photon spectrum) for both the double and the remote HXR components as well as the variation of the 17 GHz flux for both sources. There is a clear variation of the HXR spectrum of the double component from the first peak to the next as well as a much harder spectrum for the remote source. This

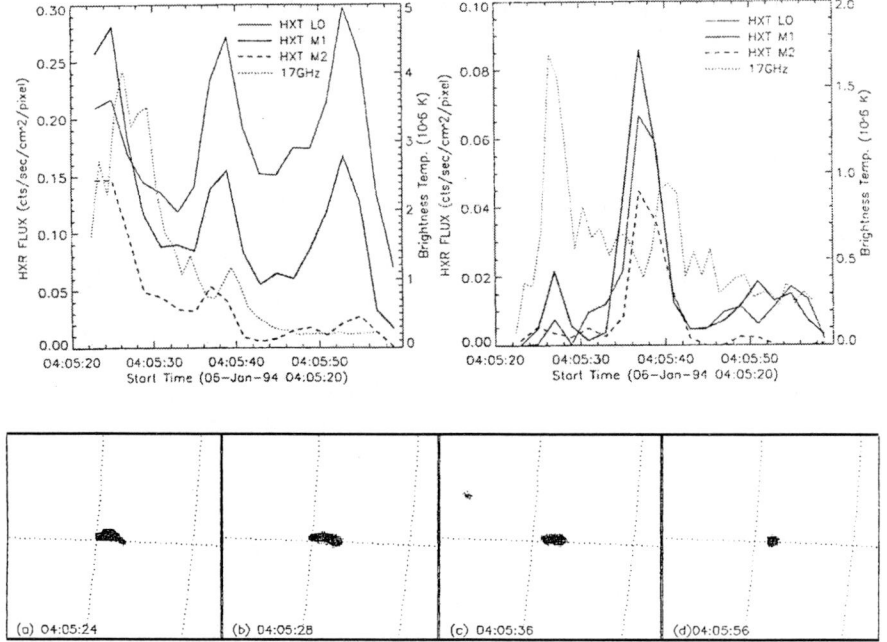

Fig. 6. Top: X-ray (L0: 13.9- 22.7 keV, M1: 22.7-32.7 keV, M2: 32.7-52.7 keV) and 17 GHz light curves for the double sources (left) and the remote source (right) of the 6 January 1994 event observed with YOHKOH/HXT and the Nobeyama Radioheliograph. Bottom: HXT M1 images at different times showing both the double and the remote sources (c) (from [104]).

is reminiscent of the observations at much higher energies [105, 100] of the production of very energetic electrons in association with remote Hα bright features appearing at the border of the main flare site. This shows that as was already suggested by e.g. [106, 93], efficient particle acceleration results from the interaction of many magnetic structures and that in addition there is a link between spatial and spectral variability of the energetic electrons.

Additional information on spatial variability of the energetic electrons on the shorter time scales than a few tens of seconds which are more representative of acceleration timescales (see Sect. 5) have been obtained using fast localization of millimeter emission. Multiple beam techniques applied at 48 GHz at Itapetinga provide the centroïd positions of bursts with a spatial accuracy of 5″ to 20″ and a time resolution down to a few seconds. In many flares this techique has shown rapid changes of the localization of time structures of a few seconds duration over a distance of the order of 10″ in the course of a flare (see e.g. [108]). Figure 7 shows such fast variations of position of the centroïd of radio pulses of a few seconds duration at 48 GHz superposed on HXR images provided by YOHKOH/HXT in two energy bands [107]. In contrast to the HXR source which shows essentially a single compact (10″), unresolved source (except for a faint

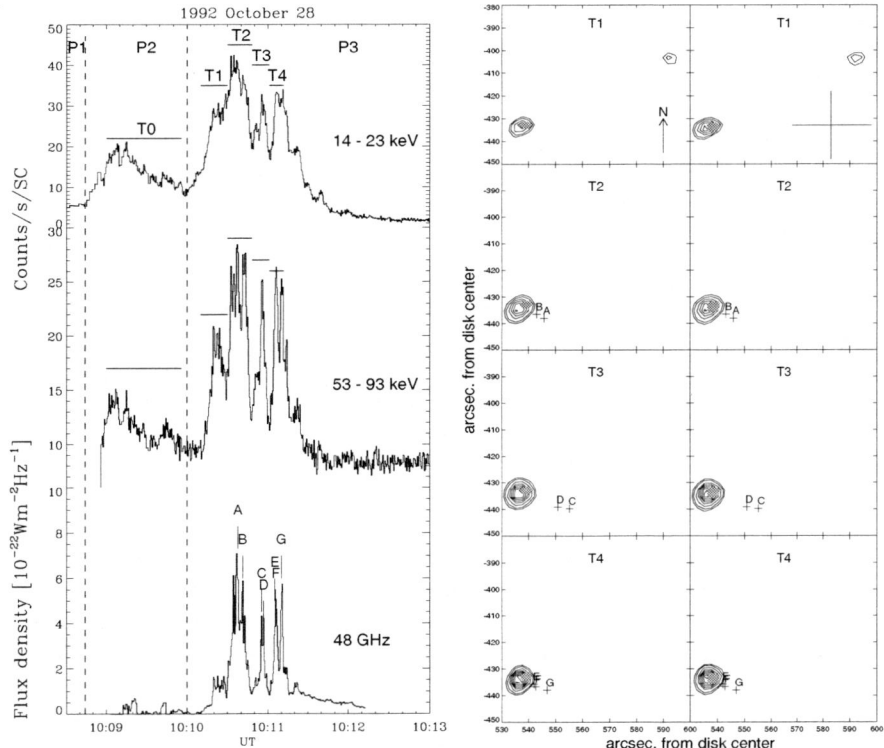

Fig. 7. Left: Temporal evolution of the HXR emission (14-23 and 53-93 keV) observed by YOHKOH/HXT and of the radio flux density at 48 GHz observed at Itapetinga. Thick horizontal bars show time intervals (T0-T4) during which HXR contour maps have been performed. Letters A to G refer to different temporal fast structures. Right: Contours of the HXR emission observed during T1-T4 in the 53-93 keV (left column) and 33-53 keV (right column). The gray patches show the HXR emission observed during T0. The crosses A to G show the position of the 48 GHz pulses during T2, T3 and T4. The small (big) cross sizes represent the relative (absolute) position of the radio pulses A-G (from [107]).

remote HXR source), the different pulses arise from positions separated by a few arcseconds. This clearly shows that the spatial resolution of the X-ray imagers is not sufficient to resolve the small scales where magnetic reconnection and energy conversion takes place. The radio observations show on the other hand that electron acceleration occurs at discrete sites, the locations of which vary on time scales of the order of a few seconds in the primary energy release volume. As for the events described above, some variations of electron numbers and spectra are found on time scales of a few tens of seconds (P1, P2, P3 in Fig. 7). On the shorter time scales ($\simeq$ few seconds) on which spatial variations are seen at millimeter wavelengths, however, no information is available on any variation of electron spectra or numbers.

As a conclusion, all the observations recalled in this section clearly show the role of the magnetic environment on the characteristics of the energetic particles and thus on the efficiency of the different acceleration processes and transport. Clearly, further probing of the link between spectral and spatial variability will be a major topic for HXR and possibly γ-ray observations from RHESSI, in combination with spatially resolved observations in the whole radio domain and at optical wavelengths.

8 Bremsstrahlung and Synchrotron Emitting Electrons

8.1 Consistency Between HXR and Microwave Radiating Electrons: Transport of Microwave Emitting Electrons and Implications for Angular Distributions of Accelerated Electrons

Complementary observations of energetic electrons are provided by the gyrosynchrotron emission that they produce in the low corona at centimeter/millimeter wavelengths. First studied by [110], the relationship between the HXR and microwave emitting electrons has been a topic with a long and controversial history. In particular, as summarised in e.g. [111], attempts to compare the numbers of X-ray and microwave emitting electrons ran into a serious discrepancy: the number of electrons necessary to produce the HXR emission was found to be 10^3 to 10^5 times larger than the number required to produce microwave emission. However, as pointed out in e.g. [112], the apparent discrepancy most likely resulted from the comparison of instantaneous electron numbers deduced separately from HXRs and microwaves, and from the assumption of a strong magnetic field in the microwave-emitting region. In practice, if electron lifetimes are short compared to instrumental resolution and electrons stop completely in the source (i.e. the source is a thick target), such a calculation underestimates the efficiency of HXR production. [111] showed that the discrepancy can in fact be removed for impulsive flares using a precipitation model for both X-ray and microwave electrons. In such a model, the same population of energetic electrons gives rise to microwave radiation in a moderate magnetic field while precipitating to the dense chromosphere, effectively a thick target, where HXR emission is efficiently produced. This study was however performed using single frequency measurements as well as flare-integrated X-ray emission, and neglected any effects of magnetic field convergence on precipitating electrons. The introduction of trap-plus-precipitation electron models allowed a more complete study of electron evolution in the trap region where microwave emission is produced (e.g. [113]), showing that both microwave and X-ray spectral evolution with time could be reproduced by the same injected (accelerated) electron population. Further developments of such models including betatron energy losses of relativistic electrons were carried out in [46] and applied to a large gradual flare, showing again that the temporal and spectral evolution of X-rays and microwaves emissions can be explained by a common source of electrons injected in trap regions. However, no constraints from HXR nor centimeter/millimeter images were available for these studies.

Combined spatially resolved observations of HXR and centimeter emissions have been obtained more recently using the SXT and HXT experiments aboard YOHKOH as well as the Nobeyama Radioheliograph (NoRH) and Owens Valley Radio Observatory Solar Array OVRO [109]. Figure 8 shows images obtained at 7 GHz with OVRO and 17 GHz with NoRH as well as HXT images for an impulsive flare. Although some diffuse component extending along the HXR source is seen at 17 GHz, a single compact radio source is found at only one of the HXR footpoints. This surprising observation is however well understood in terms of the intrinsic directionality of gyrosynchrotron radiation and the magnetic field geometry: the coronal magnetic field extrapolation shows a clear tendency for the magnetic field inclination angle to be close to 90° at the east footpoint where emission can thus be observed and close to 0° for the western footpoint where no emission is observed. Furthermore, the coronal magnetic field extrapolation reveals that the field strength does not strongly vary over the loop leading to a more or less uniform HXR loop structure with a small magnetic mirror ratio. In a trap plus precipitation model this leads to a high rate of electron precipitation losses naturally explaining the spectral evolution, in particular the steepening of the microwave spectrum in the decay phase of the emission (Fig. 8). On the other hand, an earlier flare occurring in another loop system of the same active region showed significantly different spatial and spectral characteristics. Indeed, while a single source was seen at 5 GHz with OVRO close to the top of the magnetic loops deduced from magnetic field extrapolation, double sources were seen at 10.6 and 17 GHz with respectively OVRO and NoRH on both sides of the magnetic loops. One of the sources was however found to be higher in the loops, presumably due to the higher magnetic mirror ratio found from coronal magnetic field extrapolation. The more efficient electron trapping in this magnetic loop system naturally explains the smooth microwave time profiles observed at different frequencies, together with a time delay between the peak times of the flux at different frequencies and the flattening of the optically thin part of the microwave spectrum in the decay of the event.

The modelling of the transport of the microwave emitting high energy electrons in such a loop system with magnetic mirroring was developed in [114]. The temporal evolution of the energy and pitch angle distribution of the energetic electrons together with the radiation at different frequencies can then be estimated, assuming some initial state for the electrons injected at the top of the loop, characterised by the mean and variance of an injected Gaussian pitch-angle distribution, a power-law energy spectrum and some specified, extended time profile of injection (Fig. 9). It is found that the initial pitch-angle distribution has a determining role in the temporal evolution of the microwave emitting flux at different frequencies. Applied to one of the events reported in [109], this method led to the conclusion that the accelerated electron population is produced with an initial narrow beam ($\leq 30^o$) distribution. Clearly further employment of this technique, combining spectrally and spatially resolved mi-

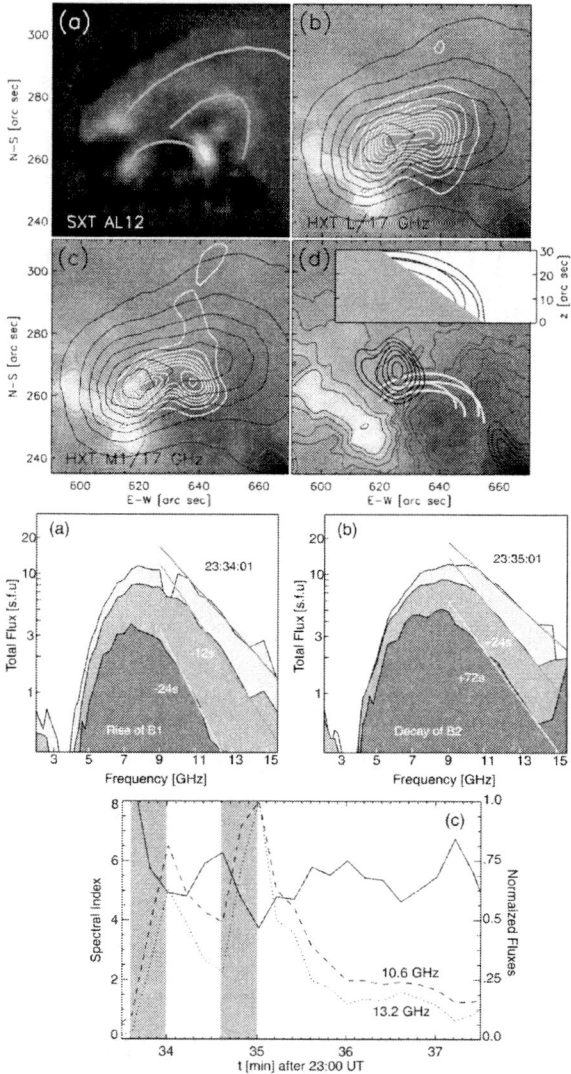

Fig. 8. Top: Radio and X-ray morphology of an event observed with YOHKOH, Nobeyama radioheliograph (NoRH) and OVRO. (a) SXT image with loops (gray solid curves). (b) 17 GHz (black contours) and HXT L0-band (white contours). (c) 17 GHz and HXT M1-band. (d) 7 GHz map (black contours) and extrapolated magnetic field lines selected to represent the loop. The inset in (d) represents a different projection of the field lines to visualise their height. Middle: Spectral variation in the radio domain. Selected times are denoted relative to time of the maximum fluxes. Note the steepening of the late spectra. Bottom: Time profiles of the microwave spectral index (solid line) and relative fluxes at 10.6 and 13.2 GHz. The shaded areas indicate the rise periods of the electron injections (from [109]).

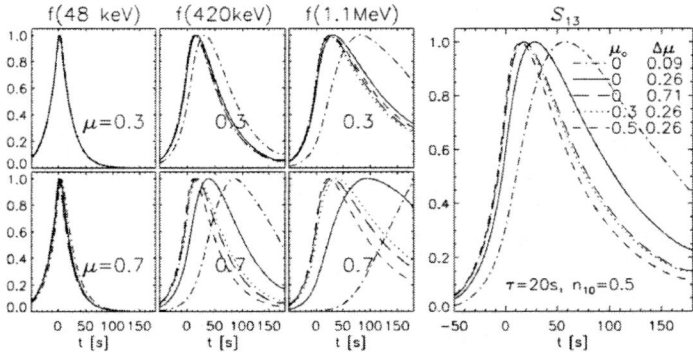

Fig. 9. Relationship between the electron injection and the radiation time profiles under a continuous injection with an exponential time profile with a half time of 20 s. Different injected beamwidths $\Delta\mu$ around a mean injection pitch angle μ_0 are considered. The ambient density is $5\ 10^{15}$ m^{-3}. Left: normalised time profiles of electrons at three energies and two pitch angles. Right: Resulting normalised 13 GHz time profile for different injections (from [114]).

crowave and HXR observations with magnetic field extrapolations to give unique information on accelerated electron angular distribution, will be important in the near future.

8.2 Electron Broken Energy Spectra

We shall focus here on results related to the comparison of bremsstrahlung and gyrosynchrotron emitting electrons in a wide energy range obtained from observations of HXR/GR spectra over a wide energy band and of centimeter/millimeter radiation in a wide frequency range. This topic has been extensively discussed in the literature (e.g. [17, 115, 116]). The relationship between the spectral information deduced from hard X-ray bremsstrahlung emitting electrons and from gyrosynchrotron emitting electrons can be best understood in the light of the spectral hardening sometimes observed at high energies (Sect. 6.2). Indeed, it had been suggested by many authors (e.g. [117, 118, 115, 17]) that millimeter wave emission (at e.g. 86 GHz) is produced by high energy electrons (above or around 1 MeV) characterised by a spectrum much flatter than the one deduced from X-ray observations around 100 keV. It is clear that the observed spectral hardening of the HXR/GR spectrum above a few hundred keV strongly supports this suggestion. Furthermore, simple attempts to relate the spectral slopes of bremsstrahlung and synchrotron emitting electrons for cases where observation in a wide frequency range and a wide energy range were available have shown that the centimeter/millimeter emitting electrons are related to the hard, high energy region of the HXR/GR spectrum [68]. More recently, the first detection of impulsive radio emission at 212 GHz coupled with the observation of the gyrosynchrotron spectrum between a few GHz and 20 GHz has also revealed that the very high frequency emission results from a population of ultra relativistic electrons with a hard energy spectrum [119]. The relationship between these electrons and any γ-ray emitting population remains an open question due to the lack of simultaneous, high-energy γ-ray observations. As evidenced by many observations in the millimeter domain, the flat part of the spectrum must also be present in the early beginning phase of the flare. The influence of the hardening of the electron spectrum at high energies on the gyrosynchrotron emission in solar flares has been modelled in detail by [120]. Observations at present have nothing to say on whether the electron populations observed at low and high energies result from the same acceleration process, or if they constitute independent accelerated populations. In the future, the combination of spatially resolved observations of the HXR/GR continuum below and above the break energy and of spatially resolved centimeter/millimeter observations will provide new clues to answer this question.

9 Where Are the Acceleration Sites?

From hard X-ray and radio observations, it is known that electron acceleration occurs in a wide range of heights in the solar atmosphere. Different studies based e.g. on the frequency drifts of decimeter type III bursts or on common starting frequencies of pairs of opposite drifting bursts indicate that electron beam acceleration arise from a medium with a density ranging from 10^{15} to 10^{17} m^{-3} (e.g. [121, 122]). Spatially resolved observations of narrowband metric radio spikes and of associated metric type III bursts also show that narrowband metric spikes are closely related to the electron acceleration region ([23]; Benz, *this volume*). All these observations suggest acceleration heights around 10^7 m. The bulk of the non-thermal electrons produced in flares is also probably produced at heights of a few 10^7 m. This is consistent with some X-ray images from YOHKOH/HXT suggesting that the emission and thus the energy release appears first at the top of magnetic loops (e.g. [123]; Scholer, *this volume*). The acceleration region possibly lies in the cusp region one would find in an arcade of flare loops (e.g. [124]). However, combined analysis of millimeter and X-ray observations also show the electron acceleration site must be close to the interaction region of magnetic loops (e.g. [125]). It is worth noting that these more recent and complete observations are consistent with earlier suggestions of energy release region at the site of interacting magnetic structures using SMM/HXIS observations (e.g. [106, 93]). As far as the ions are concerned, no direct information is available on the location of the acceleration site. It is probable that ions are accelerated in the same region as electrons, given the simultaneity of radiation from electrons and ions generally observed. Furthermore, the observation of γ-ray line radiation from thin target coronal sources (e.g. [126, 72]) also suggests that ions must be accelerated in the low corona at heights of a few 10^7 m.

Significant acceleration of energetic electrons of a few tens of keV also occurs at heights as high as a few 10^8 m. This was shown by e.g. the combined observations of an X-ray flare by experiments aboard two spacecraft [94] with different viewing angles. This is also well known from radio emission of electron beams moving downward from coronal acceleration sites [127]. In the case of large flares, these coronal acceleration sites at heights of a few 10^8 m may also be observed, delayed in time in comparison with the main acceleration episode occurring during the impulsive phase of the flare. It has been suggested by e.g. [24] that for a large flare this coronal acceleration region may also be the site of production of the relativistic protons injected in the interplanetary medium. The acceleration will result from the reconfiguration of coronal large scale magnetic loops subsequent to the propagation from the flare site of the magnetic field surrounding the eruptive filament. Such coronal acceleration sites may also result from the propagation of plasmoids ejected at the time of the flare (see e.g. [128]). They may be the site of the production of energetic electrons, but also potentially of ions in the extended phases of large flares, sometimes detected even at high photon energies (e.g. [7, 129]). A better determination of the acceleration sites of energetic electrons and potentially of energetic ions in flares is clearly one of the goals of the imaging spectroscopy provided by the RHESSI mission. This should improve our understanding of the acceleration mechanisms which provide the flare accelerated particles in the low corona as well as the energetic particles in the high coronal sites.

10 Summary of the Constraints Inferred from Multi-wavelengths Observations of Accelerated Particles and Implication for the Different Acceleration Processes

The multiwavelengths observations of flare energetic particles referred to in the previous sections provide the constraints any particle acceleration mechanism(s) must satisfy. In Table 1, the extreme properties of the accelerated electrons and ions inferred from various observations are summarised.

In addition to the constraints summarised in Table 1, energetic ions must be produced with some enrichment in α particles as well as enrichments in ^{3}He

Table 1. Extreme properties of accelerated electrons and ions

	Electrons	Ions
Number	10^{41} (> 20 keV)	$3\ 10^{35}$ (> 30 MeV)
	10^{36} (> 100 keV)	10^{32} (> 300 MeV)
Acc Times	$\simeq$ 100 ms @ 100 keV-1 MeV	< 1s @ 10 MeV
Duration(s)	10 s to hour	60 s to hour
Total Energy (J)	10^{27} (> 20 keV)	10^{25}-10^{26} (> 1 MeV)
	10^{22} (> 100 keV)	10^{23} (> 30 MeV)
Angular Distribution	Narrow beam ?(see 8.1)	Fan beam or isotropic distribution
	1 flare	2 flares (see 6.3)

and in heavy ions (Ne, Mg, Fe) compared to coronal abundances. These enhancements as well as the e/p ratio and the particle spectra vary not only from flare to flare but also in the course of a flare. Finally, there seems to be links between the charactcristics of the accelerated particles and the magnetic structures traced by the particles. Flare models must be able to reproduce this complete set of observations but we shall aim the discussion here on the (microscopic) acceleration mechanisms which can lead to the observed properties of accelerated particles.

As recalled in many reviews (e.g. [53]), only a direct or induced electric field can energise a charged particle. As usual in astrophysics, three different physical mechanisms are considered: direct electric field acceleration (the electric field being in the sub-Dreicer or super Dreicer regime (see below)), stochastic acceleration (implying e.g. wave-particle interactions) and acceleration by shocks. These different processes were reviewed in e.g. [130, 131] (cf. Dröge, Litvinenko, Schlickeiser, *this volume*). Although the ability of shocks to accelerate ions to MeV energies in the interplanetary medium is well established, their role in accelerating large electron fluxes necessary to produce the HXR emissions and also large ion fluxes to high energies has not been clarified. In particular, difficulties arise for the production of large fluxes of energetic electrons due to the electron injection energy required for the shock acceleration to proceed (see e.g. [130]).

Acceleration by direct large scale electric field has been investigated by e.g. [65] and reviewed in e.g. [66]. The characteristics of models based on acceleration by DC electric fields basically depend on the strength of the electric field with respect to the Dreicer field (i.e. field large enough to balance the drag force due to electron current and for which the entire thermal electron distribution can be accelerated). Its value in the corona is typically of the order of 1 Vm^{-1}. The energy gain is typically of 1 MeV for an electron or a proton in the case of e.g. a sub Dreicer field of 10^{-1} V m^{-1} (typical value in a reconnection region) but operating on a large scale of 10^7 m, the acceleration timescale being of the order of 0.04 s for 1 MeV electrons. Although timescales and energies can be reproduced for electrons, the existence in the coronal plasma of an electric field on such a large scale is questionable. In analogy to what occurs in the auroral arcs and based on the shape of the electron spectrum deduced from high spectral resolution X-ray measurements (see Sect. 6.1), sub-Dreicer acceleration in multiple current sheets of smaller spatial extent (10000 current sheets) has been proposed by [65] to produce simultaneously electron acceleration and heating in a way consistent with the high spectral resolution X-ray observations of [64] (Fig. 10). This produces an electron distribution that varies as $\text{E}^{-1/2}$ up to an energy determined by the magnitude of the potential drop, with an exponential accelerated tail. It is worth noting that such an exponential law behaviour for the energetic tail may be also found in the model of electron acceleration by random DC electric fields resulting from the SOC evolution of a complex and evolving inhomogeneous region in the case of a long trapping time in the acceleration region [132]. However, no significant ion acceleration is expected from the sub-Dreicer acceleration in these multiple current sheets. As discussed in [66], the acceleration of a few MeV/nuc ions becomes possible when the electric

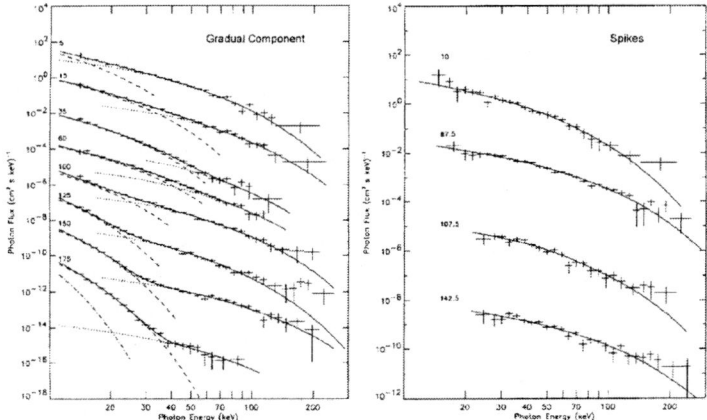

Fig. 10. Electric field acceleration model fits (from [65]) to the high resolution X-ray spectra (from [64]) from the 27 June 1980 flare. Left: for the gradual component. Right: for the spikes (from [66]).

field reaches the Dreicer limit. An interesting feature of the acceleration of ions by a given electric field is that due to the dependence of the energy gain on the charge of the ion, there is a charge dependent threshold for the acceleration of ions, the acceleration of protons requiring the highest electric field strength. An enhancement of ions with low energy thresholds (e.g. Fe, Mg, C) may thus result from such an acceleration. However due to the drag force of the electron current, the energy reached by the ions is not sufficient to produce γ-ray lines. Only the acceleration in a super-Dreicer field can lead to the production of γ-ray line producing ions as well as to relativistic electrons. Such a high electric field ($\simeq 10^3$ V m^{-1}) may be present in reconnecting current sheets. This acceleration mechanism, in which modifications to particle trajectories due to the magnetic field play an important role, has been studied in detail recently (e.g. [133, 102] and in this volume). In particular, considering a 3D magnetic field topology in the reconnecting current sheet, [102] shows that electrons can be accelerated to energies as high as 10 MeV in less than 10^{-3} s while the heavier "unmagnetised" ions (i.e. staying a shorter duration in the current sheet) have a lower energy gain. Furthermore, electrons or ions may be preferentially accelerated in sheets depending on the relative magnitude of the various field components. Such a mechanism which may reproduce a high electron-to-proton ratio and intense γ-ray continuum above 1 MeV could be the source of electron-dominated episodes in flares.

Alternatively, stochastic resonant acceleration by wave-particle interaction has also been extensively studied recently by e.g. [134] with applications to electron-dominated events, by Schlickeiser (*this volume*) and by Miller and coworkers (e.g. [135, 136]). Simultaneous acceleration of electrons and ions may occur when the ambient quasithermal particles undergo stochastic resonant acceleration by cascading Alfvén and fast mode waves initially generated at long wavelengths during the primary flare energy release phase. Using quasilinear simulations of wave-particle interactions, it is found that the particles are energised on subsecond time scales and in large numbers in good consistency with the observations. It is furthermore shown that depending on the length of the acceleration region, a varying ratio of escaping electrons and protons is to be expected with electron-dominated flares resulting from shorter acceleration regions. Stochastic resonant acceleration by wave-particle interaction on cascading Alfvén waves may also reproduce enhancements in the abundances of energetic heavy ions. Indeed, due to the resonance condition, the cascading waves will first resonate (when they still contain a large energy) with the heavier ions (having lower cyclotron frequencies) and then will go on resonating (with diminishing efficiency) with ions with higher cyclotron frequencies [137]. Although these acceleration mechanisms seem to reproduce most of the observations of solar flare accelerated particles, the main remaining question is the production of the initial waves with a large enough energy and for a sufficient duration. No direct indication of the wave level and of the wave spectrum in solar flares is available to check the consistency of these acceleration models.

11 Conclusions and Future Observations

The flare particle acceleration process is necessarily mysterious. Its coronal location (Sect. 9) places it at a remove from both the regions where we know preflare magnetic fields, and the regions where its products produce their most direct radiation signatures. Nonetheless, as we have discussed, observations at various wavelengths, alone and in combination, allow us to home in on some of its properties.

We have a good idea of the numbers and energy distribution of electrons accelerated in flares, obtained initially from HXR's and corroborated by interpretation of microwave observations in a way that sheds light on important factors in transport. There is good evidence from both radio and HXR data for an additional, hard electron population towards relativistic energies, varying both within a single flare and from one flare to the next, whose occurrence is particularly associated with changes in spatial structure. Many pieces of evidence point to a coronal site for electron acceleration, e.g.: simultaneous interpretation of microwave and HXR observations; time-of-flight interpretations of short timescale HXR spectral softening; spatial locations of metric radio spikes. The second and third of these also point to a highly fragmented acceleration process, consistent in a general way with some sort of avalanche process e.g. as in a state of SOC. Success in time-of-flight modelling also seems to point to a free streaming transport regime, albeit in a converging magnetic loop, simpler than we have any a priori right to expect. Such analyses also appear to imply an acceleration mechanism that can accelerate electrons to at least 100 keV in less than a tenth of a second or so, and ions to several MeV energies on times probably less than one second. Pitch angle information is harder to come by, although recent developments in interpretation of microwaves point to a way forward here, for electrons at least.

Some uncertainty still surrounds the number and energy distribution of fast ions. In principle, deexcitation GR lines give clear information on these, but the data so far are rather too noisy to allow, in particular, a clear discussion of the relative abundances of accelerated protons and αs. There is, however, evidence again for a coronal origin and for variation even during individual flares of fast ion relative abundances.

One major question still unresolved is the lowest energy to which the inferred ion and electron distributions hold. This is a question of major interest for flare physics generally: for these steeply declining distributions, the total energy content is governed by the values of lower energy cutoffs. In spite of this, and the uncertainties mentioned above and detailed in the relevant sections, it does appear that electrons above 20 keV energy often embody a significant fraction of the total flare energy (although definitely less than 100%); and that ions above 1 MeV/nucleon may rival or even exceed them.

It is clear that we can learn much about the workings and products of the flare particle accelerator from multi-wavelengths observations. In the final analysis, however, they must be used with other considerations in attempts to decide between competing theoretical possibilities. Particularly in view of the energy contents now mentioned many times, comparison of candidate acceleration mechanisms must involve their acceptability in terms of an overall picture of flare energy storage and release. If we invoke reconnection region electric fields, we must ensure that the total volume and rate of resupply of plasma to the reconnection region are adequate for the deduced rate of particle acceleration. If we invoke plasma turbulence or shock mechanisms, we must be able to demonstrate that a large enough fraction of the released flare energy can make its way first into the turbulence or shocks, and then into accelerated particles. We may note further that such a requirement probably drives particle acceleration modelling into a nonlinear regime where test particle discussions are no longer adequate (implying e.g. a self-consistent discussion of collisionless reconnection, along the lines sketched by [138], cf. also Biskamp, *this volume*; or a reconsideration of shock structure, e.g. [139]).

Moreover: reconnection certainly involves electric fields; strongly driven flows will certainly develop turbulence and/or solitary structure. All three main classes of candidate mechanism may well occur to varying degrees, possibly with different roles (e.g. one mechanism might be responsible for the bulk of the energy manifested as fast particles, but a different one handle the task of giving a tiny minority of particles the most extreme energies). Conditions will vary during a single flare, for instance as a result of chromospheric evaporation. It appears (some of the evidence has been given above) that several magnetic structures may be involved at various times during a single flare, each presumably with different physical conditions. The relative importance of simultaneously occurring different mechanisms might vary in consequence. The phenomenon of electron-rich events (or perhaps we should say 'episodes'), occurring briefly within single flares, may well signal such an occurrence.

If there is one general lesson to be drawn from the work reviewed here, it is the importance of taking a multi-wavelength view of flare phenomena. RHESSI will give us GR line spectra of unprecedented resolution and statistical quality, and the first ever images at $h\nu \geq 100$ keV. It should give us the first view of the interaction and acceleration regions of relativistic electrons and fast ions, and GR line shapes will yield information on fast ion pitch-angle distribution. The value of its data will be further enhanced when used together with data at radio and other wavelength ranges. The meaning of RHESSI observations will be clarified, and they will feed into the interpretation of these other wavelengths ranges. Flare physics has an exciting time ahead!

Acknowledgements

A MacKinnon thanks Mark Toner and Keith Macpherson for valuable discussions. The authors are grateful to Dr Gordon Holman for his fruitful comments and remarks on the manuscript. They would finally like to thank Dr K.-L. Klein for his careful reading of the manuscript and for valuable discussions.

References

1. J.P. Raulin, A. Kerdraon, K.L. Klein, et al.: Astr. Ap. **251**, 298 (1991)
2. N. Crosby, N. Vilmer, N. Lund, et al.: Solar Phys. **167**, 333 (1996)
3. R.D. Bentley, K.L. Klein, L. van Driel-Gesztelyi, et al.: Solar Phys. **193**, 227 (2000)
4. T. Bai, R. Ramaty: Solar Phys. **49**, 343 (1976)
5. V.V. Akimov, V.G. Afanasev, A.S. Belousov, et al.: Soviet Astronomy Letters **18**, 69 (1992)
6. N.G. Leikov, V.V. Akimov, V.A. Volzhenskaia, et al.: Astr. Ap. Supp. **97**, 345 (1993)
7. G. Kanbach, D.L. Bertsch, C.E. Fichtel, et al.: Astr. Ap. Supp. **97**, 349 (1993)
8. R. Ramaty: "Nuclear processes in solar flares", in *Physics of the Sun* Vol. 2 (1986), p. 291
9. R.J. Murphy, C.D. Dermer, R. Ramaty: Ap. J. Supp. **63**, 721 (1987)
10. E.L. Chupp: Solar Phys. **118**, 137 (1988)
11. L. Kocharov, H. Debrunner, G. Kovaltsov, et al.: Astr. Ap. **340**, 257 (1998)
12. N. Vilmer, M. Maksimovic, R.P. Lin, et al.: "Ion acceleration in solar flares: low energy neutron measurements", in *Solar Encounter: Proc. First Solar Orbiter Workshop, ESA-SP 493* (2001), p. 405
13. T.S. Bastian: "Impulsive Flares: A Microwave Perspective", in *Proc. Nobeyama Symp.*, ed. by T.S. Bastian, N. Gopalswamy, K. Shibasaki (NRO Report 479, 1999), p. 211
14. D.E. Gary, G.J. Hurford: "OVRO Solar Array Upgrades in Preparation for MAX 2000", in *Proc. Nobeyama Symp.*, ed. by T.S. Bastian, N. Gopalswamy, K. Shibasaki (NRO Report 479, 1999), p. 429

15. M. Nishio, H. Nakajima, S. Enome, et al.: "The Nobeyama Radioheliograph — Hardware System", in *Proc. Kofu Symp.* (NRO Report 360, 1994), p. 19
16. T. Takano, H. Nakajima, S. Enome, et al.: "An Upgrade of Nobeyama Radioheliograph to a Dual-Frequency (17 and 34 GHz) System", in *Coronal Physics from Radio and Space Observations*, ed. by G. Trottet (Springer, 1997), Vol. 483 of LNP, p. 183
17. S.M. White: "Millimeter Interferometer Observations of Flares", in *Proc. Nobeyama Symp.*, ed. by T.S. Bastian, N. Gopalswamy, K. Shibasaki (NRO Report 479, 1999), p. 223
18. R. Herrmann, A. Magun, J.E.R. Costa, et al.: Solar Phys. **142**, 157 (1992)
19. J.E.R. Costa, E. Correia, P. Kaufmann, et al.: Solar Phys. **159**, 157 (1995)
20. C.G. Giménez de Castro, J.P. Raulin, V.S. Makhmutov, et al.: Astr. Ap. Supp. **140**, 373 (1999)
21. P. Kaufmann, A. Magun, H. Levato, et al.: "Solar Observations at Submm-Waves", in *Recent Insights into the Physics of the Sun and Heliosphere: Highlights from SOHO and other Space Missions* (2000), IAU Symp. no. 203.
22. A. Kerdraon, J. Delouis: "The Nançay Radioheliograph", in *Coronal Physics from Radio and Space Observations*, ed. by G. Trottet (Springer, 1997), Vol. 483 of LNP, p. 192
23. G. Paesold, A.O. Benz, K.L. Klein, N. Vilmer: Astr. Ap. **371**, 333 (2001)
24. K.L. Klein, G. Trottet, P. Lantos, J.P. Delaboudinière: Astr. Ap. **373**, 1073 (2001)
25. J.C. Hénoux: "Optical View of Particle Acceleration and Complementarity with HESSI", in *High Energy Solar Physics Workshop - Anticipating HESSI*, ed. by R. Ramaty, N. Mandzhavidze (2000), Vol. 206 of ASP Conf. Ser., p. 27
26. G. Trottet, E. Rolli, A. Magun, et al.: Astr. Ap. **356**, 1067 (2000)
27. L. McDonald, L.K. Harra-Murnion, J.L. Culhane: Solar Phys. **185**, 323 (1999)
28. A.G. Emslie, J.A. Miller, E. Vogt, et al.: Ap. J. **542**, 513 (2000)
29. R.C. Canfield, C.R. Chang: Ap. J. **295**, 275 (1985)
30. B.E. Woodgate, R.D. Robinson, K.G. Carpenter, et al.: Ap. J. **397**, L95 (1992)
31. C. de Jager, G. de Jonge: Solar Phys. **58**, 127 (1978)
32. N.B. Crosby, M.J. Aschwanden, B.R. Dennis: Solar Phys. **143**, 275 (1993)
33. E.T. Lu, R.J. Hamilton: Ap. J. **380**, L89 (1991)
34. P. Bak, C. Tang, K. Wiesenfeld: PhRevA **38**, 364 (1988)
35. H.J. Jensen: *Self Organized Criticality* (Cambridge University Press, Cambridge, 1998)
36. K. Galsgaard, Å. Nordlund: Journ. Geophys. Res. **101**, 13 445 (1996)
37. K.P. MacPherson, A.L. MacKinnon: Astr. Ap. **350**, 1040 (1999)
38. S. Chapman, N. Watkins: Space Science Reviews **95**, 293 (2001)
39. H.S. Hudson: Solar Phys. **133**, 357 (1991)
40. M.J. Aschwanden, B. Kliem, U. Schwarz, et al.: Ap. J. **505**, 941 (1998)
41. P. Kaufmann, E. Correia, J.E.R. Costa, et al.: Solar Phys. **95**, 155 (1985)
42. A. Pacini, J.P. Raulin, P. Kaufmann, et al.: "Statistical study of simple and impulsive solar microwave flares", in *RAS discussion meeting on Self Organized Criticality and Turbulence in the Solar System* (2000)
43. P. Kaufmann, E. Correia, J.E.R. Costa, et al.: Solar Phys. **91**, 359 (1984)
44. T. Bai, R. Ramaty: Ap. J. **227**, 1072 (1979)
45. N. Vilmer, G. Trottet, S.R. Kane: Astr. Ap. **108**, 306 (1982)
46. G. Bruggmann, N. Vilmer, K.L. Klein, S.R. Kane: Solar Phys. **149**, 171 (1994)
47. A.G. Emslie: Ap. J. **271**, 367 (1983)
48. M.J. Aschwanden, R.A. Schwartz, D.M. Alt: Ap. J. **447**, 923 (1995)

49. T.N. Larosa, S.N. Shore: Ap. J. **503**, 429 (1998)
50. M.J. Aschwanden, R.A. Schwartz, B.R. Dennis: Ap. J. **502**, 468 (1998)
51. M.J. Aschwanden, L. Fletcher, T. Sakao, et al.: Ap. J. **517**, 977 (1999)
52. D.B. Melrose, G.A. Dulk: Ap. J. **259**, L41 (1982)
53. E.L. Chupp: "Evolution of our Understanding of Solar Flare Particle Acceleration: (1942-1995)", in *High Energy Solar Physics*, ed. by R. Ramaty, N. Mandzhavidze, X.M. Hua (1996), Vol. 374 of AIP Conf. Proc., p. 3
54. G. Trottet, N. Vilmer: "The Production of Flare-Accelerated Particles at the Sun", in *Solar and Heliospheric Plasma Physics*, ed. by G. Simnett, C. Alissandrakis, L. Vlahos (Springer, 1997), Vol. 489 of LNP, p. 219
55. N. Vilmer, G.. Trottet: "Solar Flare Radio and Hard X-Ray Observations and the Avalanche Model", in *Coronal Physics from Radio and Space Observations*, ed. by G. Trottet (Springer, 1997), Vol. 483 of LNP, p. 28
56. N. Vilmer, G. Trottet, C. Verhagen, et al.: "Subsecond Time Variations in Solar Flares around 100 keV: Diagnostics of Electron Acceleration", in *High Energy Solar Physics*, ed. by R. Ramaty, N. Mandzhavidze, X.M. Hua (1996), Vol. 374 of AIP Conf. Proc., p. 311
57. E. Rieger: Ap. J. Supp. **90**, 645 (1994)
58. F. Pelaez, P. Mandrou, M. Niel, et al.: Solar Phys. **140**, 121 (1992)
59. S.R. Kane, E.L. Chupp, D.J. Forrest, et al.: Ap. J. **300**, L95 (1986)
60. P. Kaufmann, G. Trottet, C.G. Giménez de Castro, et al.: Solar Phys. **197**, 361 (2000)
61. R.P. Lin, the HESSI Team: "The High Energy Solar Spectroscopic Imager (HESSI) Mission", in *High Energy Solar Physics Workshop - Anticipating HESSI*, ed. by R. Ramaty, N. Mandzhavidze (2000), Vol. 206 of ASP Conf. Ser., p. 1
62. C.M. Johns, R.P. Lin: Solar Phys. **137**, 121 (1992)
63. J.C. Brown: Solar Phys. **18**, 489 (1971)
64. R.P. Lin, R.A. Schwartz, R.M. Pelling, et al.: Ap. J. **251**, L109 (1981)
65. S.G. Benka, G.D. Holman: Ap. J. **435**, 469 (1994)
66. G.D. Holman: "Particle Acceleration in Large-Scale DC Electric Fields", in *High Energy Solar Physics Workshop - Anticipating HESSI*, ed. by R. Ramaty, N. Mandzhavidze (2000), Vol. 206 of ASP Conf. Ser., p. 135
67. D.A. Bryant: "Rocket studies of particle structure associated with auroral arcs", in *Physics of auroral arc formation; Proc. Chapman Conference on Formation of Auroral Arcs* (AGU, 1981), p. 103
68. G. Trottet, N. Vilmer, C. Barat, et al.: Astr. Ap. **334**, 1099 (1998)
69. B.R. Dennis: Solar Phys. **118**, 49 (1988)
70. H. Marschhäuser, E. Rieger, G. Kanbach: "Temporal Evolution of Bremsstrahlung-Dominated Gamma-Ray Spectra of Solar Flares", in *High-Energy Solar Phenomena - a New Era of Spacecraft Measurements*, ed. by J. Ryan, T. Vestrand (1994), Vol. 294 of AIP Conf. Proc., p. 171
71. M. Yoshimori: Space Science Reviews **51**, 85 (1989)
72. C. Barat, G. Trottet, N. Vilmer, et al.: Ap. J. **425**, L109 (1994)
73. N. Vilmer, G.. Trottet, C. Barat, et al.: Astr. Ap. **342**, 575 (1999)
74. G.H. Share, R.J. Murphy: "Gamma Ray Spectroscopy in the Pre-HESSI Era", in *High Energy Solar Physics Workshop - Anticipating HESSI*, ed. by R. Ramaty, N. Mandzhavidze (2000), Vol. 206 of ASP Conf. Ser., p. 377
75. A.L. MacKinnon: Astr. Ap. **226**, 284 (1989)
76. G.H. Share, R.J. Murphy: Ap. J. **452**, 933 (1995)
77. R. Ramaty, N. Mandzhavidze, B. Kozlovsky, et al.: Ap. J. **455**, L193 (1995)

78. R.J. Murphy, G.H. Share, J.E. Grove, et al.: Ap. J. **490**, 883 (1997)
79. G.H. Share, R.J. Murphy: Ap. J. **508**, 876 (1998)
80. N. Mandzhavidze, R. Ramaty, B. Kozlovsky: Ap. J. **518**, 918 (1999)
81. R.J. Murphy, G.H. Share, J.R. Letaw, et al.: Ap. J. **358**, 298 (1990)
82. R.J. Murphy, R. Ramaty, D.V. Reames, et al.: Ap. J. **371**, 793 (1991)
83. D.V. Reames, J.P. Meyer, T.T. von Rosenvinge: Ap. J. Supp. **90**, 649 (1994)
84. G. Trottet, C. Barat, R. Ramaty, et al.: "Thin Target γ-ray Line Production During the 1991 June 1 Flare", in *High Energy Solar Physics*, ed. by R. Ramaty, N. Mandzhavidze, X.M. Hua (1996), Vol. 374 of AIP Conf. Proc., p. 153
85. R. Ramaty, N. Mandzhavidze, C. Barat, et al.: Ap. J. **479**, 458 (1997)
86. R.J. Murphy, G.H. Share, K.W. Delsignore, et al.: Ap. J. **510**, 1011 (1999)
87. K.R. Bromund, J.M. McTiernan, S.R. Kane: Ap. J. **455**, 733 (1995)
88. R.P. Lin, R.A. Schwartz, S.R. Kane, et al.: Ap. J. **283**, 421 (1984)
89. S.R. Kane, K. Hurley, J.M. McTiernan, et al.: Ap. J. **446**, L47 (1995)
90. N. Mandzhavidze, R. Ramaty: American Astronomical Society Meeting **28**, 858 (1996)
91. D.V. Reames, H.V. Cane, T.T. von Rosenvinge: Ap. J. **357**, 259 (1990)
92. A.G. Emslie, J.C. Brown, A.L. MacKinnon: Ap. J. **485**, 430 (1997)
93. A.M. Hernandez, M.E. Machado, N. Vilmer, et al.: Astr. Ap. **167**, 77 (1986)
94. S.R. Kane, J. McTiernan, J. Loran, et al.: Ap. J. **390**, 687 (1992)
95. E. Rieger, H. Marschhäuser: "Electron Dominated Events during Solar Flares", in *Max '91/SMM Solar Flares: Observations and Theory*, ed. by R. Winglee, A. Kiplinger (Univ. Colorado, 1990), p. 68
96. N. Vilmer, G. Trottet, C. Barat, et al.: Space Science Reviews **68**, 233 (1994)
97. B.L. Dingus, P. Sreekumar, D.L. Bertsch, et al.: "EGRET Observation of the June 30 and July 2, 1991 Energetic Solar Flares", in *High-Energy Solar Phenomena - a New Era of Spacecraft Measurements*, ed. by J. Ryan, T. Vestrand (1994), Vol. 294 of AIP Conf. Proc., p. 177
98. M. Yoshimori, A. Shiozawa, K. Suga: "Two Types of Gamma-Ray Flares", in *Proc. Nobeyama Symp.*, ed. by T.S. Bastian, N. Gopalswamy, K. Shibasaki (NRO Report 479, 1999), p. 353
99. E. Rieger, W.Q. Gan, H. Marschhäuser: Solar Phys. **183**, 123 (1998)
100. E.L. Chupp, G. Trottet, H. Marschhäuser, et al.: Astr. Ap. **275**, 602 (1993)
101. G. Trottet, E.L. Chupp, H. Marschhäuser, et al.: Astr. Ap. **288**, 647 (1994)
102. Y.E. Litvinenko: Solar Phys. **194**, 327 (2000)
103. M. Nishio, K. Yaji, T. Kosugi, et al.: Ap. J. **489**, 976 (1997)
104. M. Nishio, T. Kosugi, K. Yaji, et al.: "Nobeyama/HXT Observations of Impulsive Flares", in *Proc. Nobeyama Symp.*, ed. by T.S. Bastian, N. Gopalswamy, K. Shibasaki (NRO Report 479, 1999), p. 235
105. J. Wülser, R. Canfield, E. Rieger: "Chromospheric Response During the Gamma Ray Flare on March 10, 1989", in *Max '91/SMM Solar Flares: Observations and Theory*, ed. by R. Winglee, A. Kiplinger (Univ. Colorado, 1990), p. 149
106. M.E. Machado, C.V. Sneibrun, M.G. Rovira: Solar Phys. **99**, 189 (1985)
107. J.P. Raulin, N. Vilmer, G. Trottet, et al.: Astr. Ap. **355**, 355 (2000)
108. R. Herrmann, E. Rolli, E. Correia, et al.: Solar Phys. **149**, 155 (1994)
109. J. Lee, D.E. Gary, K. Shibasaki: Ap. J. **531**, 1109 (2000)
110. L.E. Peterson, J.R. Winckler: Journ. Geophys. Res. **64**, 697 (1959)
111. K. Kai: Solar Phys. **104**, 235 (1986)
112. D.E. Gary: Ap. J. **297**, 799 (1985)
113. K.L. Klein, G. Trottet, A. Magun: Solar Phys. **104**, 243 (1986)

114. J. Lee, D.E. Gary: Ap. J. **543**, 457 (2000)
115. A.V.R. Silva, H. Wang, D.E. Gary: "Comparison of Microwave and HXR Spectra from Solar Flares", in *Proc. Nobeyama Symp.*, ed. by T.S. Bastian, N. Gopalswamy, K. Shibasaki (NRO Report 479, 1999), p. 255
116. V.F. Melnikov, A.V.R. Silva: "Dynamics of solar flare microwave and hard X-ray spectra", in *Ninth European Meeting on Solar Physics: Magnetic Fields and Solar Processes. ESA SP-448, ed. A. Wilson.* (1999), p. 1053
117. S.M. White, M.R. Kundu: Solar Phys. **141**, 347 (1992)
118. M.R. Kundu, S.M. White, N. Gopalswamy, et al.: Ap. J. Supp. **90**, 599 (1994)
119. G. Trottet, J.P. Raulin, P. Kaufmann, et al.: Astr. Ap. **381**, 694 (2002)
120. J. Hildebrandt, A. Krüger, I.M. Chertok, et al.: Solar Phys. **181**, 337 (1998)
121. A. Benz, M. Aschwanden: in *Eruptive Solar Flares*, ed. by Z. Švestka, B.V. Jackson, M.E. Machado (Springer, 1991), Vol. 399 of LNP, p. 106
122. M.J. Aschwanden, A.O. Benz, B.R. Dennis, et al.: Ap. J. **455**, 347 (1995)
123. T. Takakura, M. Inda, K. Makishima, et al.: Pub. Astron. Soc. Jap. **45**, 737 (1993)
124. M.J. Aschwanden: "What did YOHKOH and Compton Change in Our Perception of Particle Acceleration in Solar Flares?", in *Observational Plasma Astrophysics: Five Years of YOHKOH and Beyond*, ed. by T. Watanabe, K. Kosugi, A.C. Sterling. (Kluwer, 1998), Vol. 229 of ASSL, p. 285
125. Y. Hanaoka: "Radio and X-ray Observations of the Flares Caused by Interacting Loops", in *Proc. Nobeyama Symp.*, ed. by T.S. Bastian, N. Gopalswamy, K. Shibasaki (NRO Report 479, 1999), p. 229
126. E. Hulot, N. Vilmer, E.L. Chupp, et al.: Astr. Ap. **256**, 273 (1992)
127. K.L. Klein, H. Aurass, I. Soru-Escaut, B. Kálmán: Astr. Ap. **320**, 612 (1997)
128. M.R. Kundu, A. Nindos, N. Vilmer, et al.: Ap. J. **559**, 443 (2001)
129. V.V. Akimov, P. Ambrož, A.V. Belov, et al.: Solar Phys. **166**, 107 (1996)
130. J.A. Miller, P.J. Cargill, A.G. Emslie, et al.: Journ. Geophys. Res. **102**, 14 631 (1997)
131. A. Anastasiadis: Journal of Atmospheric and Solar-Terrestrial Physics **64**, 481 (2002)
132. A. Anastasiadis, L. Vlahos, M.K. Georgoulis: Ap. J. **489**, 367 (1997)
133. Y.E. Litvinenko: Ap. J. **462**, 997 (1996)
134. B. Park, V. Petrosian, R. Schwartz: Ap. J. **489**, 358 (1997)
135. J.A. Miller, T.N. Larosa, R.L. Moore: Ap. J. **461**, 445 (1996)
136. J.A. Miller: "Stochastic Particle Acceleration in Solar Flares", in *High Energy Solar Physics Workshop - Anticipating HESSI*, ed. by R. Ramaty, N. Mandzhavidze (2000), Vol. 206 of ASP Conf. Ser., p. 145
137. J.A. Miller, D.V. Reames: "Heavy Ion Acceleration by Cascading Alfven Waves in Impulsive Solar Flares", in *High Energy Solar Physics*, ed. by R. Ramaty, N. Mandzhavidze, X.M. Hua (1996), Vol. 374 of AIP Conf. Proc., p. 450
138. P.C.H. Martens, A. Young: Ap. J.Supp. **73**, 333 (1990)
139. L.O. Drury: Reports on Progress in Physics **46**, 973 (1983)

Transport of Energy from the Corona to the Chromosphere During Flares

Petr Heinzel and Marian Karlický

Astronomical Institute, Academy of Sciences of the Czech Republic,
CZ-25165 Ondřejov, Czech Republic

Abstract. Hard X-ray (HXR) observations frequently exhibit fast temporal variations during the impulsive phase of solar flares and this is usually ascribed to the propagation of beams of accelerated particles and to the dissipation of their energy in lower layers of the solar atmosphere. As a result of fast heating and non-thermal processes, several chromospheric lines show significant impulsive brightenings. We first review observational attempts of detecting such fast (sub-second) variations of the line intensities, namely in the Hα line, and discuss the problems associated with such observations. Second, we describe new radiation-hydrodynamical (RHD) simulations of the pulse-beam heating and show how they predict both HXR and optical-line intensity variations on very short time scales. We also discuss the effect of the return current on the energy deposit in the atmosphere. Using new spatially-resolved HXR observations (RHESSI) made simultaneously with a high-cadence detection of selected optical lines, one should be able to diagnose the properties of particle beams, provided that the response of the lower atmospheric layers to beam pulses is strong enough.

1 Introduction

Generation and transport of energy in solar flares represents a complex process which takes place on various spatial and temporal scales. Among different flare models, a scenario which starts with the energy release in the corona and considers transport of this energy down to lower atmospheric layers seems to be most viable for explaining both X-ray and optical observations. We are primarily interested in the energy transport from the corona to lower layers of the solar atmosphere, while the energy generation is discussed elsewhere in this book. Moreover, the specific aim of this review is to consider fast temporal variations of optical emission from flares which are supposed to be connected to the propagation of accelerated particle beams (electrons, protons) along the flare loops and their interactions with denser atmospheric layers. According to standard 'thick-target model', beam particles do bombard the transition region, chromosphere and in some cases may even penetrate into the photosphere. Energy dissipation takes place in two ways: Coulomb interactions of beam electrons and/or protons with the ambient plasma and direct non-thermal excitation and ionization of various atoms and ions, mainly hydrogen. Other processes of energy transport from the coronal loop, like thermal conduction, wave heating or soft X-ray heating, take place on longer time scales and we will not discuss them explicitly here.

There are two basic objectives for studying fast variations of optical emission from flares. First, we intend to develop a reliable diagnostic tool for particle beams and this can be achieved by analyzing their interaction with dense plasmas. Second, we want to understand the temporal behaviour of the flaring chromosphere, i.e. its radiation hydrodynamics. Basic data for such studies are time-resolved spectral emissions in ultraviolet (UV), optical and infrared (IR) regions, obtained simultaneously and co-spatially with hard X-ray (HXR) emissions (a particular interest represent measurements of the impact polarization due to beams [1]). A complementary information on the beam propagation can be obtained from radio spectral observations. The interpretation of such data is then based on rather complex numerical RHD-simulations which we will also briefly mention here. Particular attention is devoted to the Hα line formation in the flaring chromosphere, Hα being the representative optical line for many flare studies including observations with high temporal resolution.

2 Observations of Fast Optical Variations

Fast intensity variations, mainly in the Hα line, have been reported by several authors. These were compared with microwave and HXR fluctuations. Certain correlations were found, but in most cases on time scales larger than one second. Hα is clearly well correlated with HXR, but the fine structure of fluctuations on subsecond time scales has not been revealed until recently. Previous measurements are reported and discussed in two volumes devoted to rapid fluctuations and energetic processes in flares [2] and [3], respectively. Here we will mention some of the most recent studies.

Fast Hα variations were detected by Kiplinger et al. [4] who used fast Hα camera (0.1 s resolution) and compared the light curves with the HXRBS data. Two HXR peaks lasting about 10 s are well correlated with the Hα intensity enhancement, but the fine structure correlations are not obvious.

Multiwavelength observations of two flares, using the imaging spectrographs at Locarno-Monti (Switzerland), were reported by Rolli et al. [5, 6]. Hα spectra were obtained with the time resolution 2.3 s and the imaging spectrograph obtained Balmer Hϵ and CaII H data with resolution up to 1.1 s. Spectral data were correlated with SXR and HXR from *Yohkoh* and with radio observations. The authors found that the strongest footpoint emission in the optical lines does not coincide with the sites of the particle beam injection and these footpoints are thus heated by thermal conduction. Using density-sensitive Hϵ line (its wings are strongly Stark broadened), they derived the temporal evolution of the electron density in flaring footpoints. This electron-density variation correlates well in time with the HXR emission in one footpoint, while in another one the ionization seems to be predominantly of thermal origin.

Locarno-Monti Hα observations were also used to study fast and slow chromospheric responses to non-thermal particles during a HXR/γ-ray flare [7]. In this case the temporal resolution in Hα was 0.2 s and high temporal resolution was also achieved in HXR (full-disk flux) with the PHEBUS instrument. A new

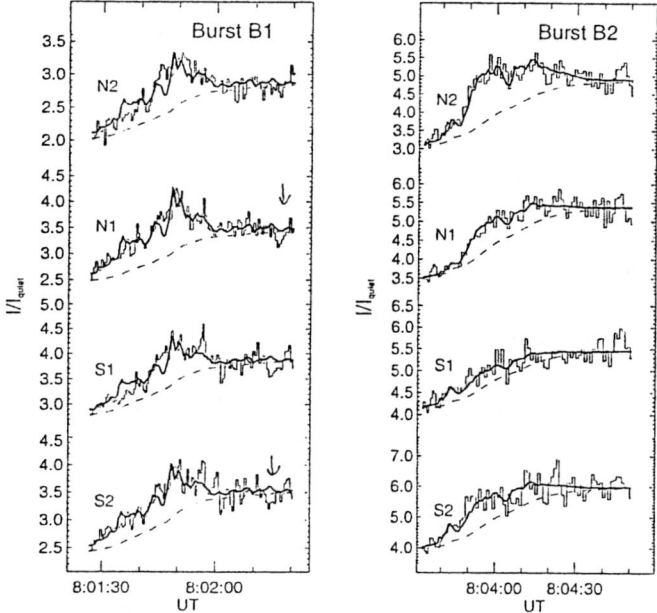

Fig. 1. Observed (thin line) and modeled (thick line) time evolution of the Hα intensity in four kernels and during two bursts. The dashed lines show the modeled slowly varying component. From [7]

idea is to model the Hα time variations using time profiles of HXR and then to compare them with the observed ones. The authors suggest the relation

$$\Delta I(t) = \frac{I(t) - I_{bg}}{I_{quiet}} = A \frac{C(t)}{C_{max}} + B \frac{\int_{t1}^{t} C(t')dt'}{\int_{t1}^{t2} C(t')dt'}, \tag{1}$$

where I is the Hα intensity, I_{bg} the backgrount intensity, I_{quiet} normalization to quiet-Sun intensity, C is the instantaneous HXR count rate (at energies larger than 73 keV), C_{max} its maximum value between times t_1 and t_2. Free coefficients A and B have been determined by χ^2-minimization. The resulting fit is shown in Fig. 1 which demonstrates generally good correlation between HXR and Hα. Fast component is represented by the first term in Eq. (1), the slow component by the second one. From the HXR time profiles, one can infer so-called *injection function* of the beam.

The energy injection function was also evaluated by [8]. Moreover, these authors have investigated correlations of fast Hα fluctuations (temporal resolution 0.27 s) between different flaring kernels. Positions of correlated kernels were compared with the magnetic connectivity between footpoints of the flaring loops.

Finally, we mention the observations of Wang et al. [9] obtained at BBSO with a narrow-band Hα filter tuned to the blue wing at 1.3 Å and with a cadence of 0.033 s. The blue wing is supposed to be more sensitive to a beam-pulse

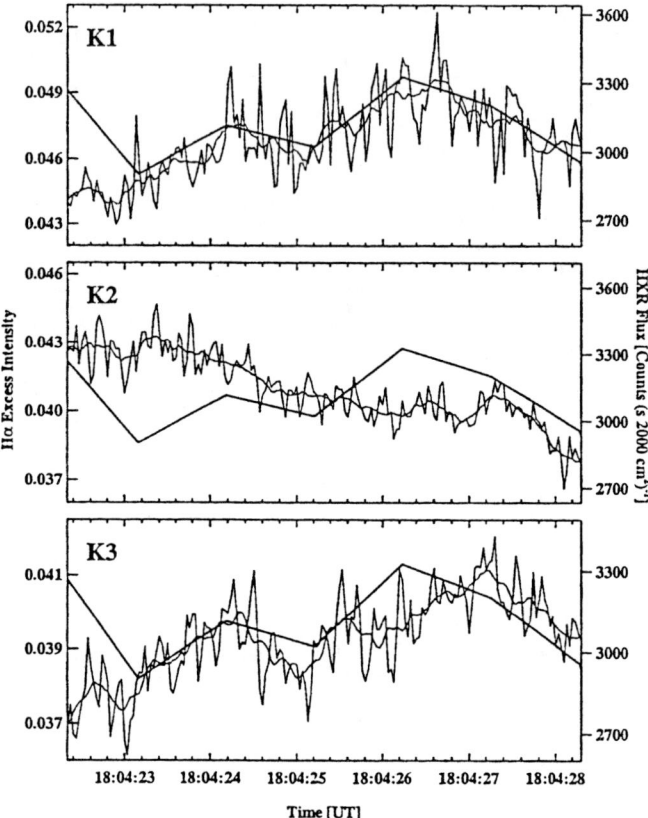

Fig. 2. Comparison of Hα-1.3 Å, intensity (thin lines) and BATSE HXR flux (thick lines) for three flare kernels during 7 s of the impulsive phase. For the Hα emission, both the raw data and a 10-point smoothed curve are plotted. From [9]

energy deposit while the red wing typically responds to the motion of the chromospheric condensation [10]. Hα time profiles were compared with HXR from BATSE, however, only 1 s resolution mode was available for the flare under study. For the flare kernel which shows a good correlation with HXR, high-frequency fluctuations on a timescale of a few tenths of a second were found. Their amplitude exceeds the noise by a factor of three. Such observations correspond only to about 7 s. In Fig. 2 we reproduce these observations taken in three different flare kernels. One can notice that indeed the highest amplitude fluctuations are seen in kernels where Hα correlates well with HXR.

The authors conclude that these fluctuations may be signatures of the Hα fine structure related to HXR elementary bursts. Based on these observations, the authors claim firmly the evidence of Hα fluctuations on the subsecond timescale which are above the noise level. However, it is not very clear what is the actual background intensity relative to which we do see such fluctuations. If the fluctuations correspond to emission enhancements only, then this level should be a curve connecting the minima. On the other hand, in their Fig. 7 Wang et al. [9] show the fluctuations after subtracting a 10-point smoothed average. Then a question arises what the negative minima actually mean (see Sect. 4).

3 Formation of the Hα Line in a Flaring Atmosphere

Before we consider the time-dependent models of the Hα line formation, we will summarize our basic knowledge about the behaviour of this strong line which appears in emission during flares. Other optical lines have similar properties and some of them (e.g. higher Balmer lines of hydrogen, CaII lines and others) can be used as complementary lines for spectral diagnostics.

In [11] one can see three Hα flare profiles which correspond to three static semiempirical models of flares as constructed by Machado et al. [12] (denoted as F1, F2 and F3 models). They are compared to a quiet-Sun absorption profile. These models, and particularly F1, were frequently used to study the behaviour of the Hα line formation in static and dynamic atmospheres and in the presence of particle beams (electrons and/or protons). When the weak-flare atmosphere F1 is perturbed by downflows with a velocity gradient simulating the onset of a chromospheric condensation, the strongly reversed Hα profile becomes asymmetrical with the blue peak being more intense (so-called 'blue asymmetry'). This was modelled by Heinzel et al. [13] and the computed profiles are consistent with those observed during initial phases of flares. Later on, during the impulsive phase, the Hα line goes to emission and the coupling between still downward velocity field and the variations of the line source function leads to a 'red asymmetry', i.e. more intense red part of the line, sometimes with pronounced red wing (see also discussion in [14]). Finally, a static flare model was perturbed by electron and proton beams and the effects of non-thermal collisional excitation and ionization of hydrogen were studied. The result is an enhancement of the Hα intensity in the presence of the beams (see reviews by Hénoux [1] and Fang et al. [15]). This has been recently confirmed by Kašparová and Heinzel [16], although their results show somewhat different behaviour. The effect of electron beams with increasing flux F_{20} on first three Balmer lines formed in the F1 atmosphere is shown in [16].

To study the beam interaction with the lower atmosphere, we must understand where the individual parts of the line profile are formed (this is normally described by so-called contribution functions) and at which layers the beam energy is deposited (energy-deposit function). Total energy loss from the beam has two components: direct Coulomb losses which convert the beam energy into the plasma heating, and the losses into hydrogen which means the non-thermal excitation and ionization. If for example the Hα line core is formed higher compared to the location of the energy deposit into hydrogen, the line brightening will take place mainly in the wings (see Fig. 3).

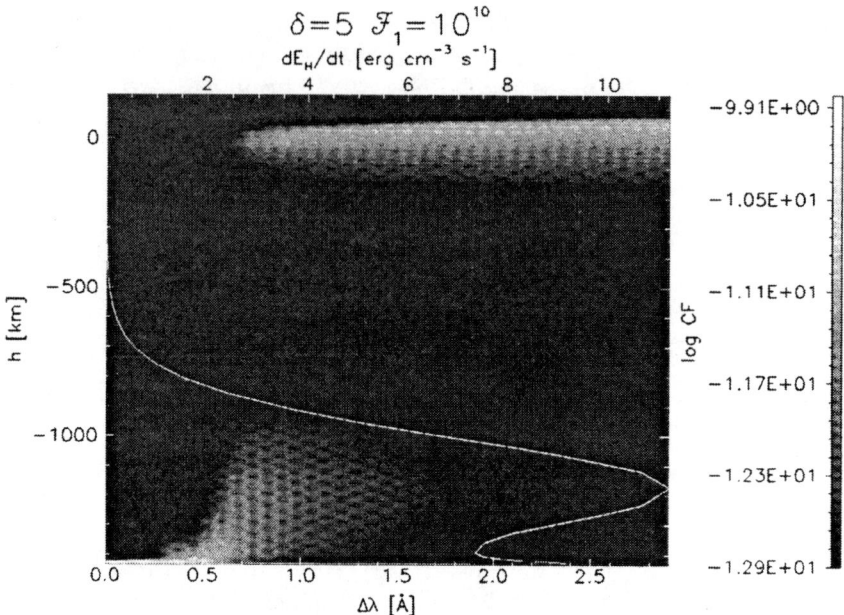

Fig. 3. Energy deposit (white profile) and the Hα contribution function CF (gray scale). Hα wings are formed at heights h where the energy deposit has the maximum for parameters indicated on top of the figure. The line core is not much affected by the beam. From [16]

4 Time-Dependent Spectral Line Formation

7 Time-dependent statistical-equilibrium equations can be written as

$$\frac{\partial n_i}{\partial t} + \frac{\partial (n_i v)}{\partial z} = \sum_{j \neq i} (n_j P_{ji} - n_i P_{ij}), \tag{2}$$

where n_i is the population of i-th level (here we consider the hydrogen atom model), v is the macroscopic flow velocity, z the vertical coordinate and P_{ij} is the total rate for transition $i \rightarrow j$. The rate P_{ij} consists of the radiative rate R_{ij} and collisional rate C_{ij}. The latter has the form

$$C_{ij} = n_e \Omega_{ij}(T) + C_{ij}^{nt}, \tag{3}$$

where n_e is the electron density, Ω is the collisional cross-section for given temperature T (see e.g. Mihalas [17]) and C_{ij}^{nt} represents the non-thermal collisional rate, i.e. the collisional excitation and ionization induced by the beam particles. According to [15], we can express the non-thermal rates as

$$C_{1j}^{nt} \simeq \frac{1}{n_1} \frac{dE_H}{dt}, \tag{4}$$

where n_1 is the ground-state population and dE_H/dt is the energy deposit from the beam into the hydrogen (by non-thermal excitations and ionization). Inverse thermal rates are obtained from the detailed balance, while inverse non-thermal rates are negligible. To obtain the radiative rates, one has to solve the radiative-transfer equation in the form

$$\mu \frac{dI_{\nu\mu}}{dz} = -\chi_{\nu\mu} I_{\nu\mu} + \eta_{\nu\mu}, \tag{5}$$

where $I_{\nu\mu}$ is the specific intensity and $\chi_{\nu\mu}$ and $\eta_{\nu\mu}$ are, respectively, the opacity and emissivity. ν is the frequency, μ the directional cosine. A coupled set of these non-LTE equations, together with additional constraint equations, is then solved using the Crank-Nicholson time-difference implicit scheme (the resulting set of equations is linearized with respect to particle number densities). An efficient technique to solve this non-LTE transfer problem is so-called MALI-method (for more details see Heinzel [18]).

It will be shown later that during the flare onset when first beam pulses start to interact with the atmosphere, the lower atmosphere is still in quasi-hydrostatic state (i.e. the flows are negligible). This is because the hydrodynamical time-scale is longer than the beam-pulse duration, if we consider very short (subsecond) beam pulses. In such a case, neglecting the velocities, we can treat the time-dependent radiation transfer in a spectral line using quite similar techniques as in the case of time-independent atmosphere. The only difference is a time-dependence of statistical-equilibrium equations, where we neglect the advection term (second term on l.h.s. of Eqn. (2)). Using such an approach, Heinzel [19] performed time-dependent simulations of the Hα line formation in an atmosphere bombarded by short-duration electron beams (as in [20]).

Three important results have been obtained: (i) for typical flare densities, the electron-density variations don't follow the temperature ones because of substantial relaxation time for hydrogen recombinations (Fig. 4); (ii) there is a significant Hα response to beam pulses as shown in Fig. 5; (iii) at the pulse onset, the Hα intensity drops down for very short sub-second period (see also Fig. 5).

Such an effect, which was also found in more sophisticated simulations, has not been observationally proved yet, however, one can deduce its signature from Fig. 6 where just before the HXR peak at time 'c', Hα drops down and this intensity drop is larger than the noise uncertainty (Trottet - private communication). In a similar way one could also interpret the negative minima in subsecond fluctuations reported in [9].

Modeling similar to that of Heinzel has been recently repeated in somewhat more complex way by Ding et al. [21], who also found instantaneous response of the Hα line intensity to the beam pulses of 0.2 s duration. Their simulations are shown in Fig. 7.

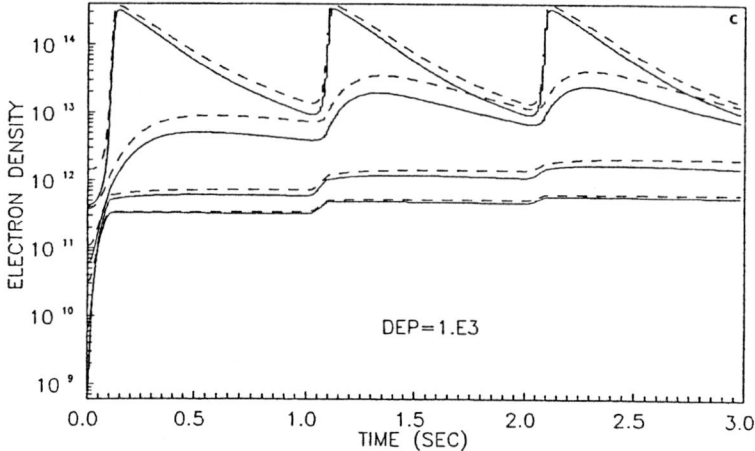

Fig. 4. Electron density evolution for multi-pulse beam heating models (for temperature variations see the following figure). Solid curves correspond to layers where the Hα line center is formed, dashed ones are for deep layers where the wings originate. Higher curves correspond to higher hydrogen densities between 10^{15} and 10^{12} cm^{-3}. DEP is the energy deposit dE/dt (erg cm^{-3} s^{-1}). Note a long relaxation at low densities. From [19]

5 RHD-Models: Basic Concepts

First complex RHD models of solar flares, which led to prediction of Hα variations, were developed by Fisher, Canfield and McClymont [22]. These models have revealed basic properties of the beam-heated atmospheric structure. However, they treated only beams with longer duration (several seconds) and the energy deposit was computed using a stationary approach of Emslie [23]. Similar models, which give variations also in other optical lines, have been published recently by Abbett and Hawley [24] who have used the new RHD code of Carlsson and Stein [25]. A kinetic solution for an electron beam precipitation is described by Zharkova [26], who also considered the effects of the return current discussed below. On the other hand, Karlický and Hénoux [20] have developed so-called 'hybrid' code which is able to treat the beam propagation and energy deposit in a non-stationary way, taking into account the finite flight-time of the beam particles. This hybrid code computes the atmospheric response to a series of short-duration beam pulses (typically sub-second pulses), the beam is represented by numerical test particles with a mono-energetic or power-law distribution function. Atmospheric response is computed by solving the hydrodynamical and energy-balance equations. This code is currently being coupled to time-dependent radiation-transfer code based on the MALI technique, as described in the previous section. The output will contain fast temporal variations of both optical (e.g. Hα) and HXR emission, we show some preliminary results in the next section.

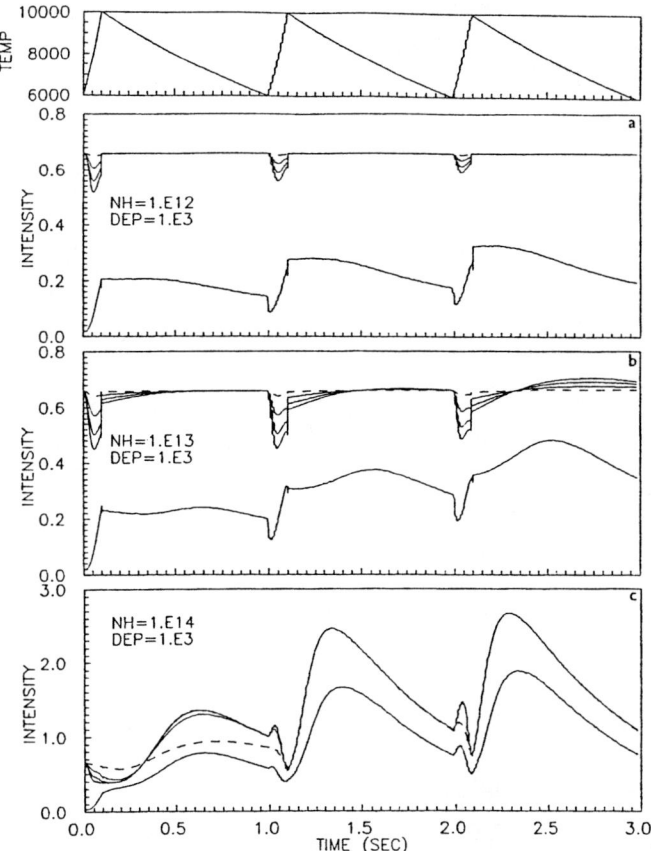

Fig. 5. Temporal variation of the Hα intensity for multi-pulse beam heating model. The line intensities are normalized to the continuum. The curves in the lower part of each figure correspond to the line center, those in the upper part belong to the wing at 1 Å. Different curves correspond to various thicknesses of the line-forming regions. Figures are labeled by the hydrogen number density NH and by the value of energy deposit DEP. The upper panel shows the kinetic temperature pulses used as the input. From [19]

In the hybrid code, particle beams are represented by a cloud of numerical particles, which are under influence of several physical processes [27]:

a) Free propagation: In presence of the magnetic field particles freely propagate along magnetic field lines and their motion can be expressed as that of their guiding centers

$$L_{new} = L_{old} + v_\parallel \Delta t, \tag{6}$$

where L_{new} and L_{old} are new and old positions of the particle, respectively, $v_\parallel$ is the particle velocity parallel to the magnetic field, and t is the time.

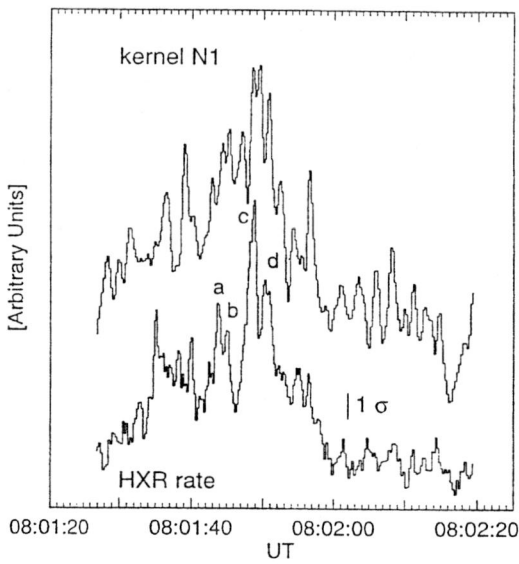

Fig. 6. Time evolution of the fast Hα intensity response to HXR rate. HXR peaks a - d have corresponding ones in Hα. Note the significant Hα intensity drop just before the pek at time c. From [7]

b) Collisional energy losses and pitch-angle scattering: In the solar atmosphere beams of particles propagate downwards into dense chromospheric layers, where the particles collisionally interact with the background plasma. Their energy E and cosine of pitch angle μ can be expressed as (see [23])

$$E = E_0 \left[1 - \left(2 + \frac{\beta}{2} \right) \frac{\gamma K \; N}{\mu_0 E_0^2} \right]^{2/(4+\beta)}, \qquad (7)$$

$$\mu = \mu_0 \left[1 - \left(2 + \frac{\beta}{2} \right) \frac{\gamma K \; N}{\mu_0 E_0^2} \right]^{\beta/(4+\beta)}, \qquad (8)$$

$$\gamma = \frac{m}{m_e} \left[x\Lambda + (1 - x)\Lambda' \right], \qquad (9)$$

where N is the column density of the background plasma, E_0 and μ_0 are initial values, $K = 2\pi e^4$, e is the electron charge, m and m_e are particle and electron masses, respectively, x is the hydrogen ionization fraction, Λ and Λ' are the Coulomb logarithms and $\beta = 2$ for the electron beam while $\beta = 0$ for the proton beam. More general models of collisional processes are based on the Monte Carlo method proposed by Bai [28].

c) Return current effects: The injection of a beam into the atmospheric plasma drives a charge and current-neutralizing plasma return current. This current is governed by Ohm's law

$$E = \left(\frac{m_e}{ne^2} \right) \frac{\partial j}{\partial t} + \eta j, \qquad (10)$$

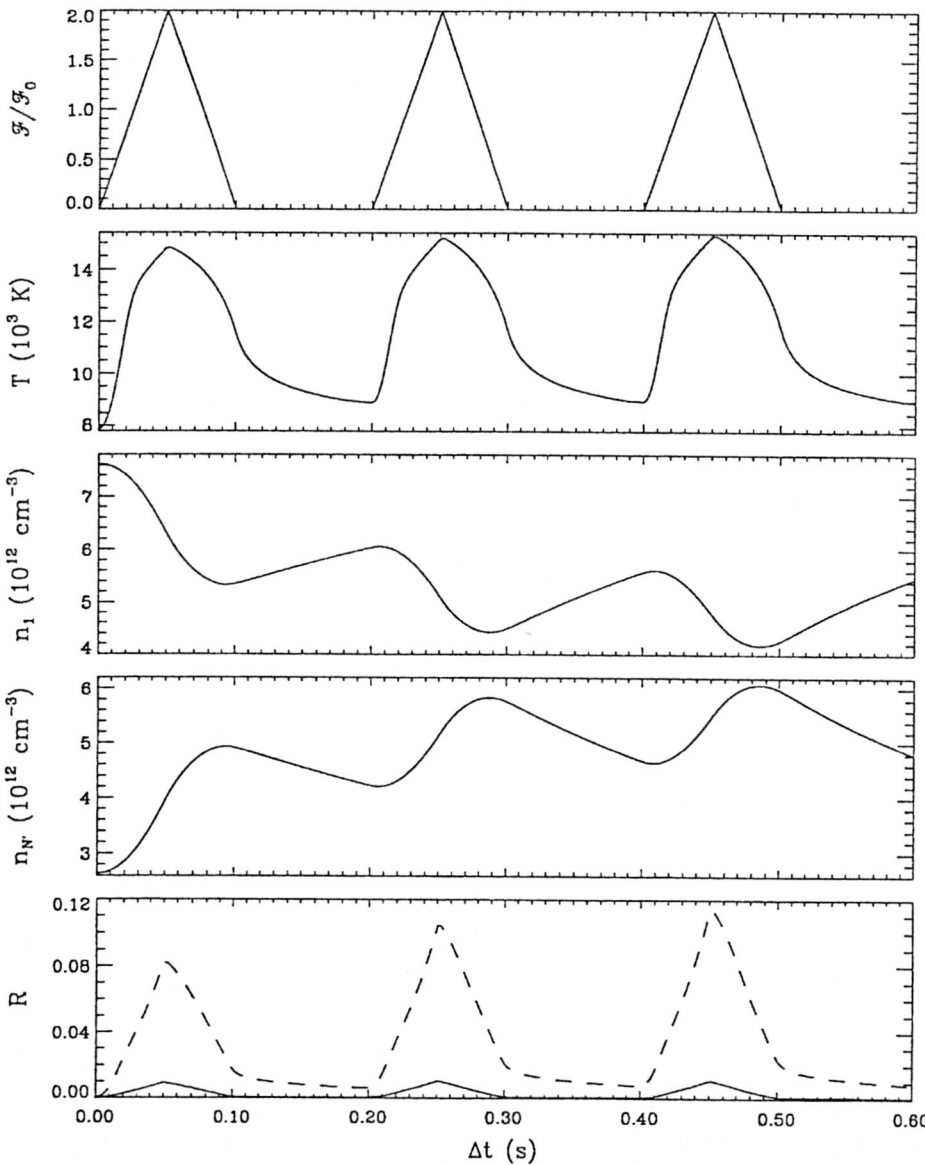

Fig. 7. Time evolution of the beam energy flux, kinetic temperature, hydrogen ground-state population, proton ($\simeq$ electron) density, and relative enhancement of the Hα line intensity at -1.3 Å. Dashed curve is obtained after convolving the emergent line profiles with a Gaussian macro-velocity of 25 km s^{-1}. Three beam pulses are used with $F_0 = 10^{11}$ ergs cm^{-2} s^{-1}. The repetition time of beam pulses is 0.2 s. From [21]

where E is now the electric field, e the electron charge, n particle density, j is the electric current and η the electric resistivity. In cases where the beam current varies slowly in comparison with the electron-ion collisional time, a steady-state form of the above equation with zero net current everywhere can be used, i.e.

$$E = \eta j_{plasma} = \eta j_{beam}. \tag{11}$$

d) Mirroring of electron beams in magnetic mirrors: Magnetic mirrors are assumed at the loop-footpoints below the transition region. But the mirroring is not simple, because in this region a relatively dense plasma is present and due to collisions the particles lose their energies. Therefore, the first adiabatic invariant $p^2 \sin^2 \alpha / B$ (p being the momentum, α the pitch angle and B the magnetic field), which is essential for computations of the particle mirroring, is not conserved. But fortunately Miller and Ramaty [29] showed that in such circumstances the conservation of $\sin^2 \alpha / B$, where α is the pitch angle, can be used.

e) The quasi-linear relaxation of particle beams: Based on in situ measurements in the heliosphere it is believed that the quasi-linear relaxation is suppressed by nonlinear processes. For details see [27].

f) Particle scattering in wave-turbulence zones: In the solar atmosphere, where many types of plasma waves are present, the scattering of particles in the plasma wave turbulence can be important.

An advantage of the particle description approach, which is used in our modeling, is that the fast transient phenomena are properly included and the hard X-ray emission can be computed directly.

6 RHD-Models: Numerical Results

We first mention complex RHD simulations of Abbett and Hawley [24]. All RHD equations are solved by linearization to the first order in all considered quantities [25]. Note however that the beam is treated within a stationary picture and the authors don't mention the inclusion of non-thermal rates into the statistical-equilibrium equations.

In most papers studying an atmospheric response to the electron beam bombardment the return-current effects are neglected. But we found that this effect plays an important role (Karlický and Hénoux [20]), as can be seen in Figs. 8 and 9, where the energy deposits without and with return-current processes are compared.

Namely, neglecting the return-current effect the beam energy is deposited in lower heights of the solar atmosphere than in the case when this effect is included. This causes essential differences in the resulting intensity of optical chromospheric lines as well as in the height distribution of plasma velocity flows. For example, the electron beam bombardment with the return-current effect can generate a downwards plasma motion in the upper chromosphere, contrary

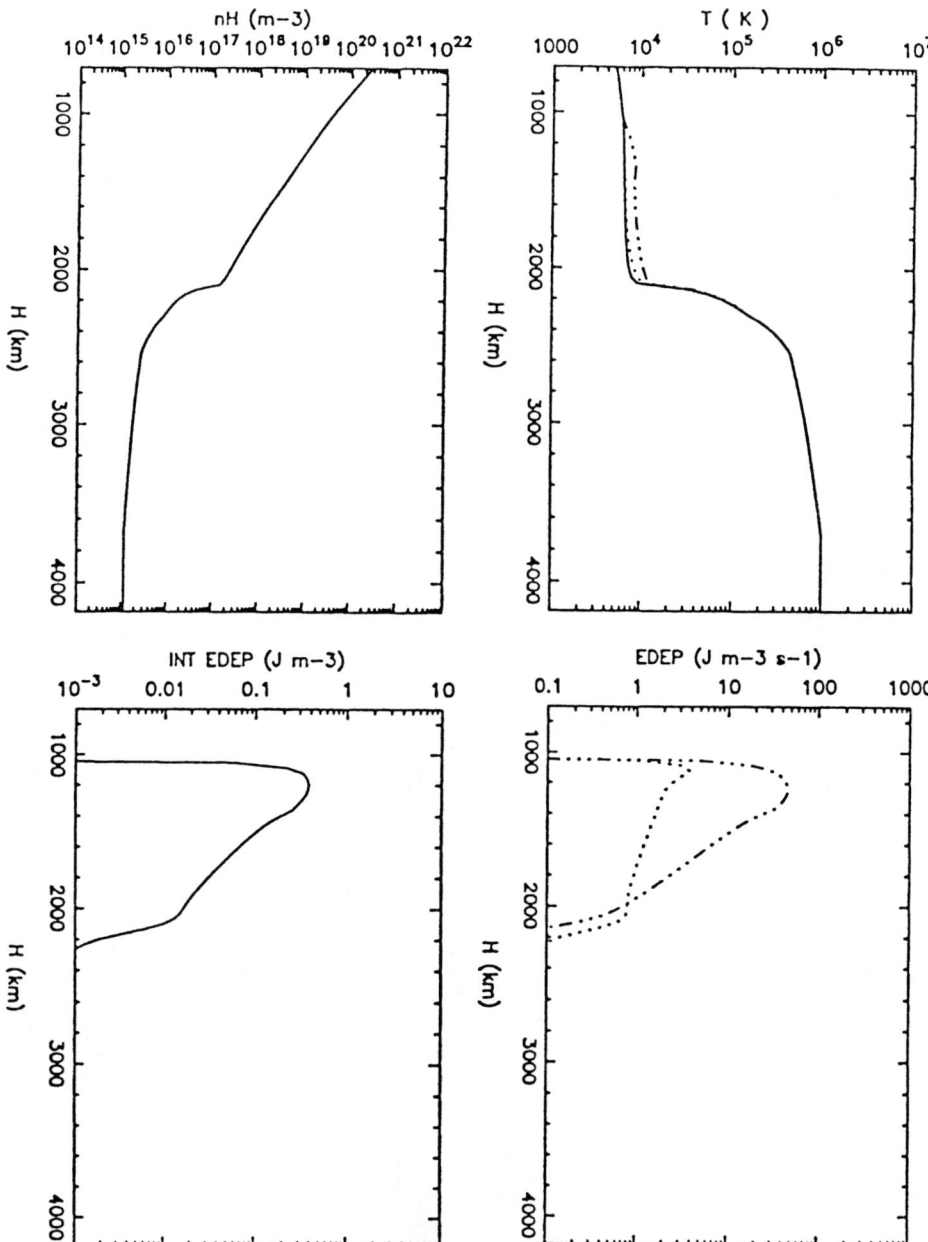

Fig. 8. The energy deposit and temperature response of the solar atmosphere when only collisional losses of beam electrons are considered. Top: temperature and energy deposit profiles in the solar atmosphere at various times: full lines - the initial state, dotted lines at 0.033 s, dashed-dotted lines at 0.043 s. Bottom: the initial hydrogen density profile of the solar atmosphere and time integrated energy deposit. From [20]

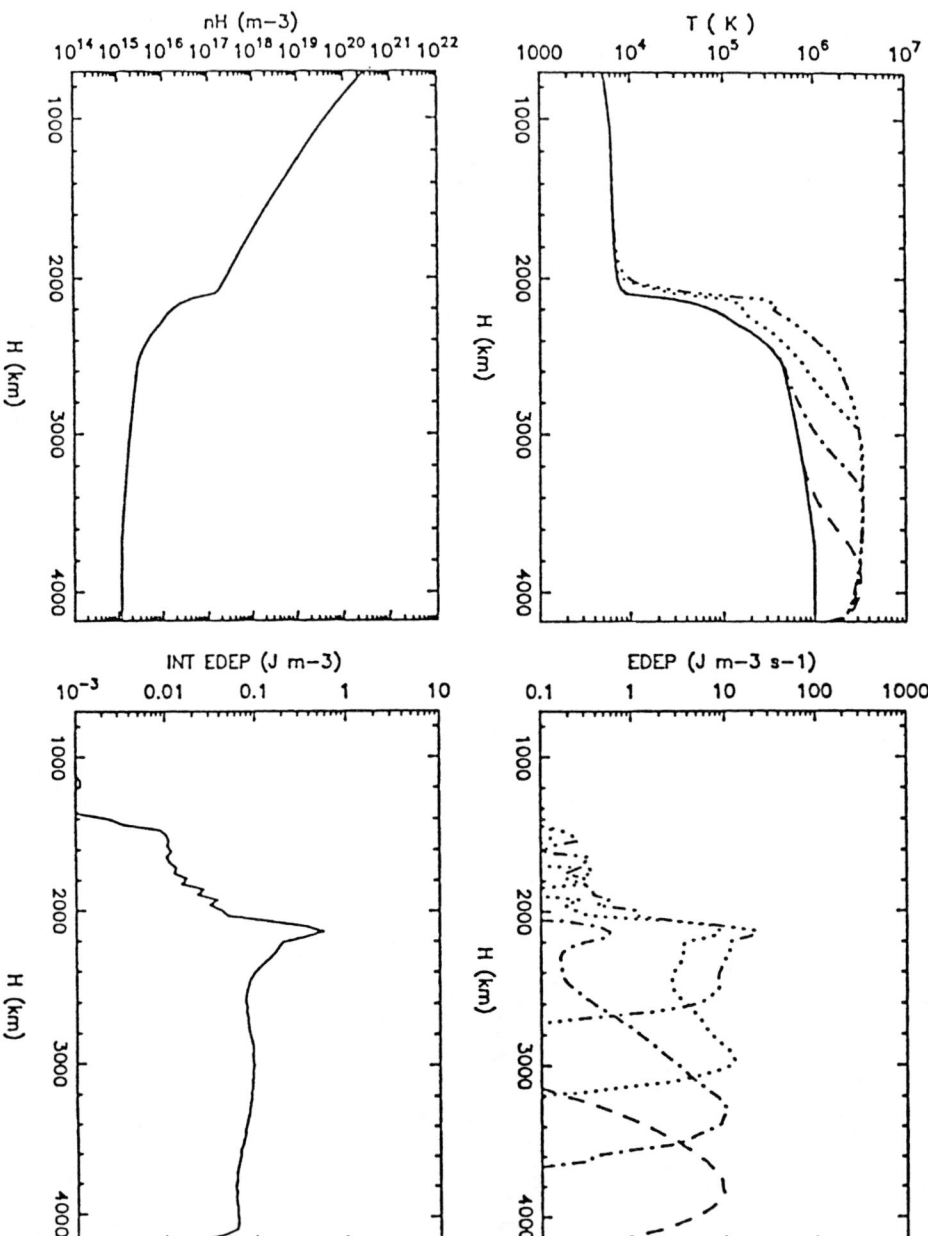

Fig. 9. The energy deposit and temperature response of the solar atmosphere when both collisional and return current losses of beam electrons are considered. Top: temperature and energy deposit profiles in the solar atmosphere at various times: full lines - the initial state, dotted lines at 0.013 s, dashed-dotted lines at 0.022 s, dotted lines at 0.031 s, and dashed-dotted (three dots per dash) lines at 0.039 s. Bottom: the initial hydrogen density profile of the solar atmosphere and time integrated energy deposit. From [20]

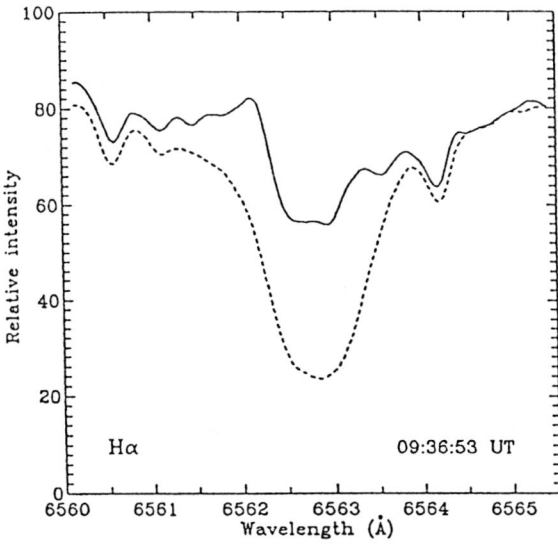

Fig. 10. Blue asymmetry observed in Hα line during the onset phase of the 4 October, 1991 flare at 9:36:53 UT. From [13]

to the case when the return current is neglected. This may then cause a blue asymmetry of Balmer lines (Fig. 10).

For a series of electron beam pulses, we have computed the time-dependent chromospheric heating and the corresponding Hα and hard X-ray flux ([30], Fig. 11).

By solving the time-dependent non-LTE problem for hydrogen, we theoretically predict the Hα-line intensity variations on sub-second time scales. Both hard X-ray fluxes and Hα wing intensities do exhibit a spiky behavior, consistent with short pulse-beam heating. However, the spikes in Hα are negative, i.e. the line intensity decreases during the beam heating (Fig. 11, see also Fig. 5). This is due to a higher rate of the second-level hydrogen population as compared to that for the third level during the fast heating process at the onset of the beam energy deposition.

New RHD calculations performed by Ondřejov group show the fast variations of the Hα line profile for one short electron beam pulse of a subsecond duration - see [31] and Fig. 12 [32].

The hydrodynamical part is described in [33], the particle code is mentioned above and the time-dependent non-LTE transfer was performed using the MALI approach. The next step is to simulate a stochastic series of bursts where the injection function can be supplied from the HXR observations.

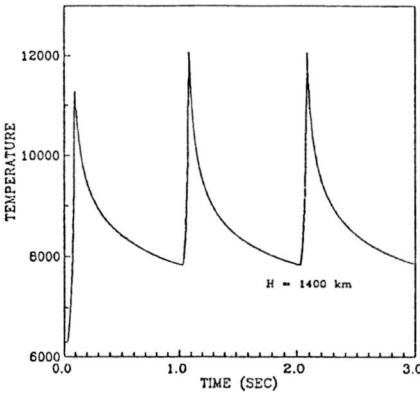
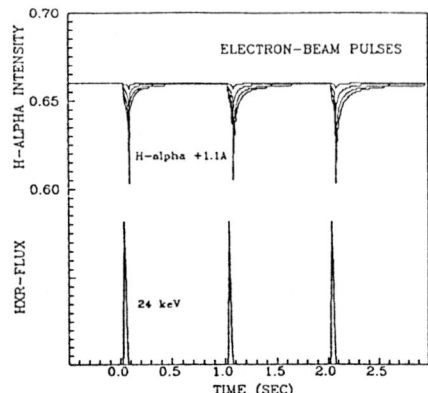

Fig. 11. (a) Temporal profile of the kinetic temperature for three beam pulses, each lasting approximately 0.1 s. H=1400 km is the chromospheric height where the heating is computed; (b) Temporal variations of HRX flux (24 keV, arbitrary units) and the corresponding Hα + 1.1 Å intensity (normalized to disk-center continuum intensity). Lower Hα intensities correspond to thicker emitting layers (from 100 km to 1500 km). From [30]

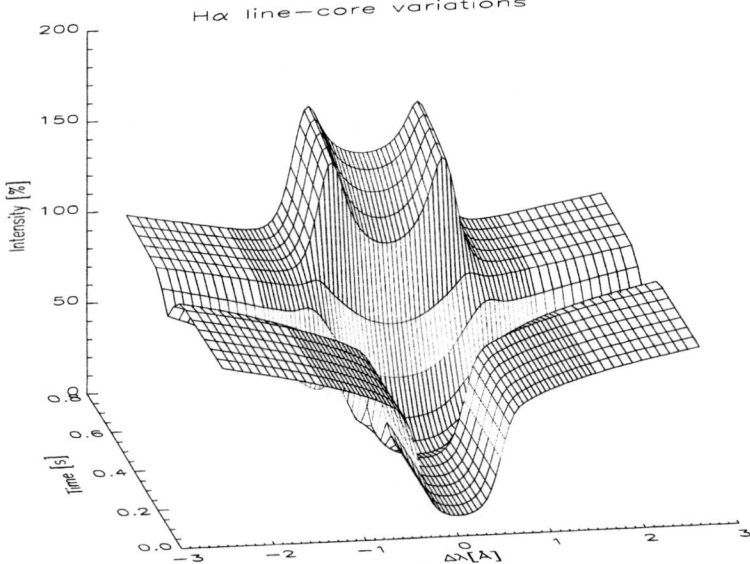

Fig. 12. Hα line-profile variations computed from RHD simulations of a sub-second electron beam puls heating. From [32]

7 Conclusions

Fast variations of the Hα line emission on very short time scales (subsecond) were detected during last decade. However, their correlation with HXR must be further confirmed. The onset of the beam interaction with chromospheric layers seems to be accompanied by a transient darkening in the Hα line, the effect which was predicted theoretically but has to be verified observationally. Currently high-cadence (0.2 s), full-disk HXR data are continuously available from the MTI satellite [34]. New spatially and temporally well resolved HXR data are expected from RHESSI. Several ground-based observatories are well prepared to observe optical line emission from flaring kernels, in Hα or in other lines of hydrogen, helium and CaII (BBSO, Ondřejov multichannel flare spectrograph, Wroclaw observatory, THEMIS on Tenerife, Locarno station and some others). Also theoretical RHD simulations do predict fast Hα response to a series of electron-beam pulses, but they are still in initial stages - some physical processes have been so far neglected, 2D or 3D MHD simulations will be needed in future. Finally, such simulations should be used to diagnose the characteristics of the electron and proton beams, which is one of the major objectives of current flare research.

Acknowledgements

The authors thank J.-C. Hénoux, J. Kašparová, G. Trottet and M. Varady for extensive discussions and the anonymous referee for several constructive comments. This work was supported by the key project K2043105 of the Academy of Sciences of the Czech Republic and by the grants A3003202 and A3003902 of the Grant Agency of ASCR. The financial support from LOC is also highly appreciated.

References

1. J.-C. Hénoux: In *High Energy Solar Physics - Anticipating HESSI*, eds. R. Ramaty and N. Mandzhavidze (ASP Conf. Ser. Vol. 206, ASP 2000) 27
2. B.R. Dennis, L.E. Orwig, A.L. Kiplinger (eds.): *Rapid Fluctuations in Solar Flares* (NASA Conf. Publ. 2449, NASA 1987)
3. M.R. Kundu, B. Woodgate, E.J. Schmahl (eds.): *Energetic Phenomena on the Sun* (ASSL, Kluwer Acad. Publ.,Dordrecht 1989)
4. A.L. Kiplinger, G. Labow, L.E. Orwig: In *Max '91 Workshop No. 3*, eds. R.M. Winglee and A.L. Kiplinger (Univ. of Colorado, Boulder 1991) 210
5. E. Rolli, J.P. Wülser, A. Magun: Solar Phys. **180**, 343 (1998)
6. E. Rolli, J.P. Wülser, A. Magun: Solar Phys. **180**, 361 (1998)
7. G. Trottet, E. Rolli, A. Magun, C. Barat, A. Kuznetsov, R. Sunyaev, O. Terekhov: Astron. Astrophys. **356**, 1067 (2000)
8. V.G. Kurt, V.V. Akimov, M.J. Hagyard, D.H. Hathaway: In *High Energy Solar Physics - Anticipating HESSI*, eds. R. Ramaty and N. Mandzhavidze (ASP Conf. Ser. Vol. 206, ASP 2000) 426

9. H. Wang, J. Qiu, C. Denker, T. Spirock, H. Chen, P.R. Goode: Astrophys J. **542**, 1080 (2000)
10. R.C. Canfield, K.G. Gayley: Astrophys. J. **322**, 999 (1987)
11. E.H. Avrett, M.E. Machado, R.L. Kurucz: In *The Lower Atmosphere in Solar Flares*, ed. D.F. Neidig (NSO, Sunspot 1986) 216
12. M.E. Machado, E.H. Avrett, J.E. Vernazza, R.W. Noyes: Astrophys. J. **242**, 336 (1980)
13. P. Heinzel, M. Karlický, P. Kotrč, Z. Švestka: Solar Phys. **152**, (1994) 393
14. W.Q. Gan, E. Rieger, C. Fang: Astrophys. J. **416**, 886 (1993)
15. C. Fang, M.D. Ding, J.-C. Hénoux, W.Q. Gan: In *Proc. of the First Franco-Chinese Meeting on Solar Physics*, eds. C. Fang, J.-C. Hénoux and M.D. Ding (World Publ. Corp., Beijing 2000) 147
16. J. Kašparová, P. Heinzel: Astron. Astrophys. **382**, 688 (2002)
17. D. Mihalas: *Stellar Atmospheres*, 2nd ed., (W.H. Freeman, San Francisco 1978)
18. P. Heinzel: In *Advances in Solar Research at Eclipses from Ground and from Space*, ed. J.-P. Zahn and M. Stavinschi (NATO ASI, Kluwer Acad. Publ., Dordrecht 2000) 201
19. P. Heinzel: Solar Phys. **135**, 65 (1991)
20. M. Karlický, J.-C. Hénoux: Astron. Astrophys. **264**, 679 (1992)
21. M.D. Ding, J. Qiu, H. Wang, P.R. Goode: Astrophys. J. **552**, 340
22. G. H. Fisher, R.C. Canfield, A.N. McClymont: Astrophys. J. **289**, 414, 425, 434 (1985)
23. A.G. Emslie: Astrophys. J. **224**, 241 (1978)
24. W.P. Abbett, S.L. Hawley: Astrophys. J. **521**, 906 (1999)
25. M. Carlsson, R. Stein: In *Physical Processes in the Solar Transition Region and Corona*, eds. P. Maltby and E. Leer (ITA, Oslo 1990), 177
26. V.V. Zharkova: In *High Energy Solar Physics - Anticipating HESSI*, eds. R. Ramaty and N. Mandzhavidze (ASP Conf. Ser. Vol. 206, ASP 2000) 227
27. M. Karlický: Space Science Rev. **81**, 143 (1997)
28. T. Bai, T.: Astrophys. J. **259**, 341 (1982)
29. J.A. Miller, R. Ramaty: Astrophys. J. **344**, 973 (1989)
30. P. Heinzel, M. Karlický: 1992, In *Eruptive Solar Flares* eds. Z. Švestka, B.V. Jackson, M.E. Machado (Lecture Notes in Physics 399, Springer-Verlag, Berlin 1992) 359.
31. P. Heinzel, M. Karlický, P. Kotrč, Yu. A. Kupryakov: In *High Energy Solar Physics - Anticipating HESSI*, eds. R. Ramaty and N. Mandzhavidze (ASP Conf. Ser. Vol. 206, ASP 2000) 289
32. J. Kašparová, M. Varady, P. Heinzel, M. Karlický: in preparation
33. M. Varady: *Observations and modeling of plasma loops in the solar corona* (PhD thesis, Charles Univ., Prague 2002)
34. F. Fárník, H. Garcia, A. Kiplinger: In *Crossroads for European Solar & Heliospheric Physics* ESA SP-417 (1998), 305.

On the Origin of Solar Energetic Particle Events

Säm Krucker

Space Sciences Laboratory, University of California, Berkeley, CA 94720-7450, USA

Abstract. The 3-D Plasma and Energetic Particles (3DP) instruments on the WIND spacecraft are providing new insights on the origin of solar energetic particles. In this paper, recent results on the solar origin of 1-300 keV electrons and 0.1-6 MeV protons are reviewed. The main findings are that one class of electron events escaping into interplanetary space are related to the impulsive phase of the flare (radio type III burst related events), but there is also a second class of events released after the impulsive phase of the flare that is possibly related to coronal shocks. In two thirds of all proton events, low energy protons (<6 MeV) are released roughly an hour later than the electrons and are most likely accelerated at the shock fronts of coronal mass ejections. The remaining third of proton events show a puzzling velocity dispersion that could be explained by a non-simultaneous release of protons at different energy.

1 Introduction

Solar particle acceleration can be studied by investigating particles escaping from the Sun into interplanetary space. Next to the quasi-steady solar wind streaming away from the Sun, there are also large, sudden increases in particle flux observed. These events detected from GeV down to keV energies are called solar energetic particle (SEP) events. In this section, new results obtained by the 3-D Plasma and Energetic Particle Instruments (Lin et al. 1995) on the WIND spacecraft are reviewed. The discussion on the origin of SEP is emphasized.

SEP events observed in interplanetary space near 1 AU generally show velocity dispersion in their onsets, indicating they are the result of sudden transient acceleration near the Sun. Two main classes of SEP are distinguished (e.g. Reames 1999): "Impulsive" SEP events (so called because of the short (<1 hr) duration of the associated soft X-ray (SXR) burst), are electron-rich and ^{3}He-rich, and show high ionic charge states. They occur frequently ($\sim 10^3$/yr over the whole Sun during solar maximum) and their escape is generally restricted to a relatively small longitudinal cone ($<50°$). Impulsive events are believed to be produced by the solar flare itself, although it is not clear how exactly the particles escape from the flare site. However, it is generally accepted that a spacecraft must be magnetically well connected to the flare site to be able to see the escaping solar energetic particles. Compared to the total energy released in a flare, the energy in escaping particles is relatively small. "Gradual" SEP events (e.g. with a long (>1 hr) duration SXR burst) occur much less frequently (several 10/yr at maximum), are proton-rich, and show normal coronal abundance and charge states corresponding to typical quiet coronal temperatures. They are typically observed over a large longitudinal cone ($\approx 180°$), and are almost always related to a fast Coronal Mass Ejection (CME).

In the last years, this classic picture of two classes has been criticized: Large gradual events have been observed with enhanced ^{3}He and high ionization states. To save the current picture of the two classes, it has been proposed that earlier impulsive events provide the seed population for later acceleration in gradual events (Mason et al. 1999) and/or that large gradual events are mixed with a central 30-50° core of flare accelerated particles surrounded by a 100° wide shock accelerated halo (Cliver 1999). Klein et al (2001 and references therein) found evidence that at least in some events neither the flare nor the CME shock are related to particle acceleration to relativistic energies; instead acceleration occurs in the high corona (0.1 to 1 solar radii). Similar ideas about particle acceleration in the high corona instead at CME shocks farer out are supported by the work of Cane et al. (2002), who pointed out that all major proton events are related to groups of type III bursts seen higher in the corona.

In this work, the temporal correlation between solar energetic particle events seen at 1 AU and events occurring near the Sun is presented. From the observed onset times of particle events at 1 AU, the solar release time is approximated and then compared with the various events occurring at the Sun.

2 Observations

The presented results are derived from observations of the 3-D Plasma and Energetic Particles (3DP) instrument on the WIND spacecraft (Lin et al. 1995). In 3DP, six double-ended telescopes (SSTs), each with a pair or triplet of closely stacked silicon semiconductor detectors, provide full 3D coverage with $36° \times 22.5°$ angular resolution for ~20-400 keV electrons and ~30 keV-6 MeV ions. One end of each SST is covered with a Lexan foil (Foil SST) which stops protons up to ~400 keV, while $\geq$20 keV electrons are essentially unaffected. The opposite end is open (Open SST) but has a magnet which sweeps away electrons below ~400 keV while leaving the ions unaffected. Thus, when no higher energy particles are present, electrons and ions below ~ 400 keV are cleanly separated. Most electrons above ~ 400 keV will penetrate the front detector and be anti-coincidenced by the rear detector. Protons from 400 keV to 6 MeV are measured by the Open SST, and their contribution to the Foil SST can be computed. Heavier ions such as He or CNO require higher energies to penetrate the foil. By comparing the response of the Open and Foil SSTs, some information on the ion species may be obtained. For some large SEP events, such as November 4 and 6, 1997, April 20, 1998, etc., the associated > 400 keV electron flux is large enough to significantly contaminate the Open SSTs. These events (~50 % of the total number of proton events) could therefore not be used to determine proton onset times. Electrostatic analyzers in 3DP additionally provide measurements of solar wind and suprathermal electrons and ions up to ~30 keV. Remote sensing solar observations from WIND/WAVES instrument (Bougeret et al. 1995), Extreme ultraviolet Imaging Telescope (EIT) on board SoHO (Delaboudinière et al, 1995), and the SoHO/LASCO coronograph (Brueckner et al. 1995) were used to search for related solar events.

3 Results of Onset Time Analysis

The arrival time at 1 AU, $t_{1AU}(E)$, of a particle with energy E is given by:

$$t_{1AU}(E) = t_{Sun}(E) + L(E)\,v^{-1}(E) \tag{1}$$

where $t_{Sun}(E)$ is the particle release time at the Sun, $v(E)$ is the velocity, and $L(E)$ is the path length. If particles at all energies are released simultaneously and travel the same path length, the observed arrival times $t_{1AU}(E)$ would be a linear function of $v^{-1}(E)$ with a slope equal to the (constant) path length, L, and an intersection at $v^{-1} = 0$ equal to t_{Sun}.

The onset times at different energies are determined by eye and then plotted versus $\beta^{-1} = c/v$ (e.g. Fig. 1). A linear fit to the observed onset times gives estimates for the path length and the release time. The error bars shown are very conservative bracketing of the onset times: Before these time periods, the event definitely has not yet started, and afterwards the flux is already clearly increased above the background (cf. Krucker et al. 1999). Hence, the shown error bars are not 1-σ error bars.

3.1 Electron Events

Using timing arguments, two different classes of ≤ 300 keV electron events have been found (cf. Krucker et al. 1999):

1. electron events related to radio type III bursts
2. electron events accelerated later than the type III burst onset, hence, events accelerated after the impulsive phase of the flare.

An example of an event of the first class is shown in Fig. 1. The derived solar release time agrees with the onset of the type III burst within the uncertainty of around ±3 minutes. The derived path length ($L_e = 1.26 \pm 0.04$ AU) is roughly in agreement with the Parker spiral length ($L_{parker} \approx 1.23$ AU) calculated from the observed solar wind speed indicating a scatterfree transport for the first arriving electrons (it is noted here that scattering is of course important in SEP events – cf. Dröge, *this volume* – however, only for the later arriving particles). In situ waves at the local plasma frequency of the spacecraft are observed about 1.5 hours after the electrons left the Sun. At this time, 2 to 5 keV electrons are arriving at the spacecraft. The speed of these electrons is in rough agreement with the velocity ($v \approx 0.1c$) derived from the drift of the type III burst. The flare site is around W70, and therefore most likely magnetically well connected to the spacecraft.

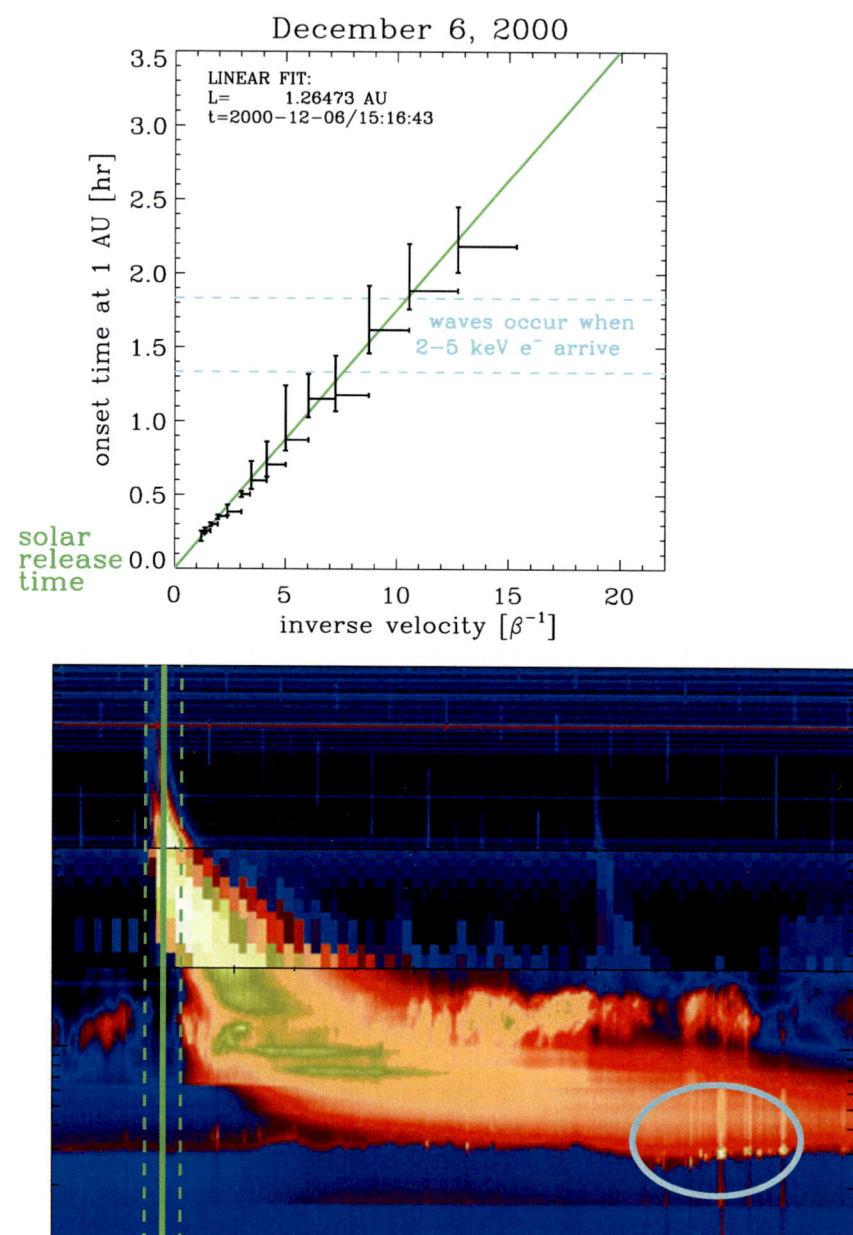

Fig. 1. (top) The electron onset times observed at 1 AU are plotted as a function of inverse velocity for the event of December 6, 2000. The green solid line is a linear fit to the data. (bottom) temporal comparison of the derived electron release time (green line) with the onset of the radio type III burst observed by WIND/WAVES. The green vertical lines give the time and uncertainty (dashed) of the release time of the electrons detected by WIND/3DP at 1 AU. The occurrence of in situ waves at the local plasma frequency of the spacecraft is marked by an ellipse. The emission seen between 0.1 and 0.2 MHz is auroral kilometric radiation.

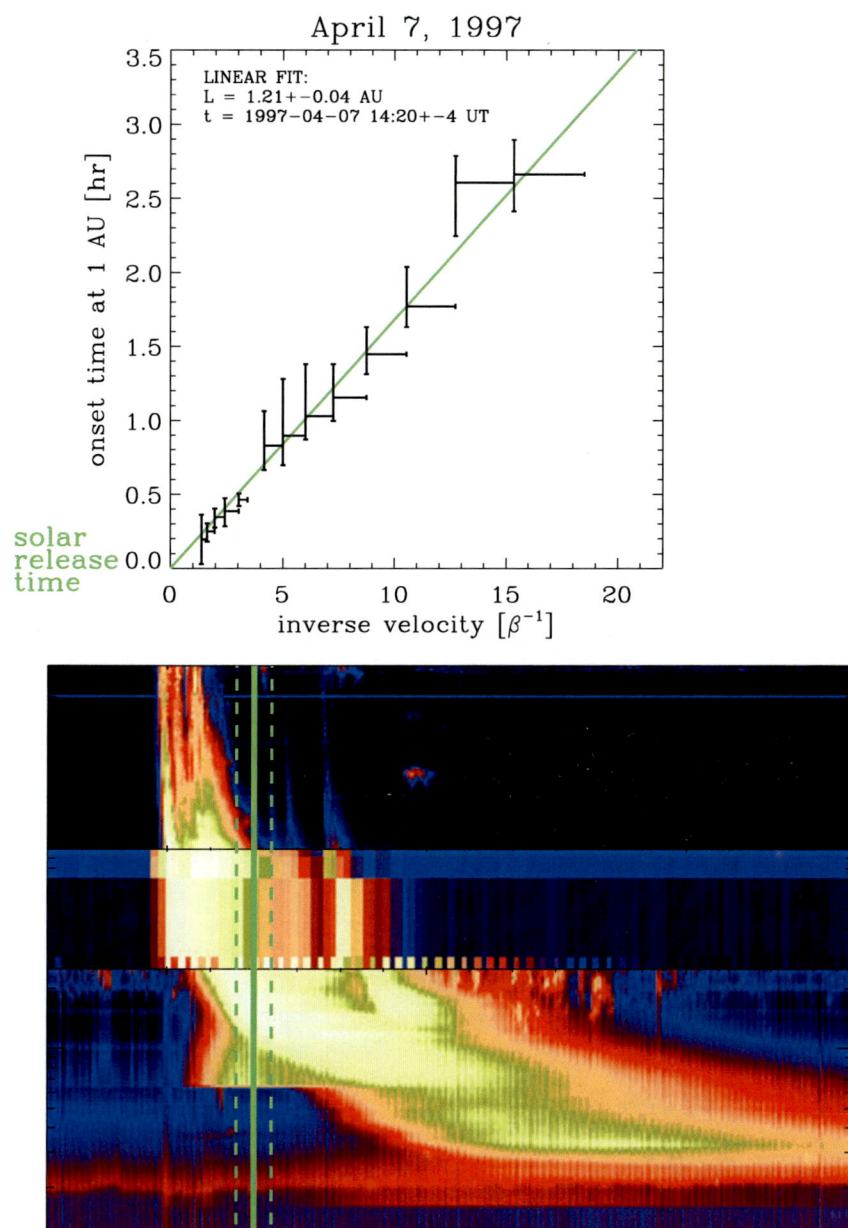

Fig. 2. Same as Fig. 1 for the event of April 7, 1997.

APRIL 7, 1997

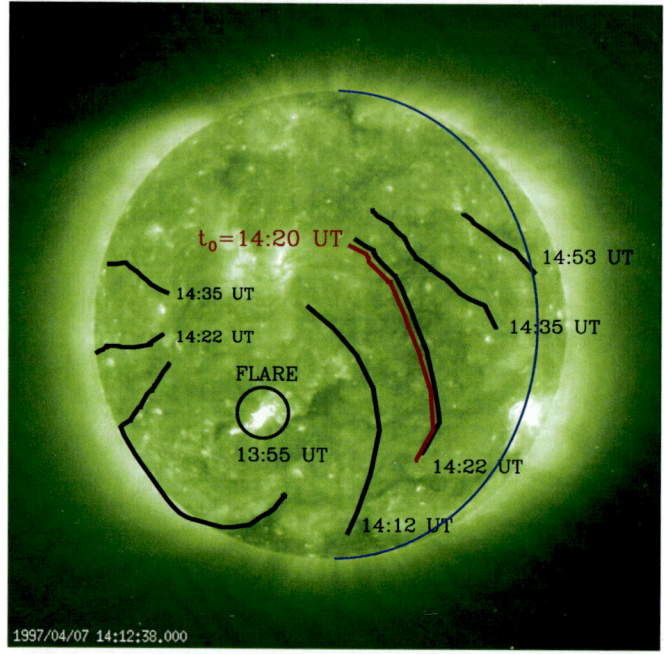

JUNE 29, 1997

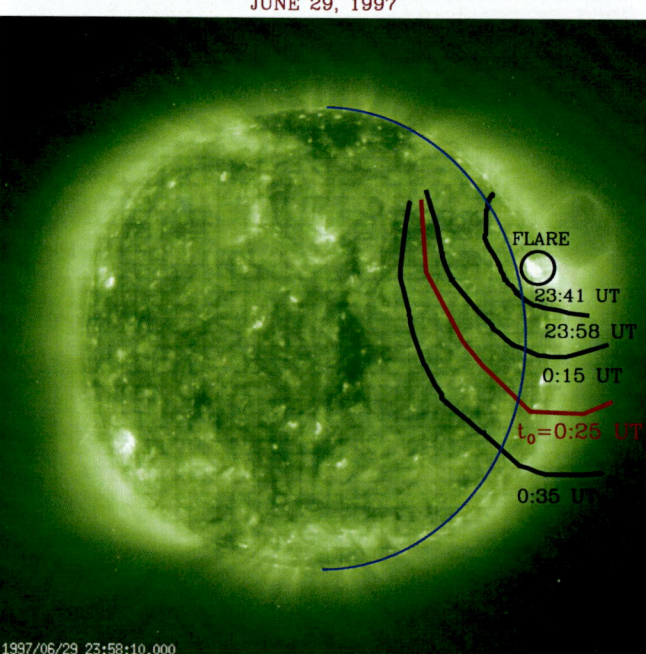

Fig. 3. Coronal EIT wave observations: The temporal evolution of the leading edge of the wave fronts outlined in black is plotted on the SoHO/EIT 195 Å image for the events of April 7, 1997 and June 29, 1997. The flare sites are marked by circles. The red curves give the interpolated wave front position at the time of the electron release. The blue curves along constant longitudes are the expected footpoint locations of field lines connecting the Sun and the spacecraft (from Krucker et al. 1999).

The observed electrons from the event of April 7, 1997 (Fig. 2, top) clearly show a later solar release time than the radio type III onset (Class 2 event). Hence, the acceleration of the electrons observed by WIND/3DP is not during the impulsive phase of the flare, but later. The derived path length of $L_e = 1.21 \pm 0.04$ AU again suggests scatterfree transport for the first arriving electrons. The flare site for this event is at eastern longitudes (cf. Fig. 3), but we expect the spacecraft to be connected to western longitudes. It is therefore not surprising that we do not see the type III producing electrons, since they travel on a field line not connected to the spacecraft. Krucker et al. (1999) speculated that electrons seen by WIND/3DP at 1 AU are related to large-scale coronal transient waves, also called EIT waves or coronal Moreton waves (Thompson et al. 1998, 2000). The timing between the wave propagation and the approximated electron release suggest that the acceleration/release takes place at higher (≥ 0.5 solar radius) altitude in the corona and could occur at numerous locations distributed over the entire solar disk (cf. Fig. 3). Detailed theoretical ideas about electron acceleration related to EIT waves and shock waves in general are discussed by Mann et al. (2002). Not all events of the second class are occurring at eastern longitudes or behind the limb, but there are also events with flare locations around W60 (cf. Krucker et al. 1999, Fig. 4, bottom). For these cases, the magnetic connection of the spacecraft to the Sun might be separated in north-south direction (cf. Fig. 3, bottom).

The second class of electron events are for some reason not or only weakly emitting type III radio bursts. In some events, like the April 7, 1997, a very faint radio type III burst is observed to occur at the time of the electron release (Fig. 4). For all class 2 events with a temporally correlating type III burst, the radio emission is faint and following a much stronger type III burst. In other

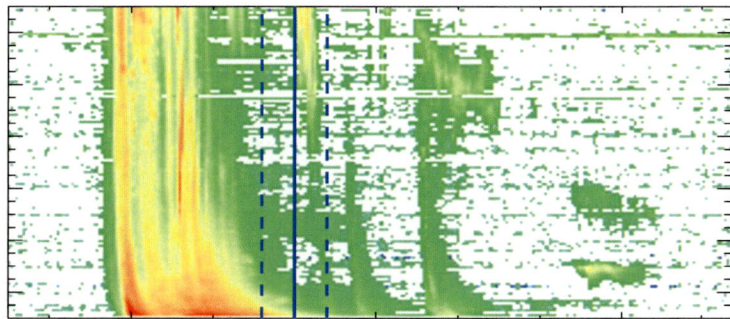

Fig. 4. WIND/WAVES radio spectrogram of the April 7, 1997 event. The pre-event background has been subtracted. The blue vertical lines mark the time and uncertainty (dashed) of the electron release (cf. Fig 2).

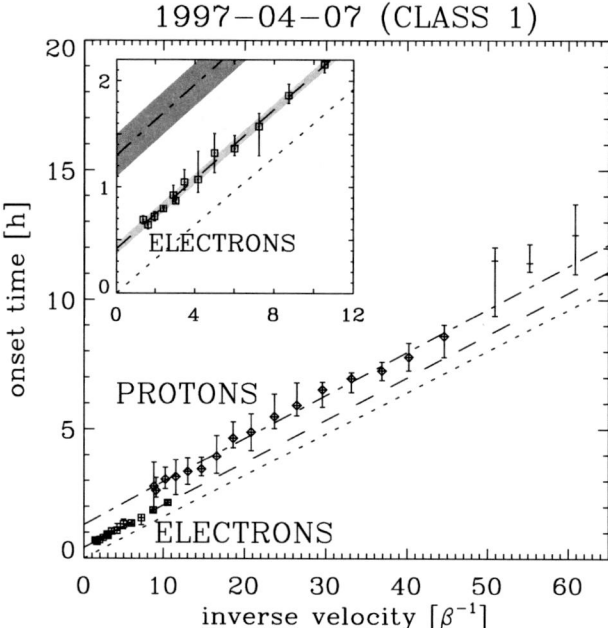

Fig. 5. Comparison of proton (diamonds) and electron (squares) onset times at 1 AU for the event of April 7, 1997. The onset time after the radio type III burst onset at the Sun as a function of the inverse velocity is shown. The dashed-dotted and dashed lines are linear fits to the proton and electron onset times, respectively. The dotted line shows the expected onset times at 1 AU for for particles traveling scatterfree along the Parker spiral assuming they are released at the time of the radio type III burst onset at the Sun. The insert shows a zoom-in of the same plot for a clearer representation of the electron onset times. The gray shaded areas represent the uncertainties in the fitted curves (from Krucker & Lin 2000).

events, however, no correlated radio emission at all is observed. The absence or faint appearance of radio type III emission could be explained by a higher energy cut off in the electron spectrum for the second class of events (i.e. the absence of ≤10 keV electrons). The beam density at higher energy might not be large enough to produce enough wave growth or might even not be large enough to start wave growth at all. It is mentioned here, that some events of the second class are speculated to be related to type III bursts (Klassen et al. 2002).

3.2 Proton Events

Krucker & Lin (2000) reported two different classes for ≤6 MeV proton events distinguished by their onset times:

1. proton events released roughly one hour after the electrons.
2. proton events that seem to have an energy dependent release time, but might originate from the same acceleration site as the electrons.

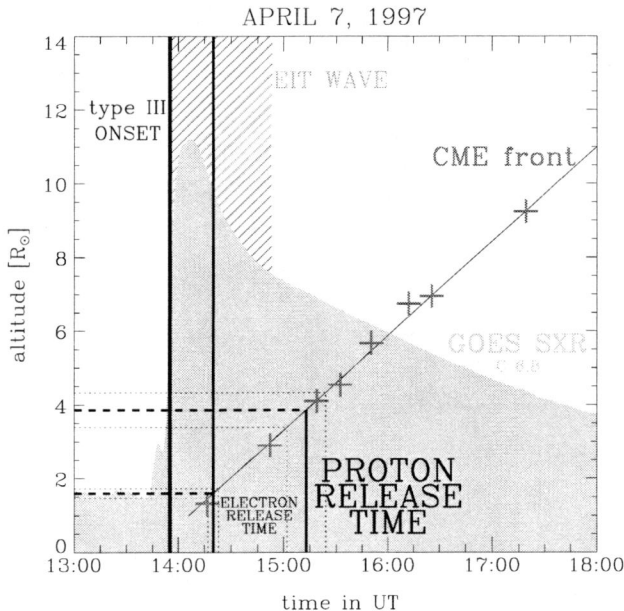

Fig. 6. The timing of the GOES soft X-rays flare (light gray), the first radio III burst (13:55 UT), the EIT wave (gray), and the CME shock front (crosses, private communication A. Vourlidas) are shown relative to the derived solar release times (8 minutes are added) of electrons (14:20 UT) and protons (15:13 UT).

It is mentioned here, that these two classes do not correlate with the two classes of electron events discussed above.

As an example for the first class of proton event, again the April 7, 1997 event is shown in Fig. 5: The linear fits show the same slope for protons (dash-dotted line) and electrons (dashed line) indicating that the first arriving protons and electrons are traveling about the same distance ($L_p = 1.20 \pm 0.05$ AU and $L_e = 1.19 \pm 0.04$ AU), a path length comparable to the Parker spiral length of 1.16 AU calculated from the averaged observed solar wind speed for this day. The intersections of the fitted lines with the vertical axis give the solar release time of the electrons, $t_{Sun} \approx 14 : 20 \pm 3$ UT (8 minutes are added to account for the time of flight of electromagnetic radiation), i.e. 25 ± 5 minutes after the type III onset at the Sun (chosen as $t = 0$), and the first protons appear to be released an additional 53 ± 12 minutes later than the first electrons. For this event, the proton onset at higher energies (24-48 MeV) derived from SoHO/ERNE (Torsti et al. 1995) is reported around 15:15 UT with no error bars given (Torsti et al. 1998). Assuming a path length of 1.2 AU and that the observed 24-48 MeV proton onset is produced by 48 MeV protons, the solar release time (plus 8 minutes) of 24-48 MeV protons is around 14:51 UT. Assuming a similar uncertainty for the SoHO/ERNE 24-48 MeV proton onset time as observed at lower energies

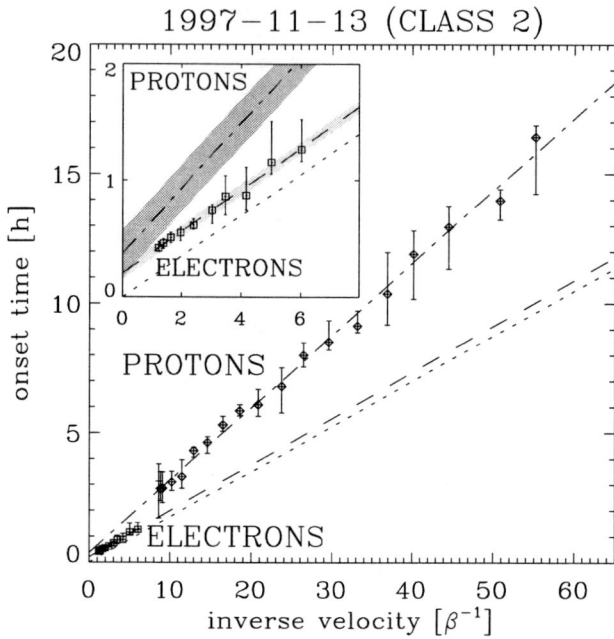

Fig. 7. Same as for Fig. 5 for the event of November 13, 1997 (from Krucker & Lin 2000).

by WIND/3DP, the release of tens of MeV protons is possibly consistent with the release of <6 MeV protons (15:13±12 UT), at least for this event. At the time of the proton release, the SXR flare, impulsive radio bursts, etc., are over, and the only ongoing solar event is the coronal mass ejection (CME) moving away from the Sun (Fig. 6). Assuming the proton acceleration/release is related to the CME shock front (Kahler 1996), LASCO CME observations suggest that this occurs when the CME reaches altitudes of several solar radii (Fig. 6). The earlier release times of electrons relative to protons suggest that electrons and protons are released at different locations. Assuming both, proton and electrons, are shock accelerated, protons would be released at higher altitude than electrons, roughly ~1-10 $R_\odot$ above the electrons.

A proton event of the second class is shown in Fig. 7. The electron onset times give again a path length ($L_e = 1.29 \pm 0.13$ AU) comparable to the Parker spiral length (1.26 AU), but the protons appear to travel a much longer distance ($L_p = 2.02 \pm 0.07$ AU). Contrary to the previously presented event, the solar release times of protons (21:36±12 UT) and the electrons (21:26±3 UT) are simultaneous within the uncertainties. For this event, the solar release time of the electrons is still significantly delayed by 12±3 minutes compared to the type III radio burst onset (~21:14 UT).

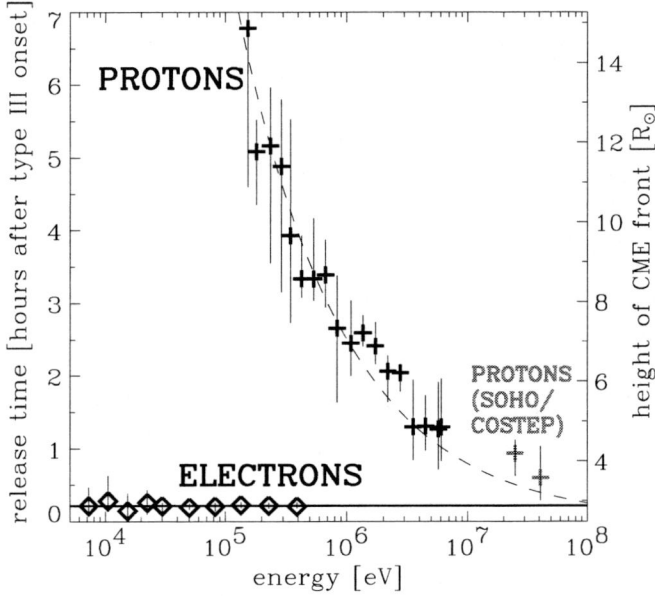

Fig. 8. The solar release times of electrons and protons for the November 13, 1997 event, assuming that the first arriving electrons as well as the first arriving protons are traveling scatterfree along the same field line. The dashed line is a fit to the proton release times assuming the release time varies with inverse velocity. The proton onset times above 10 MeV are derived from from SOHO/COSTEP observations (Müller-Mellin et al. 1995).

Since the first arriving electrons travel essentially along the Parker spiral field line, and the solar wind speed does not change significantly, it is very unlikely that the protons in Class 2 events travel along different, much longer field lines. A longer path length would result, however, if the first arriving protons had suffered much more pitch angle scattering than the electrons. However, the electron and proton pitch angle distributions during the first hours after the onset are very similar, both collimated within $\sim 40°$ of the field. While these pitch angle observations at 1 AU only provide information about scattering within a few tenths of an AU, it appears unlikely that strong proton pitch angle scattering is occurring in Class 2 events; most likely, the proton path length is the same as for electrons. Hence, to explain the observed velocity dispersion in class 2 events, the assumption that protons at all energies are released simultaneously has to be given up. Using the same path lengths for electrons and protons, the time of flight can be determined, subtracted from the observed onset time, and the solar release time as a function of energy can be approximated (Fig. 8). Under this assumption, larger than ≈ 50 MeV protons might be released simultaneously with the electrons and lower energies are released successively delayed. Assuming CME shock acceleration, the later release is translating into a release at higher altitude. If the solar release time indeed is a function of energy, it is surprising that the onset times at 1 AU are still observed to be inversely proportional to the velocity. That $t_{AU} \propto v^{-1}$ also for the second class of proton events (but with a steeper slope) suggests that the derived solar release times are also a function of inverse velocity (cf. dotted curve in Fig. 7): $t_{Sun} \propto v^{-1} \propto E^{-0.5}$, since the relativistic correction for < 6 MeV protons can be neglected.

A delayed escape of lower energy protons might be due to particle trapping or pitch angle scattering near the Sun which holds the protons back before they finally escape. Trapping can occur at a quasi-perpendicular shock (i.e. the angle θ between the shock normal and the magnetic field is close to 90°) as observed at the earth's bow shock (Anderson et al. 1979), if the particle velocity normal to the shock is slower than the shock speed. A continuous decrease of θ away from the quasi-perpendicular value, which could occur as the shock moves away from the Sun, allows the escape of particles at successively lower energies.

4 Summary

Wind/3DP observations at 1 AU have revealed that there are two classes of electron events. Next to the classic type III producing electron events, there are events accelerated up to half an hour later than the impulsive phase of the flare. These electrons are not or only weakly seen in radio emission and might therefore be called radio poor. The timing with EIT waves suggests that the second class of events could be shock accelerated. The acceleration would take place higher in the corona and numerous acceleration sites distributed over the entire solar disk could be present.

In two thirds of the <6 MeV proton events, one finds the classic gradual scenario: Protons are released about an hour after the electrons and are most likely related to the CME shock moving away from the Sun. Timing analysis suggests proton acceleration at altitude of 2 to 20 solar radii above the photosphere. The other third of proton events can be explained by an energy dependent release time, lower energies released later, higher energies are possibly released together with electrons. This is the only evidence for simultaneous (within the uncertainties of ≈ 15 min) electron and proton acceleration of later escaping SEP at the described energies.

No correlation between the two classes of electron and proton events are found. This is maybe not too surprising, since both classes of electron events can be produced by the same solar event, just separated in time. Whether a spacecraft sees a class one or two electron events, only depends on the magnetic connection of the spacecraft. If the spacecraft is well connected, the type III related electron event is observed, if there is no magnetic connection between the flare site and the spacecraft, the delayed accelerated electrons can be observed.

5 Future Observations

With the launch of the STEREO mission, major progress in understanding the origin of solar energetic particles can be achieved. The existence of two classes of electron events can be corroborated by observing both classes of events simultaneously at the different locations of the two STEREO spacecraft. The spacecraft well connected to the flare site should see the type III producing electrons. If the second spacecraft is not connected to the flare site, it should see the delayed electrons of the second class. The speculation of shock acceleration of the second class of electron events can be tested by observations when neither of the two STEREO spacecraft is connected to the flare site. Both spacecraft are then expected to see the delayed accelerated electrons. Since the shock needs time to travel from the flare site to the footpoint of the magnetic connection of the Sun and the spacecraft, the observed time delay of the release of the electrons relative to the flare onset should increase the farer away the magnetic connection of the spacecraft from the flare site. As an example, for the event of April 7, 1997 (cf. Fig. 3), one would have expected a shorter delay of around 10 minutes for the STEREO spacecraft located on earth orbit behind earth, and a longer delay around 30 minutes for the second STEREO spacecraft in front of earth.

More insights on the puzzling second class of proton events will also be provided by STEREO. The speculation about proton acceleration at quasi perpendicular shocks in the second class of events can be tested: If the one STEREO spacecraft is connected to a quasi perpendicular shock, the second one is likely connected at an angle significantly smaller than 90°. Therefore, only one STEREO spacecraft should observe the apparent longer path length but not the second one. To separate propagation effects and delayed release, observations at different radial distances would be needed. Hence, combining measurements in the inner heliosphere with STEREO observations will resolve whether the observed long path lengths are due to propagation or a delayed release.

References

1. Anderson, K.A., et al., 1979, Geophys. Res. Lett. 6, 401
2. Bougeret, J.-L. et al., 1995, Space Sci. Rev. 71, 231
3. Brueckner, G. E, et al., 1995, Solar Phys. 162, 357
4. Cane, H.V., Erickson, W.C., & Prestage, N.P., 2002, JGR, in press
5. Cliver, E.W., 1999, JGR 104, 4743
6. Delaboudinière, J.-P. et al., 1995, Solar Phys. 162, 291
7. Kahler S.W., 1996, in: 'High Energy Solar Physics', ed. by R. Ramaty, N. Mandzhavidze, X.-M. Hua, AIP Conf. Proc. 374, 61
8. Klassen, A. et al., 2002, AA 385, 1078
9. Klein, K.-L., Trottet, G., Lantos, P., & Delaboudinière, J.-P., 2001, AA 373, 1073
10. Krucker, S., Larson, D. E., Lin, R. P., & Thompson, B. J., 1999, ApJ 519, 864
11. Krucker, S., & Lin, R. P., 2000, ApJ 542, L61
12. Lin, R. P. et al., 1995, Space Sci. Rev. 71, 125

13. Mann, G.J. et al., 2002, AA, submitted
14. Mason, G.M., Mazur, J.E., & Dwyer, J. R., 1999, ApJ 525, L133
15. Müller-Mellin, R. et al., 1995, Solar Phys. 162, 483
16. Reames, D. V., 1999, Space Sci. Rev. 90, 413
17. Thompson, B. J. et al., 1998, Geophys. Res. Lett. 25, 2465
18. Thompson, B. J. et al., 2000, Solar Phys., 193, 161
19. Torsti, J. et al., 1995, Solar Phys. 162, 505
20. Torsti, J. et al., 1998, Geophys. Res. Lett. 25, 2525

Acceleration and Propagation
of Solar Energetic Particles

Wolfgang Dröge

Bartol Research Institute, University of Delaware, Newark, DE 19716, USA

Abstract. The acceleration of electrons and charged nuclei to high energies is a phenomenon occuring at many astrophysical sites throughout the universe. In the heliosphere, processes in the solar corona associated with flares and coronal mass ejections (CMEs) are the most energetic natural particle accelerators, sometimes accelerating electrons and ions to relativistic energies. The observation of these particles offers the unique opportunity to study fundamental processes in astrophysics. Particles that escape into interplanetary space can be observed in situ with particle detectors on spacecraft, and their spectra and composition can be used as diagnostic of the acceleration processes. On the other hand, energetic processes on the sun can be studied indirectly, via observations of the electromagnetic emissions (radio, X-ray, gamma-ray) produced by the particles in their interactions with the solar atmosphere. The comparison of interacting and escaping particles can provide valuable information about the question whether there is one dominant energization process in solar events, or whether particles are accelerated in multiple processes or sites. Equally important, the study of the propagation of solar cosmic rays allows to address some fundamental problems in the scattering of charged particles by magnetic fluctuations. In this article, we give an overview on models of stochastic particle acceleration and interplanetary particle transport, and discuss the question what conclusions about those models can be drawn from spacecraft observations.

1 Introduction

Energetic processes on the Sun are known to occasionally accelerate protons to GeV and electrons to tens of MeV energies. Solar energetic particles (SEPs) contain important information about the mechanisms of particle energization at astrophysical sites, as well as properties of the acceleration sites themselves. Two basic classes of solar particle events have been identified as so-called "gradual" and "impulsive" events. Impulsive particle events are associated with short timescales, large electron to proton and ^{3}He/^{4}He ratios, and high ionization states, indicating a source region with temperatures of $\sim 10^7$ K. Gradual events are associated with longer timescales, coronal and interplanetary shocks, high proton intensities, energetic particle abundances similar to the corona, and charge states corresponding to a source region with temperatures of $\sim 3 \times 10^6$ K. Whereas it was thought for a long time that all SEPs were accelerated in flares, it has been argued over the past few years [1] that only particles observed in impulsive events originate from a flare, and particles in all gradual events are instead accelerated at a shock wave in front of a CME, and not related to a flare. However, observations made by the ACE spacecraft have revealed a rich structure in the time dependence of the two classes of SEPs, much of which cannot be explained with our current understanding of solar flares, CMEs, and related phenomena. A recent analysis from ACE data [2] indicates that many events defy a simple classification, i.e., appear "impulsive" in some respects and "gradual" in others. It seems possible that gradual events possess an impulsive "core" [3] in which similar acceleration processes as in impulsive flares operate, and from which particles can escape into the interplanetary medium where they can be observed together with particles accelerated by a CME-driven shock. Observations of time-extended radio, X-ray and gamma-ray observations in gradual events indicates that particle acceleration takes place in large-scale coronal structures behind the CME, and that the CME shock is not the sole accelerator of particles in gradual events (e.g., [4]).

A complete theory of particle acceleration in solar flares must explain how electrons and ions are energized out of the thermal plasma as well as provide time scales, energy spectra, fluxes and abundance ratios of the various particle species, and the characteristics of the induced emission of radio waves, X-rays, gamma-rays, and neutrons, which are consistent with observations. However, the interpretation of spacecraft and ground-based data is often difficult because of the chain of assumptions required regarding the convolution of properties of energetic particles with properties of the photospheric, coronal and solar wind plasmas and the interaction between them. It becomes apparent that to really understand energetic processes on the sun, and to be able to distinguish between different possible scenarios, one has to address the problem of particle release from the Sun, and propagation in the inner heliosphere together with the problem of energization.

Understanding how energetic particles are transported in the turbulent solar wind from their sources to the point of observation is not only essential to study energetic processes on the sun, it is also an important problem in its own right. Considerable progress has been achieved in recent years towards a better modeling of the nature of the solar wind turbulence, and to overcome some of the deficiencies of the first, pioneering scattering theories (so-called quasi-linear theory, QLT [5, 6]) that could not be reconciled with observations. New approaches to the theory which take into account the dynamical character and the three-dimensional geometry of the magnetic field fluctuations (e.g., [7]) and, in another approach, the effects of wave propagation and thermal wave damping and resonance broadening (e.g., [8]) have shown to give better explanations for various aspects of the observations (for a recent review see [9]). These models are able to correctly account for the dependence of the scattering mean free path on the particle's rigidity, $P = pc/(Ze)$, and have improved the agreement between scattering theory and observations in a statistical sense, averaged over many events. The determination of transport parameters from the measured properties of the solar wind plasma on an event-by-event basis is still problematic, although some encouraging results have been obained recently [10].

A comprehensive representation of the current status of the observation of energetic particles from solar events, and complementary measurements of electromagnetic radiation, is well beyond the scope of this work; recent reviews, compilation of new results and re-evaluations of older results can be found in, e.g., [11, 12, 13]. Here we will focus on the interpretation of selected electron and proton observations from ISEE-3, *Helios* and *Wind*. These data are particularly well suited to study transport and acceleration processes because the spacecraft were located outside of the Earth's magnetic field, and the performance of the instruments is well understood. This work will be structured as follows: Section 2 gives an overview of the observation of solar energetic particles in the inner heliosphere. A phenomenological description of interplanetary transport of solar particles and a brief outline of the problem of wave particle interactions is given in Sect. 3. In Sect. 4 acceleration models for solar particles are discussed. Section 5 summarizes the results.

2 Observations

The Sun has long been known to be a source of energetic particles. Even before the era of spacecraft observations, in the 1950s, nearly relativistic protons from a number of solar events were detected by neutron monitors. Those early observations already revealed some information about possible acceleration mechanisms and particle propagation between the Sun and the Earth (e.g., [14]). In the decades that followed, space missions opened the window for the observation of lower energy ions and their elemental composition, electrons, solar neutral radiation, as well as properties of the solar wind. A major step forward in the understanding of solar particle events was achieved based on observations made around the maximum of solar cycle 21 from $\sim$ 1978 -1982, in particular by ISEE-3 which was positioned at the Earth-Sun Lagrangian point well outside of the Earth's magnetic field, and by *Helios 1* and *2* which operated in the inner solar system between 0.3 and 1 AU. During this time interval, electron and proton fluxes over a large energy range for many events were observed simultaneously on two or all of the above spacecraft, and complemented with detailed in situ measurements of the heliospheric plasma. For a number of events, observations of hard X-rays, gamma rays and even neutrons from the associated flares were made by the SMM satellite, as well as observations of CMEs with the NRL coronograph on the the *P78-1* spacecraft. Further progress in the study of energetic processes on the Sun was achieved with observations made by the Compton Gamma Ray Observatory and GRANAT at the maximum of the following solar cycle, especially during some very large events in the period 1989 to 1991. In the current solar cycle, space missions such as SAMPEX, *Wind* and ACE have made previously unavailable measurements of the charge states, and elemental and isotopic composition of energetic particles during solar events, and imaging instruments on board Yohkoh, SOHO, and TRACE allowed an unprecedented view on processes in the solar corona at soft X-ray, EUV and white light wavelengths.

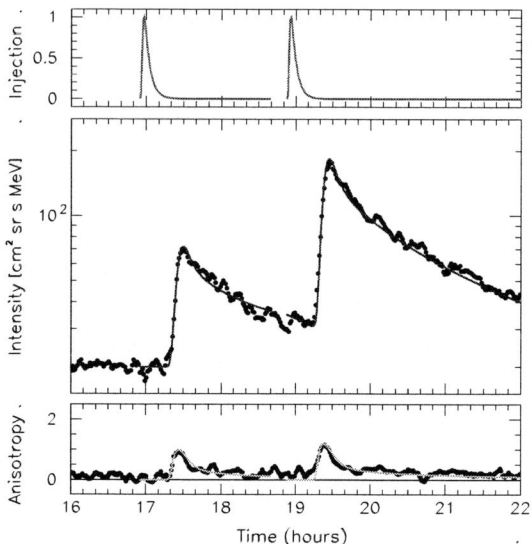

Fig. 1. Electron flux at 107 keV observed on 5 Dec 1997 by *Wind* Electron data can be fitted in intensity (upper panel) and anisotropy (middle panel) assuming a parallel mean free path of ≈ 0.4 AU.

The University of Kiel cosmic ray instrument onboard *Helios 1/2* [15] was designed to measure MeV electrons and ions from 2 to ≥ 400 MeV/n in various coincidence rate channels. Applying the results of a Monte Carlo Simulation performed with the CERN Library program GEANT 3 [16] it was possible to determine the true particle response functions of the instrument, corrected for the contamination of the electron and proton channels with particles of the other species at times when the e/p ratio was low or high, respectively. This way the useful energy range could be extended to 1-10 MeV for electrons and up to ∼ 1 GeV for protons[17]. The ISEE-3 electron observations presented here are from the ULEWAT spectrometer (MPE/University of Maryland) which measured the electron flux in the nominal energy range 0.075 − 1.3 MeV, and the University of Chicago MEH spectrometer which measured the electron flux in the energy range 5 − 100 MeV (for a full description of the instruments see [18]). Electron observations on *Wind* in the energy range ∼ 20 to 500 keV are from the Three-Dimensional Plasma and Energetic Particles (3DP) instrument ([19], see also Krucker, *this volume*), which was designed to provide full three dimensional coverage with 36° × 22.5° angular resolution.

Figure 1 shows the time history of ∼ 107 keV electrons observed on *Wind* on 5 Dec 1997. The electron event was associated with two impulsive flares which occured at ∼ 17:00 UT and 19:00 UT, respectively, in NOAA region 8113 located at N20 W50. This is not far away from the nominal beginning of the connecting interplanetary field line at ∼ 58°, as estimated from the observed solar wind speed of ≈ 380 km/s. The electron intensity and anisotropy profiles exhibit the characteristics of a typical event with a short injection at the Sun and relativeley weak scattering in the interplanetary medium: a rapid rise to a sharp maximum caused by particles arriving from the sun, followed by a smooth decay when the flux of particles scattered back from the anti-sunward direction sets in and the distribution function isotropizes. The events can be well modelled with numerical solutions of the model of focussed transport, which we will describe in more detail below. The similar overall shape of the two events indicates that the propagation conditions are constant on a time scale of several hours. Due to the nearly ideal situation found in this example, it is possible to reconstruct the injection profile of the electrons close to the Sun (shown in the upper panel of Fig. 1). In this work we have assumed that the injection takes place at 0.05 AU or 10 solar radii, where the magnetic field most likely has assumed the shape of an Archimedian spiral.

Large particle events following gradual flares and associated CMEs and interplanetary shocks usually have a much more complex structure. As a typical example we show in Fig. 2 simultaneous *Helios* 1 and 2 observations of the classic 1 Jan 1978 event (cf., [20]). The time profiles for electrons and protons, and for the various energy channels are distinctly different on the two spacecraft, but we note that this difference is much less pronounced for electrons and high energy protons. On *Helios* 1, at a coronal distance of $\sim 30°$ to the flare, the intensities show a fast rise and resemble in this respect characteristics of impulsive events, e.g., rapid acceleration close to the Sun and fast release into interplanetary space. On *Helios* 2, with its magnetic footpoint $60°$ away from the flare, particle fluxes show a slower rise and their maximum values are more than one order of magnitude smaller. Proton fluxes below ~ 20 MeV exhibit a plateau or a second rise until the passage of the shock (indicated by a dashed line in the figure). However, this effect is less pronounced on *Helios* 1, and almost totally absent for > 51 MeV protons and electrons on both spacecraft.

From many other examples that show a similar behaviour, it seems obvious that the shock has a major effect in the later phase of gradual particle events. As mentioned earlier, what exactly the role of the shock is, still remains a subject of controversial discussion. During the last decade, the view has become popular that all interplanetary particles in gradual events are accelerated by the CME-driven shock [1]. Of course, other views are also possible, and this without neccessarily having to go back to the other extreme of a 'flare' which is used as synonym for a compact acceleration region in the lower corona. It was noted some time ago [20] that the typical shapes of gradual events can be described by a superposition of a prompt component - originating from a source which is not neccessarily identical with the optical or X-ray flare and may be located higher in the corona, but likely behind the CME - and a shock-processed component late in the event. In this picture, the prompt component dominates electron and high energy proton fluxes throughout the event, wheras from medium to lower proton energies the shock component would become dominant, except in the very beginning of the event and also dependent on the azimuthal distance with respect to the particle/CME source region. Systematic studies of high energy ($>$

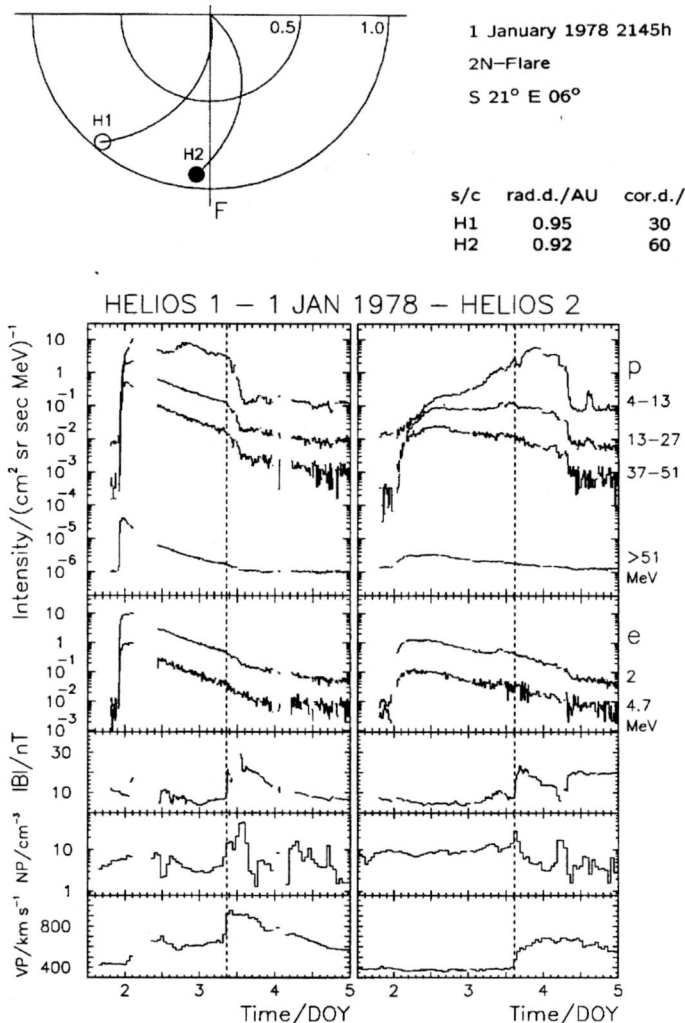

Fig. 2. The sketch shows the flare position, rotated to the central meridian, and the postition of both Helios spacecraft on 1 Jan 1978. Helios 1 (left) and Helios 2 observations of the solar particle event are shown in the panels from top to bottom: 1. protons, 2. electrons, 3. magnetic field strength, 4. solar wind proton density, 5. solar wind speed. (from [20])

1 MeV) electrons [18, 21] have revealed that their time profiles and maximum fluxes do not differ significantly in impulsive and gradual events, suggesting that electrons have little to do with the CME shock.

Besides the timeline of energetic particle events in interplanetary space and their relation to the observation of electromagnetic emission, the energy spectra of the particles also bear important information about the acceleration process.

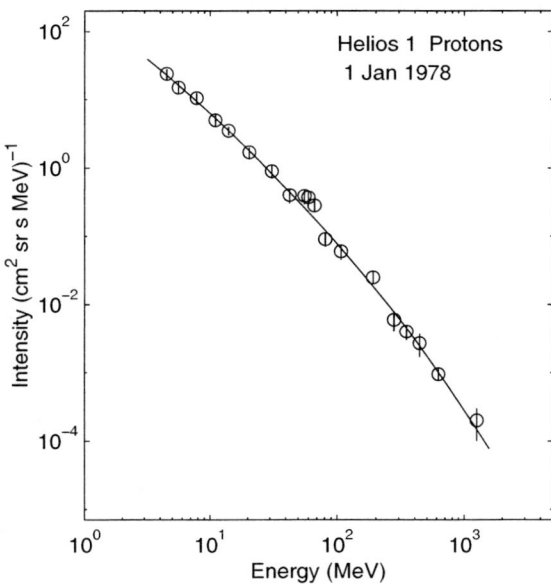

Fig. 3. Time of maximum spectrum of the prompt proton component observed on 1 Jan 1978 by the University of Kiel particle telescope onboard Helios 1.

The classic method of constructing particle spectra is to take the maximum of the observed differential flux $J(E)$ in each energy interval (cf., [22]). As will be shown below, the spectra derived with this method are representative of the source spectra injected into interplanetary space, provided that the source is located close to the Sun, that particles escape within a time period which is short compared to the time to maximum (TOM) flux at the spacecraft, and that particle propagation through the interplanetary medium can be described by pitch-angle scattering off magnetic irregularities.

Figure 3 shows the TOM spectrum of the prompt proton component observed on 1 Jan 1978 by Helios 1. The shape of the spectrum, a slightly convex curvature over the whole observed energy range, seems to be typical for the prompt proton component, and in this respect there is no significant difference between gradual and impulsive events. A detailed investigation of the 1 Jan 1978 event [20] showed that the prompt Helios 2 spectrum at energies above ~ 20 MeV, where it could be determined in a meaningful manner, had a shape quite similar to that on Helios 1. Remarkably, the proton fluxes at *Helios* 1 and 2 at the time of the shock passage were almost identical on both spacecraft, in magnitude as well as in spectral shape, but not in agreement with the spectral index diffusive shock acceleration (see below) would predict from the observed shock parameters. This and the local shock acceleration time, which can be estimated from the intensity gradient in front of the shock, indicates that the shock had stopped accelerating ions at energies above 1 MeV a long time before it reached 1 AU. Particle spectra obtained at the passage of the shock, as well as obtained by integrating the flux

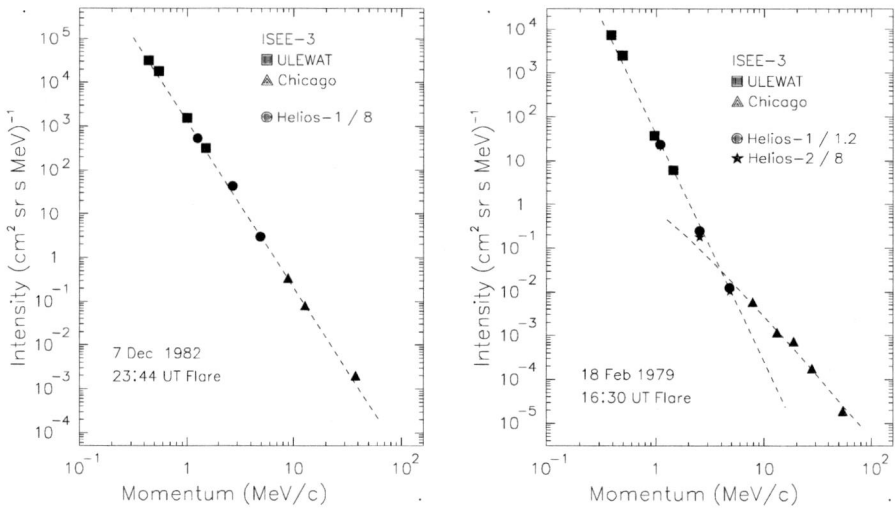

Fig. 4. Typical electron spectra from solar flares. Electron spectra from impulsive flares, such as in the right-hand panel, are much harder at high energies than those from gradual events, such as that shown in the left-hand panel (from [21])

over the event, or parts of it, are also frequently presented in the literature (e.g., [23]), but without a theory which takes into account the effects of particle injection, shock acceleration, and the propagation of particles away from the shock in a self-consistent manner they are difficult to interpret. Unfortunately, such a model is not presently available.

Energetic (> 1 MeV) electrons have not received much attention in the ongoing flare-shock controversy, and were not well measured by missions which followed Helios and ISEE-3. Nonetheless, they are as fundamental as ions to the understanding of the energy release process. Spectral shapes of electron events observed simultaneously on multiple spacecraft have been found to be in very good agreement in spite of the sometimes considerable difference in azimuthal and radial distances of the spacecraft with respect to the flare [21] and can therefore be considered characteristic for a given acceleration site at the sun. The spectral shapes also provide a nearly perfect diagnostic of the flare [18]. Gradual events (left panel of Fig. 4) produce electron spectra that are straight lines when plotted in the units of the figure (i.e., single power laws in momentum). Impulsive flares all show a spectral hardening starting at around 3-5 MeV/c in momentum (right panel of Fig. 4). Although the hardening of the spectrum is universally observed, the spectral index varies from event to event and the origin of these electrons or the reason for the hardening is still not well understood.

3 Particle Transport

3.1 Modelling

In the absence of large-scale disturbances such as CMEs and shocks, the interplanetary magnetic field can usually be described as a smooth average field, represented by an Archimedian spiral, with superimposed irregularities. In this case the propagation of charged particles consists of two components, adiabatic motion along the smooth field and pitch angle scattering off the irregularities. The quantitative treatment of the evolution of the particle's phase space density $f(z, \mu, t)$ is then given by the model of focussed transport [24]:

$$\frac{\partial f}{\partial t} + \mu v \frac{\partial f}{\partial z} + \frac{1 - \mu^2}{2L} v \frac{\partial f}{\partial \mu} - \frac{\partial}{\partial \mu} \left(D_{\mu\mu}(\mu) \frac{\partial f}{\partial \mu} \right) = q(z, \mu, t) \tag{1}$$

where z is the distance along the magnetic field line, $\mu = \cos\theta$ the particle pitch angle cosine, and t is the time. The particle velocity v remains constant in this model. The systematic forces caused by magnetic mirroring are characterized by $L(z) = B(z)/(-\partial B/\partial z)$, the focussing length in the diverging magnetic field B, while the stochastic forces are described by the pitch angle diffusion coefficient $D_{\mu\mu}(\mu)$. The injection of particles close to the Sun is given by $q(z, \mu, t)$.

If the scattering is sufficiently strong, $f(z, \mu, t)$ adjusts rapidly to a nearly isotropic distribution. In this case the particle transport can be described by a diffusion-convection equation, where the radial diffusion coefficient $K_r(r) = 1/3v\lambda_r$ is a measure for the scattering strength, with λ_r the radial mean free path. The mean free path $\lambda_\|$ which relates the pitch angle scattering rate to the spatial diffusion parallel to the ambient magnetic field is given by ([6])

$$\lambda_\| = \frac{3v}{8} \int\limits_{-1}^{+1} d\mu \, \frac{(1 - \mu^2)^2}{D_{\mu\mu}(\mu)} \tag{2}$$

with $\lambda_r = \lambda_\| \cos^2\phi$ where ϕ is the angle between the radial direction and the magnetic field. The mean free path has proven to be a convenient parameter to characterize the varying degrees of scattering from one solar particle event to another, even when it adopts values close to or larger than the observers's distance from the Sun and the transport process can not be considered spatial diffusion.

The effects of convection and adiabatic deceleration in the solar wind (moving with speed V_{SW}) can be incorporated into (1) (cf., [25]). These processes have to be taken into account when their corresponding time scale $T_C \sim r/V_{SW}$ becomes comparable with the time scale for diffusion, $T_D \sim r^2/(\lambda_r v)$ – which typically is the case for ions with energies < 1 MeV/n. Particles in this range are also heavily affected by shocks in many events, so it is difficult to determine their mean free paths. Because in this work we restrict ourselves to electrons and protons with sufficiently high energies, we can neglect the above effects. The particle events considered here were modelled with numerical solutions of

(1), obtained with a finite-difference scheme (cf., [26, 25]). Mean free paths were derived by applying simultaneous fits to the isotropic part of the distribution function and the first-order anisotropy, defined by

$$A(z,t) = \frac{3 \int_{-1}^{+1} d\mu \, \mu \, f(z,\mu,t)}{\int_{-1}^{+1} d\mu \, f(z,\mu,t)} \tag{3}$$

The transport of particles from the coronal acceleration site to, and subsequent injection at the beginning of the interplanetary magnetic field line connected to the observer can be expressed in terms of a Reid-Axford profile [27]

$$Q_R(\Phi,t) = \frac{C}{t} \exp\left\{-\frac{\Phi^2 \tau_c}{4t} - \frac{t}{\tau_L}\right\} \tag{4}$$

where Φ describes the azimuthal distance from the source to the connecting field line and τ_c, τ_L are the time scales for coronal transport and escape, respectively. We use this functional form to conveniently parameterize the source function in (1), and the actual transport has not neccessarily to be coronal diffusion as in the original Reid model. In fact, (4) can phenomenologically describe enhanced scattering close to the Sun as well, and also particle acceleration in the higher corona, even at a shock wave a few solar radii away from the Sun in the case that the acceleration process is completed and particles are released before the shock reaches interplanetary space. The injection profile (4), and fits to solutions of (1) can therefore also be used to derive mean free paths for gradual events where interplanetary particles do not neccessarily originate from an associated flare – as long as the particle profiles exhibit a fast rise to maximum, followed by a monotonic decay indicating injection close to the Sun over a finite amount of time.

The mean free path obtained from a fit to the particle data, hereafter referred to as λ_{fit}, can then be compared to the value λ_{th} based on a theoretical derivation of $D_{\mu\mu}(\mu)$ from the observed power spectrum of the magnetic fluctuations. A possible radial dependence of the mean free path can be modelled by a power law, $\lambda_r \propto r^b$. In many cases the choice of a constant λ_r is sufficient for a good fit. The fit of the 5 Dec 1997 events shown in Fig. 1 was obtained with a numerical solution of (1) assuming a constant $\lambda_r = 0.25$ AU, and a resulting $\lambda_\| = 0.55$ AU. Figure 5 shows results from a recent survey of near-earth mean free paths $\lambda_\|$ which was based on the modelling of selected solar events where electron and proton data over a large energy range were available [28].

3.2 The Pitch Angle Diffusion Coefficient

Cosmic ray pitch angle scattering is caused by irregularities of the magnetic field which violate the conservation of the first adiabatic invariant. As a cumulative result of many small random changes in its pitch angle the particle experiences a macroscopic change in direction, leading to spatial diffusion along the field line.

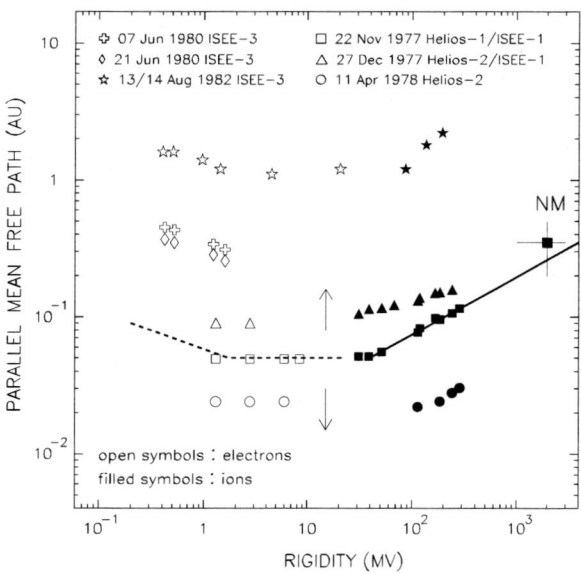

Fig. 5. Local parallel mean free path vs. particle rigidity for selected solar particle events. The form of the rigidity dependence as indicated by the curve seems to be consistent with observations from any given event, only the absolute height of the curve varies (from [28]).

A major problem in applying quasi-linear theory to comparison with observations is that in order to determine the scattering coefficient the full knowledge of the correlation tensor of the fluctuations is required. It is therefore crucial to formulate a model which describes at least the basic properties of the fluctuations correctly. Two approaches have dominated discussion of the nature of the fluctuations and subsequently, their interaction with charged particles. One model, the turbulence model, describes the observed properties of the magnetic fluctuations under the assumptions that the fluctuations interact. In this picture, there are no deterministic correlations between frequency and wave number, and the only adequate description is a statistical one. In the second model, the wave model, a much more deterministic approach is considered. The magnetic fluctuations are described as a superposition of small-amplitude plasma waves whose dispersion relations reflect the physical state of the wave-carrying background medium. Interaction between the waves is generally considered to be minimal, the time scales for excitation and damping of the waves are assumed to be large in comparison with their frequencies. The solutions of the dispersion relations lead to a variety of wave modes, such as Alfvén waves, ion and electron cyclotron waves, and magnetosonic waves. The waves can propagate parallel, or at an angle to B_0 and have different states of polarization.

Advanced approaches to QLT which take into account wave propagation effects as well as the dynamical character and the 3-dimensional geometry of

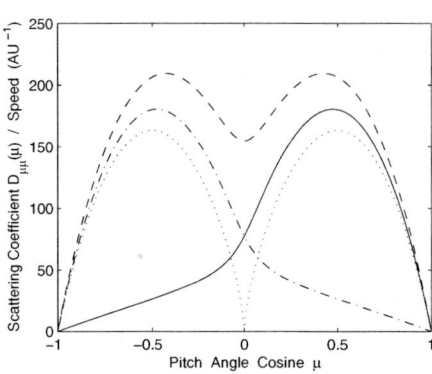

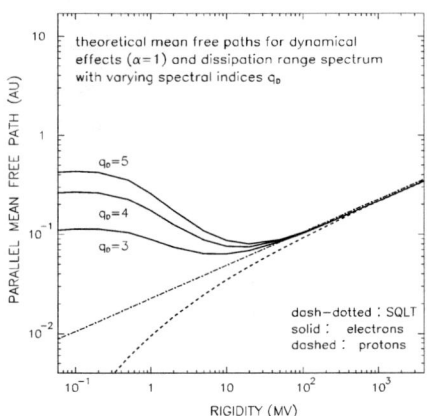

Fig. 6. Pitch angle diffusion coefficient in the case of resonance broadening for 1 MV protons and $\alpha = 1$, for positive (solid line) and negative (dash-dotted line) helicity. Also shown are the sum of both components (dashed line), and the results of standard QLT (dotted line).

Fig. 7. Theoretical mean free paths for resonance broadening and a power spectrum with spectral index $5/3$ in the inertial range, but with varying spectral indices in the dissipation range. For comparison, the result of dissipationless standard QLT is also shown. [from [10]]

the magnetic field fluctuations, and the effects of a dissipation range of the turbulence at high frequencies, predict that the mean free path has an explicit velocity dependence which, below a certain threshold, leads to larger values for electrons than for protons at the same rigidity. Both "wave" and "turbulence" approaches employ a resonance broadening of the scattering process which can be described by a resonance function $\Gamma(k) = \alpha |k| V_A$, where the parameter α allows to adjust the strength of the wave damping in the first, and the decorrelation of the fluctuations in the second case (for details, see [9, 10]). The limit $\alpha = 0$ describes the case of no wave damping, respectively magnetostatic theory, whereas for $\alpha = 1$ resonance broadening is the dominating effect for particle scattering through $\mu = 0$ (cf., Fig. 6). To reconcile the observations with the still too large absolute levels of the mean free paths, [7] suggested a composite model for the fluctuations which consists of $\approx 20\%$ slab (wave vectors parallel to $\boldsymbol{B}_0$) and 80% 2-D (wave vectors perpendicular to $\boldsymbol{B}_0$) fluctuations, the latter contributing little or not at all to particle scattering.

Figure 7 shows mean free paths calculated for a fluctuation spectrum assuming $\alpha = 1$ and for dissipation range spectral indices varying within the range of observations. The curves resulting from the combined effects of the dissipation range and resonance broadening are in good agreement with the shape of the rigidity dependence shown in Fig. 5. As can be seen from the figure, at rigidities above 100 MV standard QLT is a good approximation for the particle's mean free path, whereas below that value the free path of electrons is drastically larger, for protons even smaller compared to standard QLT.

It appears that QLT is a good approximation to particle scattering at all pitch angles if resonance broadening due to dynamic or wave damping effects is taken into account. These effects dominate scattering through $\mu = 0$ and probably make the need for investigating other additions to QLT such as non-linear corrections, mirroring, wave propagation effects and details of the dispersion relation obsolete. Resonance broadening is related to observable plasma parameters B, n, and T, so a prediction of local scattering conditions should be possible. The major remaining obstacle remains our lack of knowledge of the exact decomposition of the fluctuations, which is difficult to obtain from single spacecraft measurements, and the question whether those measurements are representative for scattering conditions in the inner heliosphere.

4 Particle Acceleration

The relatively smooth shapes of the observed spectra of solar particles, and total absence of any peaks or bumps other as produced by propagation or absorption effects at low energies, suggest that particles gain energy in a process which involves a large number of random interactions with fluctuating electromagnetic fields, or which has at least some sort of diffusive nature. To explain the sometimes very short timescale of a few seconds to reach relativistic energies, the energy transfer in these interactions must be efficient, and to explain the large numbers of particles which are sometimes released impulsively from the Sun, the mechanism must be able to provide an efficient containment during its course, so that a large number of particles can undergo many interactions.

Current theories of acceleration mechanisms in solar flares focus on direct acceleration by electric fields, acceleration at shock waves, and stochastic acceleration in turbulent magnetic fields. For the former two mechanisms the reader is referred to the articles by Litvinenko and Schlickeiser, respectively, (*this volume*) and references therein. Here we will discuss stochastic acceleration and a comparison of predictions of this theory with interplanetary particle observations.

There are several basic mechanisms that lead to stochastic acceleration of particles by magnetohydrodynamic turbulence. In the original stochastic Fermi mechanism [29], the process was reflection from randomly moving clouds, but stochastic acceleration can also result from scattering off magnetized fluid elements in a plasma, resonant pitch-angle scattering from Alfvén waves, and interaction of particles with magnetosonic and Langmuir waves. In general, these mechanisms can be regarded as systematic acceleration of particles plus a diffusion in momentum space. The usual scenario for stochastic acceleration of solar particles is a closed magnetic configuration filled with hydromagnetic turbulence, e.g., a flare loop or a complex field structure higher in the corona which is possibly related to an evolving CME. The magnetic field confines the particles long enough that the acceleration process reaches an equilibrium, and at some point opens up in the course of the reconfiguration of the coronal magnetic field.

In order to compare the results of acceleration models with TOM spectra in interplanetary space it is convenient to perform a separation ansatz using the scattering time method (cf., [30]) and get a set of particle spectra depending only on momentum and the eigenvalues of the spatial problem. Because the particle spectra are observed far away from their sources, it is possible to study the problem under the assumption that the transport equation with the first eigenvalue, which can be expressed as an escape time, describes the momentum dependence of the distribution function reasonably well. This leads to a transport ("leaky-box") equation for the spatially averaged phase space density $f(p,t)$

$$\frac{\partial f}{\partial t} - \frac{1}{p^2}\frac{\partial}{\partial p}\left(p^2 D(p)\frac{\partial f}{\partial p}\right) + \frac{1}{p^2}\frac{\partial}{\partial p}\left(p^2(\dot{p}_G + \dot{p}_L)f\right) + \frac{f}{T(p)} = Q(p,t) \quad (5)$$

where $\dot{p}_G$ and $\dot{p}_L$ are momentum gain and loss rates, respectively, and $Q(p,t)$ represents sources and sinks of the particles.

The phase space density is related to the particle flux $J(E)$, which is the quantity usually derived from spacecraft observations, by $J(E) = p^2 f(p)$. The momentum diffusion coefficient $D(p)$ depends on the nature of the stochastic process and can be written the form

$$D(p) = \frac{\alpha}{3\beta}p^2 \quad (6)$$

where $\alpha=V^2/lc$ describes the acceleration efficiency, $c\beta$ is the particle's speed, and l is an effective mean free path against interaction with waves of speed V which energize the particles.

Particle escape is described by $f/T(p)$. If the escape is due to diffusion out of the acceleration region, the escape time becomes

$$T(p) \approx L^2/K = 3L^2/(v\ell) \quad (7)$$

where $K = v\ell/3$ is the spatial diffusion coefficient and L the size of the physical region. In the case that particles escape from the acceleration site by convective motion (with speed V_{con}) of the plasma they are coupled to the escape time is

$$T(p) \approx L/V_{con} \quad (8)$$

Additional energy changing processes of some relevance, which can also be incorporated into (5), can be adiabatic deceleration due to expansion of the acceleration region, and energy loss due to Coulomb collisions. For electrons of ~ 1 MeV energy the loss rate in a fully ionized hydrogen plasma and for reasonable flare parameters ($T \approx 10^7 K, n_e = n_p = 10^{10} cm^{-3}$) can be written as

$$-\dot{p}_{Coul} = \frac{mc}{\tau_L \beta^2} \quad (9)$$

where $\tau_L = 0.146 \times 10^{13} \ (n/cm^{-3})^{-1}s$ is a characteristic time scale of the loss process.

In general, the coefficients in (5) are functions of momentum, and also might vary with time during the acceleration process, so that solutions have to be obtained with numerical methods. If the acceleration time scale is short compared with the time scales of subsequent injection of the particles into interplanetary space and transport therein, it is probably not unreasonable to assume that a steady state over the energy range of interest can be reached. For this case a number of classes of analytic solutions of (5) are known. The shape of the spectrum at energies high enough above the energy of the source is then solely determined by the interplay of the three time scales of stochastic acceleration, $\tau_F \sim \alpha^{-1}$, the escape time T, and the respective loss time τ_L. An overview of general solutions and the appropriate boundary conditions of (5) was given by [31]. In the following, we will discuss two analytic solutions which are suited for modelling proton and electron spectra, respectively, and compare them with observed particle spectra.

4.1 Protons

A special model of stochastic acceleration, the scattering of particles by hard spheres moving with speed V, has to some extent been compared with interplanetary particle observations (e.g., [32]) and also been employed to calculate gamma-ray and neutron production in solar flares (e.g., [33]). In this model α and T are assumed to be constant. A steady-state solution of (5) for non-relativistic energies and impulsive monoenergetic injection at low energies is $J(E) \sim p\, K_2(x)$, where $K_2(x)$ is a modified Bessel function, and $x = (12/(\alpha T) \cdot p/(Mc))^{1/2}$. The shape of the particle spectrum is solely determined by the constant parameter αT, so that a larger value gives a flatter spectrum. Although a number of solar proton spectra have been successfully modelled with the above solution, other spectral shapes, ranging from near power laws to spectra which exhibit either weaker or stronger curvature as predicted by the K_2 Bessel function solution, are observed as well.

Realistically, one would not expect that the turbulence in the acceleration region always evolves in the same manner. The energy density of the fluctuations which the particles interact with, and their wave number spectra, will differ from one event to the next, and the acceleration efficiency will not likely be momentum-independent in every event. A natural extension of the above model which allows the modelling of a larger variety of spectral shapes can be achieved with the generalization

$$\alpha = \frac{V^2}{l_c\, c}\left(\frac{p}{p_c}\right)^{-s} = \alpha_c \left(\frac{p}{p_c}\right)^{-s} \tag{10}$$

Here l_c and α_c denote the interaction length with respect to a particular wave mode and the acceleration efficiency, respectively, at a characteristic momentum p_c, and V is the speed of the wave.

The variable s allows one to relate the momentum dependence of $D(p)$ to the spectral index of the turbulence spectrum of certain wave modes (for details cf., [34]). Similarily, we can introduce a momentum dependent escape time, $T(p) = T_c\,(p/p_c)^{-b}$, so that its product with the acceleration efficiency now reads

$$\alpha T = \alpha_c T_c (p/p_c)^{-(s+b)} = \alpha_c T_c (E/E_c)^{-(s+b)/2} \tag{11}$$

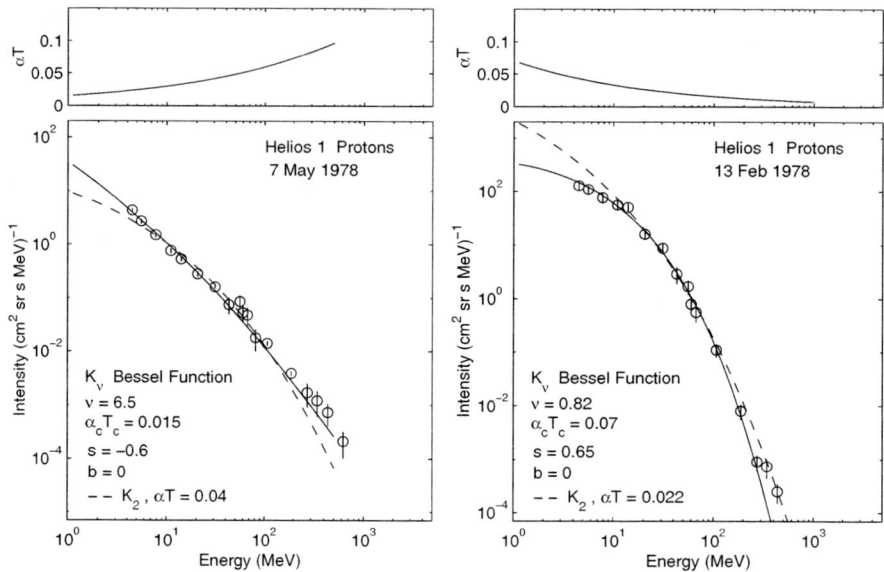

Fig. 8. Proton spectra of the 13 Feb 1978 (left) and the 7 May 1978 (right) solar particle events measured by the University of Kiel particle telescope onboard Helios 1, fitted by two different Bessel function solutions (details see text). Upper rows show the energy dependence of αT which carries imortant information about the acceleration mechanism.

where the latter holds for the case of non relativistic energies, $p \approx \sqrt{2ME}$. With the above assumptions, the solution of (5) is

$$J(E) \sim E^{-(2+s)/4} \mathrm{K}_\nu(z) \tag{12}$$

The index of the Bessel function is here given by $\nu = |(2-s)/(1+s+b)|$, and its argument, choosing $E_c = 1$ MeV, by

$$z = \frac{1}{|1+s+b|} \left(\frac{1}{3.26\,\alpha_c^2\,T_c^2} \right)^{1/4} \left(\frac{E}{\mathrm{MeV}} \right)^{(1+s+b)/4} \tag{13}$$

According to the properties of the Bessel function, a larger index will result in a weaker curvature of the spectrum. Miller et al. [35] numerically calculated solutions of (5) for a constant αT in the transrelativistic regime and noted that they start to become flatter compared to the K_2 solution for protons with energies above ~ 300 MeV. It is also possible that due to a finite duration of the acceleration process the steady state solution is not reached at high energies, resulting in a comparatively steeper spectrum.

In the following, we will present some results from a recent study [36] which analyzed a total of ten proton spectra observed on Helios 1. Fits with solution (12) to the observed proton spectra up to 500 MeV were performed. In order to keep the number of free parameters as small as possible, only a constant escape time was considered. Figure 8 (left panel) shows data for the 7 May 1978 event (gradual, also observed by neutron monitors) and gives an example for a spectrum which is close to a power law in energy up to above 600 MeV. This event can be well fitted with (12), assuming that αT increases with energy, i.e., $s < 0$. The proton spectrum of the 1 Jan 1978 event (cf., Fig. 3) shows a similar behaviour in this respect. The spectrum of an event associated with the gradual flare of 13 Feb 1978 is shown in the right panel of Fig. 8. Neither a power law in energy nor the K_2 solution give good fits. The generalized Bessel function solution, now with αT decreasing towards higher energies, models the spectrum well.

It was found that all proton spectra analyzed in the study (and seemingly most of the published flare spectra which have been obtained with the selection criteria applied here) can be well modelled with the generalized Bessel function solution. From the small number of spectra analyzed at present a statistical significance cannot be claimed, and an ordering of spectral shapes with respect to associated gradual/impulsive flares, if it exists, definitely is not as strong as is the case for electron spectra. Recent *Wind* observations of H, He, C, O, and Fe spectra in the range of 20 to 100 MeV/amu [37], and observations of heavy ion spectra at higher energies [23] seem to reveal a similar ordering although those studies used somewhat different methods to construct particle spectra and draw different conclusions on the acceleration mechanisms.

4.2 Electrons

Electron acceleration in a high-density environment where Coulomb losses can be important was investigated by Steinacker et al. [38]. For energetic ($v \sim c$) electrons, a constant αT, the inclusion of the the energy loss rate (9), and for monoenergetic injection $Q(p) \sim q_0 \delta(p - p_0)$ an analytic solution of (5) for the steady-state case of the form

$$J(E) \sim p^{1/2-\mu} M(\mu - \frac{1}{2}, 1 + 2\mu, \frac{4p_c}{p}) \tag{14}$$

was derived, where M is the Kummer function, or confluent hypergeometric function [39], $\mu = \sqrt{9/4 + 3/\alpha T}$, and $p_c = 3mc/(4\alpha \tau_L)$ is a characteristic momentum where the gain rate due to stochastic acceleration and the loss rate are equal. With the asymptotic expansions of the confluent hypergeometric functions, J(E) can at high energies be approximated as a power law in momentum

$$J(E) \sim p^{1/2-\mu} \qquad p \gg p_c \tag{15}$$

with a spectral index which is determined by αT only. Figure 9 (left panel) shows the electron spectrum observed after the impulsive 18 Feb 1979 flare (cf., Fig. 4), fitted by the solution (14) of the stochastic model with ionization losses. The fit requires three parameters, the absolute normalization, αT, and τ_L/T, and improves the fit significantly compared to a single power law. To highlight details of the spectrum, and the quality of the fit, the right panel of the figure

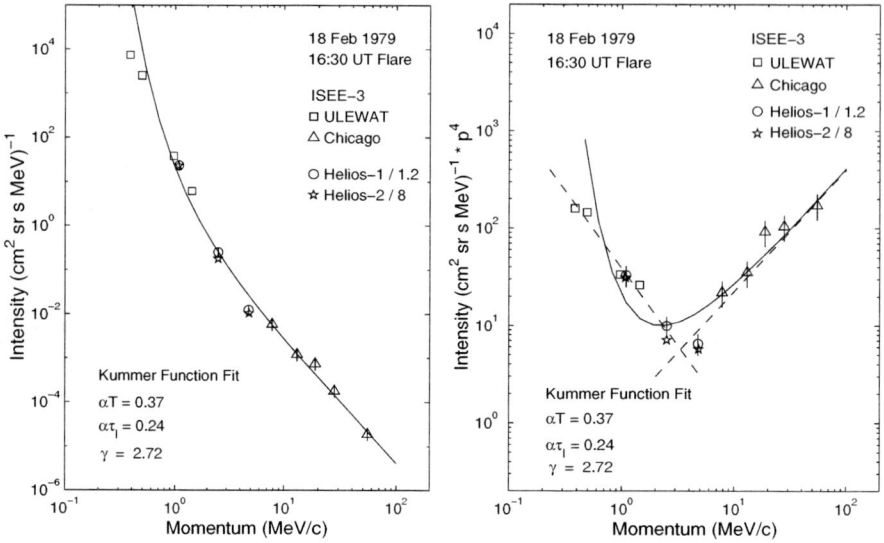

Fig. 9. Left panel: Electron spectrum observed after the impulsive 18 Feb 1979 flare, fitted by (14). Right panel: same spectrum multiplied by a power law in momentum, to highlight details. The ionization loss model cannot describe the spectrum, in particular the sharp break at ~ 3 MeV/c, as well as a double power law in momentum (dashed lines).

shows the spectrum multiplied by a power law in momentum. It is obvious that the ionization loss model cannot describe the spectrum, in particular the sharp break at ~ 3 MeV/c, as well as a double power law in momentum. All spectra from the survey [21] of simultaneous ISEE-3/Helios observations show a similar behaviour in this respect. It seems that the strongly diffusive nature of the stochastic acceleration mechanism, which prevents any sudden changes in the spectral shape, makes a single acceleration process for $\sim 0.2 - 50$ MeV electrons observed in impulsive events unlikely. In summary, we find some evidence that in gradual events electron acceleration above 200 keV occurs in or is dominated by a single stage mechanism which produces a power law in momentum whereas in impulsive events electrons are accelerated by two different mechanisms or in two (or more) different locations.

5 Conclusions

The propagation of energetic particles in the solar wind, under conditions when no large scale disturbances are present, can be well described with improved transport models which recently have become available. This allows one to relate in situ measurements of solar particles to acceleration processes at or close to the Sun in a meaningful way. Particle events observed in energetic electrons, and ions with energies above a few MeV/n, often follow a pattern which can roughly be characterized as a three-stage process: energization on a timescale of tens of seconds to minutes, release/injection into the interplanetary medium which lasts tens of minutes to hours, and interplanetary propagation which, at 1 AU, has time scales of hours to days. The release/injection process can be mathematically well described with the model of diffusion on a sphere - whether this process actually is propagation in turbulent fields or azimuthal distribution of particles related to the large-scale reconfiguration of the coronal magnetic field after the flare/CME remains unclear.

The relatively simple versions of stochastic acceleration models which we have briefly outlined in this review give a good phenomenological description of observed features of the above mentioned particles in interplanetary space, and also those derived from particles interacting at the Sun: spectral shapes, absolute numbers, acceleration time, spatial scales, and special properties such as the enhanced acceleration of ^{3}He in some events due to resonance effects in the wave-particle interaction. They can be used as input for models which consider the spatial dependence of the acceleration process in more detail. The source of the plasma turbulence necessary for particle acceleration may be related more to "flares" in some events, and more to "CME-related" processes, taking place in the higher corona, in others. Additional acceleration processes in the corona, such as by a coronal shock wave or magnetic reconnection cannot be ruled out, but at present their modelling seems to make less specific predictions which can be compared with interplanetary particle observations. Whether the CME shock is able to accelerate particles to high energies, and not merely re-process particles that can originate from other known sources, can only be decided if models which describe wave generation and wave-particle interactions in the vicinity of the shock, and resulting energization and transport processes in a self-consistent way become available, as well as in situ measurements in the inner heliosphere to test this hypothesis.

Acknowledgements

I wish to thank the organizers of the CESRA Workshop for inviting me to participate in this meeting. Thanks are also due to R.P. Lin and S. Krucker for providing Wind 3DP particle data, and to C.W. Smith and N.F. Ness for providing ACE magnetic field data. I benefitted from discussions with K.-L. Klein, J.W. Bieber, W.H. Matthaeus, and from the comments of an anonymous referee. This work was supported by NASA grants NAG 5–7142 and NAG 5–8134.

References

1. D. V. Reames: Rev. Geophys. (Suppl.) 33, 585, (1995)
2. E. Moebius et al.: Proc. 26th Internat. Cosmic Ray Conf. (Salt Lake City), **6**, 87 (1999)
3. E. Cliver: in *High Energy Solar Physics*. ed. by R. Ramaty, N. Mandzhavidze, X.-M. Hua, (American Institute of Physics, 1996) 45

4. K.-L. Klein, G. Trottet: Space Sci. Rev. **95**, 215 (2001)
5. J. R. Jokipii: Ap. J. **146**, 480 (1966)
6. K. Hasselmann, G. Wibberenz: Z. Geophys. **34**, 353, (1968)
7. J. W. Bieber, W. H. Matthaeus, C. W. Smith, W. Wanner, M.-B. Kallenrode, G. Wibberenz: Ap. J. **420**, 294 (1994)
8. R. Schlickeiser, U. Achatz: J. Plasma. Phys. **49**(1), 63 (1993)
9. W. Dröge: Space Sci. Rev. **93**, 121 (2000a)
10. W. Dröge: Proc. 27th Internat. Cosmic Ray Conf. (Hamburg) **8**, 3285 (2001)
11. R. Ramaty, N. Mandzhavidze, X. M. Hua: *High Energy Solar Physics*, (AIP, New York 1996)
12. D. V. Reames: Space Sci. Rev. **85**, 327 (1998)
13. R. Ramaty, N. Mandzhavidze: *High Energy Solar Physics – Anticipating HESSI*, (ASP, San Francisco 2000)
14. P. Meyer, E. N. Parker, J. A. Simpson: Phys. Rev. **94**, 1017 (1956)
15. H. Kunow et al.: J. Geophys. **42**, 615 (1977)
16. R. Brun et al.: GEANT3, CERN Data Handling Division, DD/EE/84-1 (1987)
17. M. Bialk, W. Dröge, B. Heber: Proc. 22nd Internat. Cosmic Ray Conf. (Dublin) **3**, 764 (1991)
18. D. Moses, W. Dröge, P. Meyer, P. Evenson: Ap. J. **346**, 523 (1989)
19. R. P. Lin et al: Space Sci. Rev. **71**, 125 (1995)
20. M. Bialk, W. Dröge: Proc. 23rd Internat. Cosmic Ray Conf. (Calgary) **3**, 278 (1993)
21. W. Dröge: in: *High Energy Solar Physics*. ed. by R. Ramaty, N. Mandzhavidze, X.-M. Hua, (American Institute of Physics, 1996) 78
22. R. P. Lin, R. A. Mewaldt, M. A. I. van Hollebeke: Ap. J. **253**, 949 (1982)
23. A. J. Tylka, W. F. Dietrich, P. R. Boberg: Proc. 25th Internat. Cosmic Ray Conf. (Durban) **1**, 101 (1997)
24. E. C. Roelof: in: *Lectures in High Energy Astrophysics*. ed. by H. Ögelmann, J. R. Wayland, NASA SP-199, 111 (1969)
25. D. Ruffolo: Ap. J., **382**, 688 (1991)
26. W. Schlüter: Ph.D. Thesis, University of Kiel (1985)
27. G. C. Reid: J. Geophys. Res. **69**, 2659 (1964)
28. W. Dröge: Ap. J. **573**, 1073 (2000b)
29. E. Fermi: Phys. Rev. **75**, 1169 (1949)
30. Y. M Wang, R. Schlickeiser: Ap. J. **313**, 200 (1987)
31. B. T. Park, V. Petrosian: Ap. J. **446**, 699 (1995)
32. R. E. McGuire, T. T. von Rosenvinge: Adv. Space Res. **4**, 117 (1984)
33. R. Ramaty, R. J. Murphy: Space Sci. Rev., **45**, 231 (1987)
34. J. A. Miller et al.: J. Geophys. Res. **102**, 14631 (1997)
35. J. A. Miller, N. Guessum, R. Ramaty: Ap. J. **361**, 701 (1990)
36. W. Dröge: in: *High Energy Solar Physics – Anticipating HESSI*, ed. by R. Ramaty, N. Mandzhavidze, (ASP, San Francisco 2000) 191
37. D. V. Reames et al.: Ap. J. **483**, 515 (1997)
38. J. Steinacker, W. Dröge, and R. Schlickeiser, Sol. Phys. **115**, 313 (1988)
39. M. Abramowitz, I. A. Stegun: *Handbook of Mathematical Functions*, (Dover, NewYork 1965)

Particle Acceleration by Magnetic Reconnection

Yuri E. Litvinenko

Institute for the Study of Earth, Oceans, and Space, University of New Hampshire, Durham, NH 03824-3525, USA

Abstract. This is a review of theoretical models for particle acceleration by DC electric fields in reconnecting current sheets during solar flares. Particular emphasis is placed on models for collisionless acceleration in a large-scale reconnecting current sheet with a nonzero magnetic field and a highly super-Dreicer electric field of order a few hundred V m^{-1}. Theoretical arguments and observational evidence for such electric fields are also discussed. An approximate analytical approach is employed to identify the effects of the electric and magnetic fields on particle orbits. The magnetic field structure in the sheet is shown to determine both the electron to proton ratio for the accelerated particles and their typical energies and spectra. Formulae for the particle energy gains and acceleration times are presented. Recent numerical calculations of particle orbits are described, stressing the use of exact MHD solutions for the magnetic fields and plasma flows in the sheet. The analytical and numerical results form the basis for electric field acceleration models in solar flares. In particular, physical conditions can be identified that lead to either electron acceleration to gamma-ray energies of a few tens of MeV in electron-rich flares or the generation of protons with energies up to several GeV in large gradual events.

1 Introduction

Particle acceleration to super-thermal energies is an important signature of energy release in solar flares. Electrons with energies above 20 keV, which are responsible for the flare hard X-ray and gamma-ray continuum emissions, contain a significant fraction of the flare energy, up to 10^{24} J[1]. This corresponds to the rate of particle production of order 10^{35} s^{-1} and higher. Protons can be accelerated to several GeV, although the bulk of accelerated protons, which are responsible for flare gamma-line emission, have energies within the range of 0.1 MeV to 10 MeV. The production rate for protons above 1 MeV can reach 10^{34} s^{-1}, and their energy content can exceed 10^{23} J. There are indications that the typical energy content of protons above 1 MeV is in fact comparable to that of electrons with energies above 20 keV (see [33] for a review, and [46]).

---

[1] At the insistence of the publisher, SI units are used throughout the paper. The use of eV $\approx 1.6 \times 10^{-19}$ J is still accepted by the International Committee for Weights and Measures (see http://physics.nist.gov).

The current consensus is that flare energy is released through rapid dissipation of magnetic energy in the corona by virtue of magnetic reconnection in current sheets [35]. Many important properties of the reconnection process, however, remain poorly understood. Of particular interest is the fact that rapid reconnection in a large-scale current sheet is associated with a quasi-static super-Dreicer electric field in the corona, which provides the most direct way to accelerate particles in solar flares. The goal of this paper is to describe a model for charged particle acceleration by the direct electric field associated with magnetic reconnection.

The reconnection electric field in the current sheet is determined by the plasma inflow speed to the sheet and the local magnetic field. For inflow speeds of order a few km s^{-1} and magnetic fields of order $10-100$ mT, the electric field can be as strong as a few hundred V m^{-1} and hence can lead to relativistic energies of charged particles, provided their acceleration length—the particle displacement along the electric field in the current sheet—is large enough. The acceleration length itself is controlled by the structure of both electric and magnetic fields inside the sheet.

A large body of research has been devoted to the question of charged particle orbits in reconnecting current sheets in the context of particle acceleration on the Sun and in the geotail. Various solutions have been obtained that describe particle orbits both in two-dimensional [41, 28, 15] and three-dimensional current sheets [49, 26, 21, 22] as well as in magnetic X-point geometries [6, 4]. The principal point in these studies is that although the magnetic field cannot change the particle energy, it can change the orbit, determining the displacement along the electric field and hence the energy gain. It is conceptually important that particle acceleration in this approach is considered as an inherent part of the flare energy release in a large-scale current sheet (or several current sheets) in the solar corona. It appears possible, in particular, to relate the electron to proton ratio of accelerated particles to the magnetic field geometry in the reconnection region [11].

The paper is organized as follows. Section 2 presents arguments for highly super-Dreicer electric fields in solar flares and delineates some general aspects of particle acceleration by the electric field. Section 3 gives basic results on charged particle orbits in the current sheet far from a singular line of magnetic field in the sheet. The acceleration of hard X-ray generating electrons in solar flares is discussed. Section 4 analyzes the energy gains for electrons near the singular line. It is shown that impulsive electron acceleration to tens of MeV is possible in the flare reconnecting current sheet undergoing the tearing instability. Conclusions are presented in Sect. 5, which also describes recent results on the acceleration of protons to GeV energies in long-duration gamma-ray flares.

2 Electric Fields and Particle Acceleration in Solar Flares

The particle acceleration model discussed below is based on the idea that magnetic reconnection in solar flares corresponds to electric fields E of order several kV m^{-1} in the corona. This is a few orders of magnitude larger than the Dreicer field, which means that the acceleration process is essentially collisionless, and the usual expression for electrical conductivity is not applicable in the current sheet. Hence either plasma turbulence or the particle escape itself provide an effective plasma resistivity. In the latter case, the acceleration time, which is the lifetime of the particle in the system, replaces the mean collision time in an expression for the electrical conductivity [27, 22].

Electric fields in the solar corona are extremely difficult to detect. Hence it is instructive to review the available estimates for direct electric fields in solar flares. The simplest argument is based on the flare energy requirements. The electrodynamic power dissipated in a flare, which is up to $(0.1 - 1) \times 10^{22}$ W, is determined by the free magnetic energy in the corona and the corresponding electric current I. The rate of work of an electromagnetic field on a system of electric currents and charges is $\mathbf{j} \cdot \mathbf{E}$ per unit volume, where $\mathbf{j}$ is the electric current density. Hence the flare energy release rate

$$P = IU = IEl, \tag{1}$$

where from observations $I \leq 10^{12}$ A and the active region length scale is $l \leq 100$ Mm. This leads to $E \geq 100$ V m^{-1} [32].

Other approaches lead to essentially the same value of the electric field in the reconnection region $E = 0.1 - 1$ kV m^{-1}. For example, the analysis of the current sheet structure [40], which uses balance equations based on the conservation laws and the Maxwell equations, shows that for the flare energy requirements to be satisfied, reconnection inflow has to be fast: $v_{\mathrm{in}} \geq 10$ km s^{-1}. The corresponding motional electric field is $v_{\mathrm{in}} B_0 = 0.1 - 1$ kV m^{-1}, where the coronal magnetic field $B_0 = 10 - 100$ mT, and continuity dictates that the same electric field be present in the reconnection region itself. Note that the reconnection inflow of up to a few km s^{-1} has been recently discovered in a well-observed flare, giving direct evidence of fast reconnection [47]. Finally, this theoretical estimate is confirmed by the measurements based upon the Stark effect. In particular, $E \approx 3.5$ kV m^{-1} was reported in a flare surge [12] (see also [13]).

It is immediately clear from the extremely large value of the total potential $U = El$ that the particle acceleration length l_{acc} in current sheets in solar flares has to be much less than the total length of the sheet l [28, 21]. The relation $l_{\mathrm{acc}} \ll l$ is a salient feature of particle acceleration in current sheets with a nonzero magnetic field, which prevents electrons from gaining unreasonably large energies of order eU. It is also important that a small acceleration length limits the total electric current I through the sheet:

$$I = e\dot{N}(l_{\mathrm{acc}}/l), \tag{2}$$

where $\dot{N}$ is the number of particles flowing into and out of the sheet per unit of time. The factor $l_{\mathrm{acc}}/l \ll 1$ appears because the particles that had left the sheet cannot contribute to the current inside it.

A simpler particle runaway model that ignores the magnetic field in the sheet altogether provides an alternative approach to the particle acceleration problem, in particular in application to the hard X-ray generating electrons in solar flares. This model envisions the formation of an electron beam by postulating that $l_{acc} = l$ for all particles since they are presumed to move from one end of the sheet to the other. The correspondingly small value of E also has to be postulated. This model encounters a difficulty, however. In order to be consistent with observational estimates of the number of energetic electrons ($10^{34} - 10^{36}$ s^{-1}), the electric current associated with the beam would have to be so large that its magnetic field would exceed typical coronal values by several orders of magnitude [16]. To avoid this contradiction, one would have to postulate the existence of at least 10^5 acceleration regions producing oppositely directed electron beams.

An advantage of the former model is that charged particles in the current sheet are accelerated locally all along its length because $l_{acc} \ll l$. This approach not only drastically decreases the electric current through any cross-section of the sheet but also has the advantage of treating both particle acceleration and global flare energy release as parts of a single physical process—magnetic reconnection in a large-scale current sheet. One should not forget, however, that both the current sheet model and the particle runaway model are likely to require unreasonably large electric currents outside of the current sheet. Hence the return current of thermal electrons appears to be inevitable in the solar atmosphere [20]. The return current has to neutralize the direct current of accelerated electrons, thus avoiding the problems of the huge magnetic field of the electron beam and the enormous charge displacement [45].

3 Particle Orbits in the Reconnecting Current Sheet

Given the complicated three-dimensional nature of the magnetic and velocity fields in the solar corona, not only the reconnecting component of the magnetic field, say $B_x = B_x(y)$, but also the longitudinal (along the electric field) and the transverse (perpendicular to the plane of the sheet) magnetic field components are likely to be present in the current sheet. The problem of charged particle motion in a current sheet is greatly simplified by the fact that typical acceleration length and time scales under solar flare conditions turn out to be very small compared with typical global parameters. This is why the usual approach is to approximate the field by the first nonzero terms in the Taylor expansion inside the sheet located at $y = 0$:

$$\mathbf{B} = -(y/a)B_0\hat{\mathbf{x}} - B_\perp\hat{\mathbf{y}} + B_\parallel\hat{\mathbf{z}}. \tag{3}$$

Here the minus signs correspond to the electric current in the positive z-direction, and a is the current sheet half-thickness (Fig. 1). The reconnection electric field inside the sheet is

$$\mathbf{E} = E\hat{\mathbf{z}}. \tag{4}$$

Both E and the nonreconnecting component $B_\parallel$ may be assumed locally constant. The transverse field $B_\perp = B_\perp(x)$ changes sign at the center of the sheet

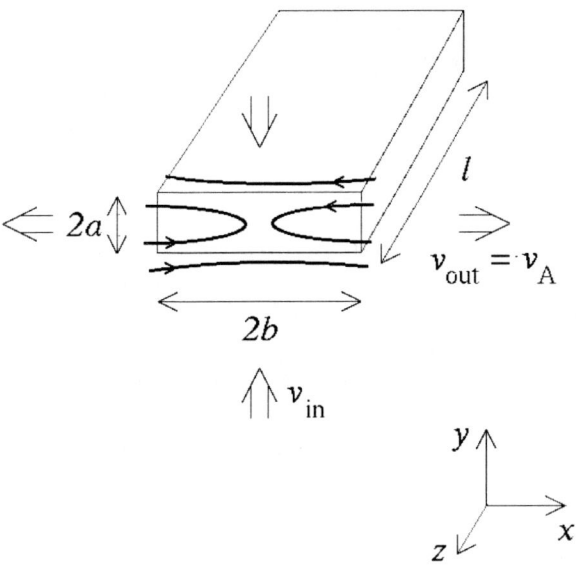

Fig. 1. Projection of the magnetic field in the reconnecting current sheet (length l, thickness $2a \ll l$, width $2b \approx l$) on the xy plane. The electric field E and the longitudinal magnetic field $B_\parallel$ are along the z-axis.

and reaches a maximum at its edges $x = \pm b$. When the half-width of the sheet $b \gg a$, this component is a very slowly varying function of x. Hence $B_\perp$ is often also assumed constant on a given particle trajectory. The fact that $\mathbf{E} \cdot \mathbf{B} \neq 0$ indicates the presence of a significant resistive term in Ohm's law, required for magnetic reconnection:

$$\mathbf{R} = \mathbf{E} + \mathbf{u} \times \mathbf{B}, \tag{5}$$

where $\mathbf{u}$ is the reconnection flow velocity.

It should be stressed that the magnetic field in the sheet is neither uniform nor static. The reconnecting field lines move into the sheet with speed u_{in} and out of the sheet with the characteristic speed of order the Alfvén speed

$$u_{\mathrm{out}} = v_{\mathrm{A}} = \frac{B_0}{(\mu_0 m_{\mathrm{p}} n)^{1/2}} \tag{6}$$

and carry the magnetized particles with them. Here n is the particle density and m_{p} is the proton mass. This familiar "sling-shot" effect causes the reconnected field lines to straighten out so that $B_\perp$ increases from zero at $x = 0$ to the maximum value $\pm B_{\perp,\mathrm{max}}$ at $x = \pm b$ for each reconnected field line, leading to a dependence $B_\perp = B_\perp(x)$ in a steady state and to the corresponding temporal evolution of each reconnected field line. It is only when the length scale of particle acceleration in the sheet is small enough that the spatial dependence of the field lines can be ignored with the exception of the variation of $B_x \sim y/a$ across the current sheet thickness $2a$. This simplification appears to be justified for the hard X-ray generating electrons in flares but will have to be relaxed below when considering MeV electrons in the electron-rich flares.

Even under the simplifying assumption $B_\perp = $ const, the character of the charged particle motion for various relative values of the magnetic field components in the current sheet is nontrivial [21]. In the limit $B_\perp = 0$, whether $B_\parallel = 0$ or not, the motion consists of the acceleration along the electric field $\mathbf{E} = E\hat{\mathbf{z}}$ and finite oscillations along the y-axis caused by the Lorentz force component $F_y \sim v_z B_x$ [41, 49]. This idealized, highly symmetric situation, however, is unlikely to occur. In fact any sheet model requires a nonzero $B_\perp$ as a result of reconnection itself [28].

Particle orbits in current sheets with $B_\perp \neq 0$ are very complex in general, but the situation is simpler in two limiting cases.

If the longitudinal field $B_\parallel$ is small enough, then the maximum displacement along the electric field and the energy gain are determined by the particle gyroradius in the transverse field $B_\perp$:

$$\mathcal{E} = 2m \left(\frac{E}{B_\perp} \right)^2 , \tag{7}$$

where m is the particle mass [41]. Self-consistent treatment based on test-particle orbits shows that this energy gain is equal to $m_\mathrm{p} v_\mathrm{A}^2 / 2$ for protons [27]. Thus the test-particle approach is in agreement with the basic prediction of the MHD treatment, namely that the reconnection outflows are characterized by the Alfvén speed [3, 38]. The corresponding energy gain can explain the acceleration of protons to MeV energies. Because $\mathcal{E} \sim m$, however, it is too small to explain the electron acceleration to energies above 1 keV in solar flares [26].

Since the magnetic field in the solar corona is known to have a significant axial component along the coronal loops, the other limit of a strong longitudinal field $B_\parallel$ on the order of the main reconnecting field B_0 could be appropriate for flaring current sheets. The strong longitudinal field $B_\parallel > B_{\parallel,\mathrm{c}}$ magnetizes the particles and makes them follow the field lines. The critical field $B_{\parallel,\mathrm{c}}$ that leads to the transition to this new type of motion is given by

$$B_{\parallel,\mathrm{c}} = \left(\frac{mEB_0}{eaB_\perp} \right)^{1/2} , \tag{8}$$

where m and e are the particle mass and electric charge [21].

This result is easy to understand from the physical viewpoint. If $B_\parallel = 0$, a typical time scale for particle ejection out of the sheet is of order

$$t_\perp = \left(\frac{am}{F} \right)^{1/2} . \tag{9}$$

The Lorentz force component perpendicular to the sheet is evaluated as

$$F \approx ev B_0 = \frac{B_0}{B_\perp} eE, \tag{10}$$

assuming some typical value of $v = E/B_\perp$. The particle escape from the sheet is prevented if the longitudinal field is sufficiently strong:

$$t_\perp > t_\parallel, \tag{11}$$

so that the particle is effectively tied to a field line in the sheet. Here $t_\parallel$ is the time scale introduced by the longitudinal field:

$$t_\parallel = \frac{m}{eB_\parallel}. \tag{12}$$

Equation (11), which is equivalent to Equation (8), defines when the particle becomes magnetized in the sheet.

For electrons in solar flares, $B_{\parallel,c} \leq 0.1B_0$ for typical parameters of current sheets in the solar corona [26]. Therefore electrons are magnetized efficiently by the longitudinal field $B_\parallel$ in the sheet. The effect, however, is harder to achieve for much heavier protons that will still follow the Speiser-type orbits [41].

A potential complication in the problem is the charge separation electric field that arises in the sheet because electrons and protons follow different orbits. This effect is present in particle simulations of collisionless reconnection [17]. The simulations, nevertheless, are in a surprisingly good agreement with the results of the test-particle approach [22].

Thus the magnetized electrons will mainly move along the magnetic field lines in the current sheet. The adiabatic particle motion in principle can be described by drift theory. The main effect though is acceleration along the field lines that will cease when the particles leave the sheet. Integrating the magnetic field line equations

$$-\frac{a}{y}\frac{dx}{B_0} = -\frac{dy}{B_\perp} = \frac{dz}{B_\parallel} \tag{13}$$

defines the acceleration length l_{acc} as the displacement δz along the electric field, which corresponds to $|\delta y| = a$ when the magnetized electrons initially inside the sheet at $y = 0$ leave the sheet along the field lines:

$$l_{acc} = \frac{B_\parallel}{B_\perp} a \tag{14}$$

(see [21] for a detailed discussion of particle orbits in this case). The displacement perpendicular to the electric field is given by a similar formula $\delta x = \frac{1}{2}(B_0/B_\perp)a$. The predicted energy gain for the magnetized particles

$$\mathcal{E} = \frac{B_\parallel}{B_\perp} eEa \tag{15}$$

and the corresponding acceleration time

$$\Delta t = \left(\frac{B_\parallel}{B_\perp} \frac{2am}{eE} \right)^{1/2} \tag{16}$$

are the main results of the local analytical approach to DC electric field acceleration in current sheets in solar flares.

It is worth noting that the magnetic field structure in the current sheet is reflected not only in the dynamics of a single particle but also in the velocity distribution function. A "ridge" in velocity space is of particular interest as a signature of reconnection in the magnetotail of the Earth. A nonzero $B_\parallel$ should have an observable effect on the ridge signature as well [30].

In the numerical estimates that follow, the reconnecting and nonreconnecting components of the field are assumed to be of the same order, $B_\parallel = B_0 = 10$ mT. This choice ensures that $B_\parallel \gg B_{\parallel,c}$ for electrons. Using $B_0 = 10$ mT and the numbers below leads to $B_{\parallel,c} \approx 0.4$ mT for electrons whereas $B_{\parallel,c} \approx 18$ mT for protons.

Models for fast reconnection in solar flares suggest that the average transverse field is of order $\langle B_\perp \rangle = 0.1$ mT [40]. The particle density in the current sheet is $n = 10^{16}$ m^{-3}. The sheet dimensions are as follows: $a = 1$ m, $b = 1$ Mm, $l = 10$ Mm. These are typical estimates used in studies of collisionless acceleration processes in flares [28]. In accordance with the arguments of the previous section, the reconnection electric field is taken to be $E = 300$ V m^{-1}, implying fast reconnection as conventionally measured by the Alfvén Mach number $M = u_{in}/v_A$. The inflow speed is determined by the plasma electric drift speed into the sheet:

$$u_{in} = \frac{E}{B_0}. \tag{17}$$

The numbers adopted above imply that the reconnection regime considered is fast since $u_{in} \approx 30$ km s^{-1} and $u_{out} = v_A \approx 2000$ km s^{-1}, leading to $M \approx 10^{-2}$. Fast reconnection is necessary to ensure the large magnetic energy release rate P implied by observations of the flare impulsive phase. The power output is determined by the Poynting flux into the sheet:

$$P = \frac{4}{\mu_0} u_{in} B_0^2 bl. \tag{18}$$

The choice of parameters above leads to a reasonable value of 10^{20} W. A larger power output would be possible in a larger current sheet.

Using the formulae and numbers above, it is straightforward to see that the hard X-ray generating electrons in solar flares can indeed be accelerated in a large-scale current sheet. Equation (15) gives the electron energy gain of about 30 keV, which would lead to X-rays in the same energy range. The acceleration time defined by Equation (16), which is simply the time spent by a particle inside the sheet, can be as small as $\Delta t \approx 20$ μs. The particle influx to the sheet is defined as

$$\dot{N} = 4lbu_{in}n, \tag{19}$$

which is about 10^{34} s^{-1} for the parameters above. This is clearly enough to cover the needed supply of electrons in small impulsive flares [10]. The influx can be much larger in gradual flares that correspond to large-size current sheets. For example, assuming the length scale of the current sheet to be of order the size of an active region, $l \approx b \approx 100$ Mm, leads to the electron flux $\dot{N} \approx 10^{37}$ s^{-1}. The flux can be even higher when reconnection is faster and u_{in} is larger.

Recall that the magnetizing longitudinal field $B_{\parallel,\text{c}}$ in Equation (8) is proportional to the square root of the particle mass. This is why even $B_{\parallel} = B_0$ is not high enough to make the proton motion adiabatic. The typical proton energy is found from Equation (7) to be about 200 keV. Martens and Young [29] presented several arguments in support of proton beams in solar flares. A simple continuity argument shows that the inhomogeneity in $B_y(x)$ results in a power-law energy distribution. In general, though, the problem of proton energization in the sheet is more complicated (see Sect. 5).

Allowing for a nonzero magnetic field inside the sheet has important consequences for electron orbits. First of all, the particle escape is much more efficient across the sheet than along it. If electrons could simply move along the electric field direction through the total current sheet length, their typical energy would be determined by the total potential drop

$$eU = eEl \approx 3\,\text{GeV}. \tag{20}$$

Electrons with GeV energies are hardly ever observed in solar flares. The effect of the magnetic field is to decrease the average acceleration length by about five orders of magnitude, resulting in $\mathcal{E} \ll eU$. This makes the electric field acceleration a local acceleration mechanism that can occur throughout the reconnection region. In contrast to electron runaway models, the particles leave the current sheet sideways—perpendicular to the electric field—rather than at its top or bottom.

Equation (15) with $B_{\perp} = \langle B_{\perp} \rangle$ leads to electron energies in the range of tens of keV. The next section investigates the possibility that the electric field acceleration in the current sheet can lead to MeV energies in those parts of the sheet where $B_{\perp}$ is much less than its average value.

4 Acceleration at Singular Lines and Electron-Rich Events

"The particle acceleration problem of electron-rich γ-ray flares is perhaps the most challenging" [18]. These flares are defined by an unusually intense gamma-ray continuum above 1 MeV, which can be fit well by a hard power law [18, 36]. Nuclear gamma-line radiation in electron-rich events is dominated by electron bremsstrahlung. This implies the presence of large fluxes of relativistic electrons with energies up to a few tens of MeV and a high electron to proton ratio in this energy range, although the total energy content of > 1 MeV ions may be comparable to that of > 20 keV electrons [44, 46].

Because the transverse field $B_{\perp}$ varies along the current sheet, going through zero at least at one point (its center), the magnetic field projection onto the xy-plane has the geometry of a standard magnetic X-point. In other words, this point is a projection of the singular magnetic field line with $B_x = B_y = 0$ and $E \neq 0$ onto the xy-plane. This is where the field lines are "cut" and "reconnected." More complicated geometries with multiple singular lines are also possible due to the tearing instability in the sheet, for example. Since the particle acceleration length scale is typically much less than the length scale of $B_\perp$ variation that can be of order b, Equations (14) and (15) derived for $B_\perp = $ const remain valid for $B_\perp = B_\perp(x)$, unless $B_\perp \to 0$. Now, however, the energy gain depends on the location in the sheet as a parameter with $B_\perp = B_\perp(x)$.

Syrovatskii [42, 43] suggested that the impulsive phase of a flare corresponds to tearing instability of a current sheet and to the formation of multiple singular lines. Charged particles are accelerated to very high energies in the vicinity of the singular lines where the electric and magnetic field lines are coaligned. Fast particles themselves, however, create the electric current that corresponds to a new current sheet growing at the original singular line of the magnetic field. This eventually leads to a nonlinear stabilization of the instability. The process repeats itself at the newly formed singular lines. Under these conditions the magnetic energy accumulated in a pre-flare current sheet is primarily converted to the kinetic energy of nonthermal particles rather than bulk flows as in standard MHD reconnection models [3].

It should be remembered that the electron motion remains adiabatic even at the singular line itself because of a strong longitudinal field $B_\parallel$. In spite of this, integrating the equations of particle motion, whether directly or in the guiding center approximation, is a very complicated task. Moreover, the simplified representation of the electric and magnetic fields in the sheet by the first nonzero terms in the Taylor expansion may not be sufficient to determine rigorously the time spent by the particles near the singular line and their energy gain. The following simple approach, based on the picture of the magnetic line motion, can be used instead [24].

The electrons are assumed to follow the reconnected field lines that move out of the current sheet with the speed of order v_A, much like beads sliding on a moving wire. The effect of the reconnection electric field is to accelerate the electrons along the magnetic field lines until the particles leave the singular line vicinity. Goldstein, Matthaeus, and Ambrosiano [14] were evidently the first to use this approximation to estimate the particle energy gains in two-dimensional current sheets ($B_\parallel = 0$) in various astrophysical environments. Taking the magnetic field dynamics into account is what makes this approach different from the studies of charged particle acceleration at singular lines with a static magnetic field [6, 5, 4].

For simplicity, assume that the reconnected field lines move out of the sheet with a constant speed v_A along the x-axis (Fig. 1):

$$B_\perp(t) \approx \frac{2}{b}\langle B_\perp \rangle v_A t. \tag{21}$$

Here $\langle B_\perp \rangle = B_{\perp,\mathrm{max}}/2$, and the scale of the field variation from zero to $B_{\perp,\mathrm{max}}$ is half of the current sheet width, b. The effect of multiple singular lines on the motion of a given particle is also ignored for simplicity. The approximation of a constant speed v_A corresponds to a linear dependence $B_\perp \sim x$ in the steady state when $x = v_A t$.

It is clear from Equation (15) that higher particle energies can be reached close to the singular line where $B_\perp \to 0$. One might think that arbitrarily large energy gains (up to those given by Equation (20)) could be possible near the singular line. This is not the case though because the acceleration time t_{acc} is finite. Eventually the magnetized electrons are carried with the reconnected field lines away from the singular line and acceleration ceases. This effect limits the electron energy.

The maximum electron energy is estimated as follows. The magnetized electrons move almost along $\mathbf{B}$ inside the sheet, and their relativistic kinetic energy increases with time as

$$\mathcal{E}_{\mathrm{e}}(t) \approx ceEt. \tag{22}$$

The maximum energy is defined by Equation (15). The maximum energy itself, though, is a function of time because the particles move out of the sheet together with the magnetic field lines. Therefore the time-dependent $B_\perp$ given by Equation (21) should be substituted into Equation (15):

$$\mathcal{E} = \frac{B_\parallel}{\langle B_\perp \rangle} \frac{eEab}{2v_{\mathrm{A}}t}. \tag{23}$$

Now, since the actual electron energy is given by Equation (22) and the maximum energy is given by Equation (23), equating the two expressions gives an equation for the electron acceleration time, which is solved to give [24]

$$t_{\mathrm{e,acc}} \approx \left(\frac{B_\parallel}{\langle B_\perp \rangle} \frac{ab}{2cv_{\mathrm{A}}} \right)^{1/2} \approx 0.3\ \mu\mathrm{s}. \tag{24}$$

Substituting this result back into Equation (22) gives the sought-after maximum electron energy

$$\mathcal{E}_{\mathrm{e,max}} \approx ceE \left(\frac{B_\parallel}{\langle B_\perp \rangle} \frac{ab}{2cv_{\mathrm{A}}} \right)^{1/2} \approx 30\ \mathrm{MeV}. \tag{25}$$

The same numerical values as before have been employed in these estimates. The derived maximum energy of the accelerated electrons in the current sheet is compatible with the observations that imply electron energies of a few tens of MeV in impulsive electron-rich solar flares. These electrons create the strong bremsstrahlung that dominates the nuclear gamma-line radiation.

Recall that it was possible to ignore the variation of $B_\perp$ in Equations (14) and (15) when the acceleration time is much less than the time scale of the field variation. It is the increasing acceleration time near the singular line that limits the particle energy gain. Physically, the energy of a magnetized electron increases with time but the maximum possible energy decreases as the particles move out of the sheet and the transverse magnetic field "felt" by the particles becomes larger, which makes it easier for them to escape the current sheet.

Clearly the maximum acceleration length is still much less than the total length of the current sheet:

$$l_{\mathrm{e,acc}} = \frac{\mathcal{E}_{\mathrm{e,max}}}{eE} \approx 0.1 \text{ Mm} \ll l. \tag{26}$$

Thus even for the highest energies, the strong DC electric field acceleration remains a local acceleration mechanism. In other words the maximum energy is still much less than the total potential drop given by Equation (20). This confirms the result of the previous Section that the acceleration by strong electric fields in the reconnection region is a local mechanism acting all along the length of the sheet. The acceleration length remains small enough to ignore the Coulomb losses as well. Thus the particle energization process in impulsive flares is essentially collisionless as assumed throughout this paper. It is also interesting to note that $t_{\mathrm{e,acc}}$ in Equation (24) does not depend on the reconnection electric field E.

For time intervals and energy gains smaller than those given by Equations (24) and (25) the particle motion with the magnetic field lines can be ignored, so that the electron acceleration far from the magnetic singular line can be studied assuming a constant instantaneous value of $B_{\perp}$ that depends on $x = v_{\mathrm{A}}t$ as a parameter. This leads to a continuous electron spectrum extending to high energies. Calculation of the detailed spectrum and the total number of accelerated electrons is a complicated problem that requires numerical simulations including the effects of nonuniform electric and magnetic fields, particle escape from the sheet, and possibly the electric field due to charge separation. Nevertheless, it can be demonstrated that a power-law spectrum may result [21, 24]. As before, consider acceleration in the case of a linear magnetic X-point in the xy-plane: $B_{\perp} \sim x$. The energy spectrum $f(\mathcal{E}_{\mathrm{e}})$ below $\mathcal{E}_{\mathrm{e,max}}$ follows from the continuity equation $f(\mathcal{E}_{\mathrm{e}})\mathrm{d}\mathcal{E}_{\mathrm{e}} = f(x)\mathrm{d}x$ with $\mathcal{E}_{\mathrm{e}} \sim B_{\perp}^{-1} \sim x^{-1}$ from Equation (15). Assuming a spatially uniform inflowing distribution $f(x) = \mathrm{const}$ leads to

$$f(\mathcal{E}_{\mathrm{e}}) \sim \frac{\mathrm{d}x}{\mathrm{d}\mathcal{E}_{\mathrm{e}}} \sim \mathcal{E}_{\mathrm{e}}^{-2}. \tag{27}$$

Simulations of charged particle acceleration in the vicinity of a singular line in the solar corona indeed demonstrate the formation of a power-law spectrum with the index of about $2 - 2.2$ for a wide range of parameters [34]. It appears, however, that a somewhat steeper spectrum would be necessary to interpret observations of the continuous gamma-ray radiation in electron-rich solar flares [48]. The discrepancy is not suprising given the simplifying assumptions used in the estimate above. A more complicated geometry, for example, could make the actual spectrum steeper. Equation (27) is valid for a linear magnetic X-point. This result is easily generalized for a singular line of any order. Assuming $B_{\perp} \sim x^{\alpha}$ leads to the energy spectrum

$$f(\mathcal{E}_{\mathrm{e}}) \sim \mathcal{E}_{\mathrm{e}}^{-(1+\alpha)/\alpha}, \tag{28}$$

which gives $f(\mathcal{E}_{\mathrm{e}}) \sim \mathcal{E}_{\mathrm{e}}^{-3}$ for $\alpha = 1/2$. Yet another possibility is to relax the assumption of the electric field homogeneity. Particle acceleration by inhomogeneous electric fields in reconnection regions appear to result in observed energy spectra [39].

Other properties of the electron-rich events can be addressed in the context of the model. The event duration, in particular, should be determined by the Alfvén transit time along the current sheet width:

$$t_A = \frac{b}{v_A} \approx 0.5 \text{ s}, \tag{29}$$

which is in agreement with the typically observed electron-rich event durations of a few seconds. As far as the number of energetic electrons produced in an "elementary" acceleration pulse is concerned, it can be estimated as the number of particles accelerated at one singular line:

$$\delta N = 4 l a v_A t_{e,acc} n \approx 2 \times 10^{26}, \tag{30}$$

where the acceleration time $t_{e,acc}$ is given by Equation (24). This estimate appears to be consistent with the electron fluence data for the energy range of a few MeV in the electron-rich events [36, 46]. It is of interest that a typical length scale δb associated with each acceleration site is given by the Alfvén transit time:

$$\delta b = v_A t_{e,acc} \approx 600 \text{ m}. \tag{31}$$

Because $\delta b \ll b$, each singular line can indeed be treated independently, justifying the notion of multiple acceleration sites in a single current sheet.

The question remains whether or not the considered mechanism of particle acceleration at singular lines of three-dimensional magnetic field is relatively inefficient for proton acceleration. This indeed appears to be so [24], explaining a high electron to proton ratio in the energy range of a few tens of MeV, which is the essential property of the electron-rich flares. Recall that substituting the numerical values given in the previous section into Equation (8) gives the magnetizing field $B_{\parallel,c} \approx 18$ mT for protons. Because $B_\parallel < B_{\parallel,c}$, the protons are not magnetized by the longitudinal field in the current sheet. Hence the protons move rapidly out of the singular line vicinity, and their energy gains are smaller than those for electrons.

5 Discussion and Future Research

Analytical studies of charged particle motion and acceleration in reconnecting current sheets have demonstrated very interesting effects associated with the three-dimensional magnetic field structure in the sheet. Values of the field components determine the character of particle orbits and acceleration efficiency. Direct electric field acceleration leads to electron energies and fluxes required to explain the generation of hard X-rays in solar flares. The acceleration time is very short, just a few milliseconds, which may correspond to the observed strong variability of the hard X-rays. The deka-keV electron beams produced in the reconnecting current sheet can also be responsible for the generation of various waves that later interact with ions and create heavy ion anomalous abundances in flares.

The electric field acceleration model is in nice agreement with studies of hard X-ray impulsive flares [1]. These studies strongly suggest that electron acceleration in impulsive flares occurs in the cusp region above the flare loop, leading to the formation of the coronal hard X-ray sources [31]. The inferred geometry is that of a reconnecting current sheet outside the flare loop. This paper envisions particle acceleration in such a current sheet in the corona. Electrons in particular are accelerated at the reconnection site and ejected from the current sheet sideways into the flare loop where they produce the observed hard X-rays. The acceleration of relativistic electrons occurs in the vicinity of singular lines of magnetic field in the current sheet with a strong longitudinal component of the field. The physical mechanism that may underlie the electron-rich impulsive solar flares is the tearing instability of a pre-flare current sheet in the corona, which leads to effective electron acceleration at the singular magnetic field lines where the electric and magnetic fields are coaligned.

Highly super-Dreicer electric fields associated with magnetic reconnection should also lead to proton acceleration. Observations of solar flares indicate that the bulk of accelerated protons have energies within the range of 0.1 MeV to 10 MeV. The simple analytical model described in Sect. 3 can explain the generation of proton beams with these energies. In some flares, however, protons are accelerated up to GeV energies, which presents an important theoretical problem for future research. Long-duration solar gamma-ray flares are of particular interest in this regard since they indicate the presence of continuously accelerated ions for several hours after the impulsive phase [37]. The particles with the highest energies typically come from large gradual events that are most likely a consequence of magnetic field relaxation following a coronal mass ejection (CME). The CME generates a shock wave that contributes to particle acceleration. Significant observational evidence, however, suggests that at least some particles are energized in the current sheet formed in the wake of the CME [19].

Protons are much heavier than electrons, hence simple arguments based on particle magnetization in the current sheet cannot be used. It is necesary to investigate particle orbits in realistic magnetic field configurations. Exact flux pile-up solutions are now available that describe steady-state magnetic reconnection in both two and three dimensions [8, 7], which display many of the characteristics required for the mechanism of flare energy release by reconnection. In particular the exact solutions explicitly demonstrate the appearance of small length scales (defined by either anomalous or classical resistivity η), magnetic sling shots, and Alfvénic outflows [25].

The method used to describe proton acceleration in realistic current sheets is straightforward. Given a solution for the plasma velocity $\mathbf{u}$ and magnetic field $\mathbf{B}$, the electric field responsible for particle acceleration is calculated:

$$\mathbf{E} = \frac{\eta}{\mu_0} \nabla \times \mathbf{B} - \mathbf{u} \times \mathbf{B}. \tag{32}$$

Test particles are traced numerically, using the equation of motion

$$\dot{\mathbf{p}} = e\left(\mathbf{E} + \mathbf{v} \times \mathbf{B}\right), \tag{33}$$

where $\mathbf{p} = \gamma m \mathbf{v}$, $\mathbf{v}$, e, m are the momentum, velocity, charge and mass of the particle, and γ is the relativistic Lorentz factor.

As mentioned above, the reconnecting magnetic field configuration follows from an exact MHD solution for the magnetic and velocity fields (normalized below by a typical field B_0 and the Alfvén speed $v_{A0} = B_0/\sqrt{\mu_0 \rho_0}$):

$$\mathbf{B} = \frac{\beta}{\alpha}\mathbf{P} + \mathbf{Q}, \tag{34}$$

$$\mathbf{u} = \mathbf{P} + \frac{\beta}{\alpha}\mathbf{Q}, \tag{35}$$

where $\mathbf{P}$ is a global background field, $\mathbf{Q}$ is a reconnection-associated field, α and β are numerical constants [7]. For the simplest case of a planar potential background, the uniform electric field is directed along the z-axis, and $\mathbf{P}$ and $\mathbf{Q}$ components are as follows:

$$\mathbf{P} = \alpha\Big[x, -y, 0\Big], \tag{36}$$

$$\mathbf{Q} = \Big[\frac{Q_0}{S\mu}\mathrm{daw}(\mu y), 0, \frac{\sqrt{\pi}}{2\mu}a_1\mathrm{erf}(\mu y) + a_0\Big], \tag{37}$$

where $\mu^2 = (\alpha^2 - \beta^2)/2\alpha S$, $S = \mu_0 v_A l/\eta$ is the Lundquist number, $\mathrm{erf}(x)$ is the error function, and $\mathrm{daw}(x) = \int_0^x \exp(t^2 - x^2)\mathrm{d}t$ is the Dawson function.

Simple local models for the electric and magnetic fields, which are used in this paper, provide reasonable results. These local models, however, contain a number of arbitrary factors that are difficult to interpret in terms of the physical properties of reconnecting current sheets. For example, it is not clear how the field strength should be normalized, or what value of the transverse field should be used. The use of an exact global reconnection model, described by Equations (36) and (37) in the simplest two-dimensional case, should lead to detailed quantitative predictions that are unhindered by extraneous parameterizations. In the exact analytic MHD solution for the current sheet structure, the local values of the electric and magnetic fields in the sheet are determined unambiguously in terms of the electric resistivity and boundary conditions.

Preliminary study of proton orbits in a two-dimensional current sheet with $a_0 = a_1 = 0$ has already lead to several nontrivial results that could provide a framework for the interpretation of the accelerated particle data without the introduction of numerous free parameters. For example, the maximum kinetic energy scaling with resistivity,

$$\mathcal{E}_{\mathrm{max}} \approx 3 \times 10^{12}\, S^{-1/2}\, B_{\mathrm{max}}^{3/2}\, \mathrm{eV}, \tag{38}$$

suggests that inverse Lundquist numbers as small as 10^{-10} could be sufficient (taking the dimensionless magnetic field $B_{\mathrm{max}} = 10$ corresponding to localized sheet fields of 1 kG) to produce GeV protons [15]. Even the simplest exact MHD reconnection solutions involving strictly two-dimensional fields appear well suited to rapid energy release and significant particle acceleration [9]. New interesting results should be expected for more realistic three-dimensional magnetic fields, in particular for the "separator" solutions with $a_0, a_1 \neq 0$ as well as "fan" and "spine" reconnection solutions based on a three-dimensional global field **P**. Observations strongly suggest that such three-dimensional null points are involved in active phenomena in the solar corona [2].

Finally, super-Dreicer electric fields should be expected when rapid magnetic reconnection occurs in nearly collisionless space plasmas. Hence the results described in this paper may be applicable not only to solar and stellar flares but also to other phenomena in space, characterized by efficient particle acceleration. For example, electron acceleration in a current sheet was demonstrated to lead to the observed synchrotron radiation by relativistic electrons in extragalactic jets [23].

Acknowledgements

I am grateful to CESRA for the opportunity to attend the workshop on energy conversion and particle acceleration in the solar corona at Ringberg Castle, Bavaria. Grateful acknowledgement is due in particular to K.-L. Klein and A. L. MacKinnon. This work was supported by NSF grant ATM-9813933, NASA grants NAG5-7792 and NAG5-8228, and the PPARC Short Term Visitor grant to the Astronomy and Astrophysics Group at the University of Glasgow.

References

1. M. J. Aschwanden, T. Kosugi, H. S. Hudson, M. J. Wills, R. A. Schwartz: Astrophys. J. **470**, 1198 (1996)
2. G. Aulanier, E. E. DeLuca, S. K. Antiochos, R. A. McMullen, L. Golub: Astrophys. J. **540**, 1126 (2000)
3. D. Biskamp: *this volume*
4. D. L. Bruhwiler, E. G. Zweibel: J. Geophys. Res. **97A**, 10825 (1992)
5. S. V. Bulanov: Soviet Astron. Lett. **6**, 206 (1980)
6. S. V. Bulanov, P. V. Sasorov: Soviet Astron. **19**, 464 (1976)
7. I. J. D. Craig, R. B. Fabling, S. M. Henton, G. J. Rickard: Astrophys. J. **455**, L197 (1995)
8. I. J. D. Craig, S. M. Henton: Astrophys. J. **450**, 280 (1995)
9. I. J. D. Craig, Y. E. Litvinenko: Astrophys. J. **570**, 387 (2002)
10. B. R. Dennis: Solar Phys. **100**, 465 (1985)
11. L. Fletcher, P. C. H. Martens: Astrophys. J. **505**, 418 (1998)
12. P. V. Foukal, B. B. Behr: Solar Phys. **156**, 293 (1995)
13. P. V. Foukal, S. Hinata: Solar Phys. **132**, 307 (1991)
14. M. L. Goldstein, W. H. Matthaeus, J. J. Ambrosiano: Geophys. Res. Lett. **13**, 205 (1986)
15. J. Heerikhuisen, Y. E. Litvinenko, I. J. D. Craig: Astrophys. J. **566**, 512 (2002)
16. G. D. Holman: Astrophys. J. **293**, 584 (1985)
17. R. Horiuchi, T. Sato: Phys. Plasmas 4, 277 (1997)
18. H. S. Hudson, J. M. Ryan: Annual Rev. Astron. Astrophys. **33**, 239 (1995)
19. K.-L. Klein, G. Trottet: Space Sci. Rev. **95**, 215 (2001)
20. J. W. Knight, P. A. Sturrock: Astrophys. J. **218**, 306 (1977)
21. Y. E. Litvinenko: Astrophys. J. **462**, 997 (1996)

22. Y. E. Litvinenko: Phys. Plasmas **4**, 3439 (1997)
23. Y. E. Litvinenko: Astron. Astrophys. **349**, 685 (1999)
24. Y. E. Litvinenko: Solar Phys, **194**, 327 (2000)
25. Y. E. Litvinenko, I. J. D. Craig: Astrophys. J. **544**, 1101 (2000)
26. Y. E. Litvinenko, B. V. Somov: Solar Phys. **146**, 127 (1993)
27. L. R. Lyons, T. W. Speiser: J. Geophys. Res. **90A**, 8543 (1985)
28. P. C. H. Martens: Astrophys. J. **330**, L131 (1988)
29. P. C. H. Martens, A. Young: Astrophys. J. Suppl. **73**, 333 (1990)
30. R. F. Martin, T. W. Speiser, K. Klamczynski: J. Geophys. Res. **99A**, 23623 (1994)
31. S. Masuda, T. Kosugi, H. Hara, S. Tsuneta, Y. Ogawara: Nature **371**, 495 (1994)
32. D. B. Melrose: Aust. J. Phys. **43**, 703 (1990)
33. J. A. Miller, P. J. Cargill, A. G. Emslie, *et al.*: J. Geophys. Res. **102A**, 14631 (1997)
34. K.-I. Mori, J.-I. Sakai, J. Zhao: Astrophys. J. **494**, 430 (1998)
35. E. R. Priest, T. G. Forbes: *Magnetic Reconnection: MHD Theory and Applications* (Cambridge University Press, Cambridge 2000)
36. E. Rieger, W. Q. Gan, H. Marschhäuser: Solar Phys. **183**, 123 (1998)
37. J. M. Ryan: Space Sci. Rev., **93**, 581 (2000)
38. M. Scholer: *this volume*
39. R. Schopper, G. T. Birk, H. Lesch: Phys. Plasmas **6**, 4318 (1999)
40. B. V. Somov: *Physical Processes in Solar Flares*, (Kluwer, Dordrecht 1992)
41. T. W. Speiser: J. Geophys. Res. **70**, 4219 (1965)
42. S. I. Syrovatskii: Comments on Astrophys. and Space Phys. **4**, 65 (1973)
43. S. I. Syrovatskii: Bulletin Acad. Sci. USSR **39**, 359 (1975)
44. G. Trottet, N. Vilmer, C. Barat, *et al.*: Astron. Astrophys. **334**, 1099 (1998)
45. G. H. J. van den Oord: Astron. Astrophys. **234**, 496 (1990)
46. N. Vilmer, A.L. MacKinnon: *this volume*
47. T. Yokoyama, K. Akita, T. Morimoto, K. Inoue, J. Newmark: Astrophys. J. **546**, L69 (2001)
48. M. Yoshimori, Y. Takai, K. Morimoto, K. Suga, K. Ohki: Publ. Astron. Soc. Japan **44**, L107 (1992)
49. Z. Zhu, G. Parks: J. Geophys. Res. **98A**, 7603 (1993)

Particle Acceleration Processes in Cosmic Plasmas

Reinhard Schlickeiser

Institut für Theoretische Physik, Lehrstuhl IV: Weltraum- und Astrophysik
Ruhr-Universität Bochum, 44780 Bochum, Germany

Abstract. Three acceleration processes of charged particles in cosmic plasmas are reviewed: resonant stochastic acceleration, diffusive shock acceleration and conversion of bulk motion to individual charged particle energies by relativistic pick-up. All three processes rely on the interactions of charged particles with partially random electromagnetic fields which are theoretically described within quasilinear theory. We demonstrate that in all three cases the modeling of the dynamics of the electromagnetic plasma turbulence is most crucial, and we discuss the relevant wave-particle and wave-wave interaction processes that control the turbulence dynamics.

1 Introduction

The detection of nonthermal radiation from many astrophysical objects requires the acceleration of charged particles to relativistic energies in these plasma systems. The acceleration mechanisms, whatever they may be, are remarkably efficient, converting a major fraction of the total energy into fast particles. The principal limitation on particle acceleration theories has been the realisation (e.g. Parker (1976)) that the universe is not filled with a perfect vacuum, but rather is pervaded everywhere by tenous ionized gases with high electrical conductivity quite able to short circuit any large-scale electric fields that occur under ordinary circumstances. But electric fields are needed for particle acceleration as the scalar product of the equation of motion of charged particles (charge q, mass m) in electromagnetic fields,

$$\frac{d\boldsymbol{p}}{dt} = q\left[\boldsymbol{E} + \frac{1}{\gamma mc}\boldsymbol{p} \times \boldsymbol{B}\right] ,\tag{1}$$

with the particle momentum $\boldsymbol{p}$ demonstrates,

$$\boldsymbol{p} \cdot \frac{d\boldsymbol{p}}{dt} = \frac{1}{2}\frac{dp^2}{dt} = q\boldsymbol{p} \cdot \boldsymbol{E}\tag{2}$$

Eq. (2) readily yields for the change of the particle energy, $W = mc^2\sqrt{1 + \frac{p^2}{m^2c^2}}$, the expression

$$\frac{dW}{dt} = \frac{qc^2}{W}\,\boldsymbol{p} \cdot \boldsymbol{E}\tag{3}$$

Apart from unipolar inductor sources like pulsars and the electric field in perpendicular collisionless shock waves, the large electrical conductivity of most space plasmas prevents to sustain stationary and steady large-scale electric fields. Electric fields in space plasmas then only appear as (i) transient phenomena (as in magnetic reconnection situations), or as (ii) fluctuating fields (plasma turbulence) in magnetized gases

$$E = 0 + \delta E \, , B = B_0 + \delta B \tag{4}$$

with vanishing ensemble averages $\langle \delta E \rangle = 0$ and $\langle \delta B \rangle = 0$.

As a consequence, four basic types of particle acceleration processes in space plasmas are relevant:

(1) acceleration by magnetic reconnection,
(2) stochastic acceleration by resonant particle interactions with plasma waves,
(3) (diffusive) shock acceleration,
(4) conversion of bulk motion to relativistic particles.

Particle acceleration in magnetic reconnection situations is reviewed in the lectures by Litvinenko (2002) and Scholer (2002) and will not be considered here. We will restrict our discussion to processes (2)-(4), which all rely on interactions with random electric fields. Because of lack of space we will limit our discussion to the basic physics of these three acceleration processes and some recent achievements in their theoretical description. A more detailed discussion of processes (2) and (3) is given in my monograph (Schlickeiser (2002)).

2 Gas Component Coupling in Cosmic Plasmas

As a result of efficient acceleration processes the gas of energetic particles (commonly referred to as cosmic ray gas) coexists with the background gas of slow and often thermal particles. The fluctuating electromagnetic fields (δE, δB) are the agency that provides the dynamical coupling between these two gases because both of them are collision-free plasmas, i.e. the plasma parameter $g = (\nu_{e,e}/\omega_{p,e})$, defined as the ratio of the scattering rate for elastic electron-electron collision ($\nu_{e,e}$) to the plasma frequency ($\omega_{p,e}$), in each gas is much smaller than unity. Only by interacting with the same electromagnetic plasma turbulence the energetic cosmic ray gas is tied to the background plasma.

There are at least five important interaction processes of these two gases and the electromagnetic turbulence that also determine the properties of fluctuating fields:

(a) Because of its normally much larger density the background thermal plasma is the wave-carrying agency. In a theoretical description this means that the real part of the plasma wave dispersion relation $\omega_R = \omega_R(k)$ is determined solely by this background plasma. Moreover, large-scale irregular motions of the background plasma, arising in galaxies from effects due to e.g. stellar winds, stellar explosions, galactic inflows, and in the solar wind due to e.g.

coronal mass ejections, solar flares, fast solar wind streams overrunning small solar wind streams, serve as input of turbulent energy at large turbulence scales corresponding to small wavenumbers.

(b) On the other hand, the background thermal plasma gains energy from the turbulent electromagnetic field by collision-less Landau and/or cyclotron damping of plasma waves. In a theoretical description this gives rise to a negative imaginary part of the plasma wave dispersion relation $\Gamma_d = \Gamma_d(\mathbf{k}) < 0$ and a corresponding heating term in the change of entropy equation in a hydrodynamical description of the background gas.

(c) The cosmic ray gas serves as important source of plasma turbulence at nearly all wavenumbers by efficient instabilities driven by pitch-angle anisotropies, streaming instabilities, loss-cone distributions in converging magnetic field lines or inverted energy distributions. In a theoretical description this gives rise to a positive imaginary part of the plasma wave dispersion relation $\Gamma_c = \Gamma_c(\mathbf{k}) > 0$.

(d) On the other hand, the cosmic ray gas gains energy from the turbulent electric field components by stochastic resonant acceleration processes. In a theoretical description these enter as momentum diffusion and momentum convection terms in the transport equations for the cosmic ray particles. Moreover, the turbulent magnetic field components provide the scattering of particles along the ordered background magnetic field which is crucial for cosmic ray confinement in an astrophysical system. In a theoretical description these enter as spatial diffusion and spatial convection terms in the transport equations for the cosmic ray particles.

(e) At appreciable wave intensities plasma waves at different wavenumbers interact by various nonlinear wave-wave-interaction processes that are responsible for the cascading (up in wavenumber) and/or inverse cascading (down in wavenumber) in wavenumber space.

To a large extent, our progress in understanding cosmic ray acceleration in cosmic plasmas depends on our understanding of the dynamical evolution of power spectra of the electric field fluctuations, which is a difficult subject in the light of the various coupling ((a)-(d)) and interaction (e) processes.

3 Theoretical Methods to Study Cosmic Particle Acceleration

Because of the complicated nonlinear equations of motion of charged particles in partially random electromagnetic fields there are only two methods to study theoretically particle acceleration: (i) numerical simulations of highly idealized configurations, (ii) quasilinear theory. Both have their advantages and shortcomings, and they complement each other.

Obviously, besides limited computer power numerical simulations require the precise knowledge of many important input plasma parameters as well as the specification of initial and boundary conditions which at least for the more distant cosmic objects are not known. By chosing the wrong input plasma quantities one may end up in an irrelevant range of solution space. Of course, when all these input quantities are known and given, the simulations result in a very accurate and detailed description of the acceleration processes on all spatial, momentum and time scales of interest.

On the other hand, the discussion of the accuracy of quasilinear theory is legendary in the literature of theoretical plasma physics especially in its application to plasma fusion devices. However, it seems that up to factor 2 uncertainties it gives a reasonable description of the ongoing fundamental physical processes if its basic assumptions are fulfilled. After its original developments for longitudinal plasma waves (Vedenov, Velikhov and Sagdeev 1962) the application of quasilinear theory to astrophysical plasmas has turned out to be very fruitful in explaining the dynamics of energetic charged particles in these plasmas. Quasilinear transport equations for magnetohydrodynamic plasma waves were formulated by Kennel and Engelmann (1966), Hall and Sturrock (1967) and Lerche (1968).

The quasilinear approach to the interaction of charged particles with partially random electromagnetic fields $(\boldsymbol{B}_0 + \delta\boldsymbol{B}, \delta\boldsymbol{E})$ is a first-order perturbation calculation in the ratio $q_L = (\delta B/B_0)^2$ and requires smallness of this ratio with respect to unity. In most cosmic plasmas this requirement is well satisfied as has been established by direct in situ measurements in interplanetary plasmas, or due to saturation effects in the growth of fluctuating fields. The standard quasilinear approach also requires incoherent mode coupling of the fluctuating electromagnetic fields described as the superposition of individual plasma wave modes. It happens in some circumstances that the instabilites responsible for plasma wave growth be narrow band and that the wave trains develop in a rather coherent way, so that the energetic particles interact with them in the trapping mode. Such situations occur for example in geophysical environments but not only. Coherent wave particle interaction processes are discussed for eample in Le Quéau and Roux (1987a,b).

Another aspect not detailed here concerns the issue of particle acceleration by turbulence when the turbulence takes the form of an ensemble of coherent structures like double layers or soliton structures, as is presumably the case in auroral situations. For more details on these structures the interested reader is referred to the work of Pelletier et al. (1988), Treumann et al. (1996), Treumann and Pottelette (1999), and Lerche and Schlickeiser (2002).

We review here some recent results of standard quasilinear cosmic ray transport theories assuming incoherent mode coupling. We show that modelling the plasma turbulence beyond the magnetostatic approximation is most important for a correct description of cosmic ray particle dynamics. In particular, the effects of finite plasma wave speeds play a crucial role both for spatial diffusion and the acceleration of cosmic ray particles.

4 The Quasilinear Diffusion–Convection Transport Equation

Due to the high conductivity of most cosmic plasmas large-scale steady electric fields are absent, so that the interest concentrates on magnetized plasma. Linear stability calculations show that these systems contain low-frequency magneto-hydrodynamic turbulence such as shear Alfvén waves and fast and slow magnetosonic waves. Because for these plasma waves the magnetic part of the Lorentz force is much larger than the electric part of the Lorentz force, the time scale for rapid pitch angle scattering of energetic charged particles is much shorter than the time scale for energy changes. As a consequence, the cosmic ray particles' gyrotropic distribution function adjusts rapidly to quasi-equilibrium, which is close to the isotropic distribution function, in excellent agreement with the observational fact of the isotropy of galactic cosmic ray particles. The diffusion–convection transport equation for the isotropic part of the phase space density $F(z, p, t)$ can be derived from the quasilinear Fokker–Planck equation. While the derivation of the quasilinear Fokker–Planck equation from the particles' equations of motion (1) can be found in standard plasma physics textbooks (e.g. Swanson 1989, Ch. 7) the reduction of the diffusion–convection equation for the isotropic part of the phase space density $F(z, p, t)$ is less well-known. We therefore detail in Appendix A this derivation for nonrelativistic ($u \ll c$) bulk speeds of the turbulence-carrying background plasma based on the approximation scheme by Jokipii (1966), Hasselmann and Wibberenz (1968), Earl (1973) and Schlickeiser (1989). The corresponding derivation for relativistic bulk speeds has been given by Kirk et al. (1988).

For non-relativistic bulk speeds the diffusion-convection equation reads

$$\frac{\partial F}{\partial t} - S_0 = \frac{\partial}{\partial z}\left[\kappa\frac{\partial F}{\partial z}\right] - V\frac{\partial F}{\partial z}$$
$$+ \frac{p}{3}\frac{\partial V}{\partial z}\frac{\partial F}{\partial p} + \frac{1}{p^2}\frac{\partial}{\partial p}\left[p^2 A\frac{\partial F}{\partial p} - p^2\dot{p}_{\text{Loss}}F\right] - \frac{F}{T_c}, \qquad (5)$$

where the spatial diffusion coefficient κ, the cosmic ray bulk speed V and the momentum diffusion coefficient A are determined by pitch-angle averages of three Fokker–Planck coefficients

$$\kappa = \frac{v^2}{8}\int_{-1}^{1} d\mu\frac{(1-\mu^2)^2}{D_{\mu\mu}(\mu)},$$

$$V = u + \frac{1}{3p^2}\frac{\partial}{\partial p}(p^3 D), \quad D = \frac{3v}{4p}\int_{-1}^{1} d\mu(1-\mu^2)\frac{D_{\mu p}(\mu)}{D_{\mu\mu}(\mu)}$$

$$A = \frac{1}{2}\int_{-1}^{1} d\mu\left[D_{pp}(\mu) - \frac{D_{\mu p}^2(\mu)}{D_{\mu\mu}(\mu)}\right]. \qquad (6)$$

In Eq. (5) S_0 is the source term, and $\dot{p}_{\text{Loss}}$ and T_c describe continuous and catastrophic momentum loss processes. The three Fokker–Planck coefficients entering the averaging in equations (6) are calculated (Hall and Sturrock 1967, Krommes 1984, Achatz et al. 1991) from ensemble-averaged first-order corrections to the particle orbit,

$$D_{\mu\mu} = \Re \int_0^\infty d\tau \, \langle \dot{\mu}(t)\dot{\mu}^*(t+\tau) \rangle \,, \quad D_{\mu p} = \Re \int_0^\infty d\tau \, \langle \dot{\mu}(t)\dot{p}^*(t+\tau) \rangle \,,$$

$$D_{pp} = \Re \int_0^\infty d\tau \, \langle \dot{p}(t)\dot{p}^*(t+\tau) \rangle \,, \tag{7}$$

with $\boldsymbol{p} = (p\sqrt{1-\mu^2}\cos\phi, p\sqrt{1-\mu^2}\sin\phi, p\mu)$ and the equation of motion

$$\dot{\boldsymbol{p}} = q\Big[\delta\boldsymbol{E} + \frac{1}{mc\gamma}\boldsymbol{p} \times (\boldsymbol{B}_0 + \delta\boldsymbol{B})\Big] \tag{8}$$

In its general form the diffusion–convection transport equation contains spatial diffusion and spatial convection terms as well as momentum diffusion and momentum convection terms. Since the pioneering work of Fermi (1949, 1954) it has become customary to refer to the latter two as Fermi acceleration of second and first order, respectively. Note, however, that the momentum convection term only leads to acceleration for converging bulk flow (i.e., $dV/dz < 0$) but to deceleration for expanding flows (i.e., $dV/dz > 0$). The converging bulk flow condition $dV/dz < 0$ is fulfilled at cosmic shock waves and leads to diffusive shock acceleration which is discussed in Sect. 6.

The value of the three quasilinear transport parameters (6) depends on the nature and the statistical properties of the electromagnetic turbulence and the turbulence-carrying background medium. Idealized physical situations can be constructed where some of the three transport parameters (6) do not occur, e.g., in the magnetostatic approximation of the turbulence the parameters $A = D = 0$, so that the transport equation (5) in particular would contain no momentum diffusion term. Despite its frequent use, such a truncated transport equation is unrealistic and its applicability therefore rather limited.

Because of the smallness of the quasilinear parameter $q_L = (\delta B/B_0)^2 < 1$ (also referred to as weak turbulence limit), it is justified to adopt the plasma wave approach (Schlickeiser and Achatz 1993) to describe the electromagnetic turbulence as the superposition of $j = 1, ..., N$ individual plasma modes with definite dispersion relation $\omega_j = \omega_j(\boldsymbol{k})$ so that

$$\delta\boldsymbol{B}(\boldsymbol{x},t) = \sum_{j=1}^N \int d^3k \int d\omega \, \boldsymbol{B}_1(\boldsymbol{k},\omega) \exp[\imath(\boldsymbol{k}\cdot\boldsymbol{x} - \omega_j(\boldsymbol{k})t)],$$

$$\delta\boldsymbol{E}(\boldsymbol{x},t) = \sum_{j=1}^N \int d^3k \int d\omega \, \boldsymbol{E}_1(\boldsymbol{k},\omega) \exp[\imath(\boldsymbol{k}\cdot\boldsymbol{x} - \omega_j(\boldsymbol{k})t)]. \tag{9}$$

Augmented with terms representing perpendicular spatial diffusion, for which yet no rigorous theory is available, the diffusion–convection transport equation

(5) has passed all tests provided by the wealth of solar modulation and in situ interplanetary cosmic ray data with flying colours (Jokipii 1983) reproducing qualitatively and quantitatively the observations if the required input turbulence data are accurately known. Therefore we have high confidence that this equation also can be applied to cosmic ray dynamics in more distant objects.

Because they often operate on longer time scales than those of many interplanetary processes, the effects of continuous momentum losses ($\dot{p}_{\mathrm{Loss}}$), catastrophic losses (T_c) and momentum diffusion (A) are less exhibited in some interplanetary phenomena, but clearly play a crucial role in solar flares as proven by the successful modelling of particle acceleration in impulsive solar flares (e.g., Steinacker et al. 1988, Schlickeiser and Steinacker 1989, Park et al. 1997).

5 Relevant Plasma Modes

Most cosmic plasmas have a small value of the plasma beta $\beta = c_S^2/V_A^2$, which is defined by the ratio of the ion sound to Alfvén speed, and thus indicates the ratio of thermal to magnetic pressure. For low-beta plasmas the two relevant magnetohydrodynamic wave modes are the

(1) *shear Alfvén waves* with dispersion relation

$$\omega_R^2 = V_A^2 k_\parallel^2 \tag{10}$$

at parallel wavenumbers $|k_\parallel| \ll \Omega_p/V_A$, which have no magnetic field component along the ordered background magnetic field δB_z ($\parallel \boldsymbol{B}_0$) = 0,
(2) the *fast magnetosonic waves* or *fast mode waves* with dispersion relation

$$\omega_R^2 = V_A^2 k^2, \quad k^2 = k_\parallel^2 + k_\perp^2 \tag{11}$$

for wavenumbers $|k| \ll \Omega_p/V_A$, which have a compressive magnetic field component $\delta B_z \neq 0$ for oblique propagation angles $\theta = \arccos(k_\parallel/k) \neq 0$.

In the limiting case (commonly referred to as slab model) of parallel (to $\boldsymbol{B}_0$) propagation ($\theta = k_\perp = 0$) the shear Alfvén waves become the left-handed circularly polarised Alfvén-ion-cyclotron waves, whereas the fast magnetosonic waves become the right-handed circularly polarised Alfvén-whistler-electron-cyclotron waves.

Schlickeiser and Miller (1998) investigated the quasilinear interactions of charged particles with these two plasma waves. In case of negligible wave damping the interactions are of resonant nature: a cosmic ray particle of given velocity v, pitch angle cosine μ and gyrofrequency $\Omega_c = \Omega_{c,0}/\gamma$ interacts with waves whose wavenumber and real frequencies obey the condition

$$\omega_R(k) = v\mu k_\parallel + n\Omega_c, \tag{12}$$

for integer $n = 0, \pm 1, \pm 2, \ldots$.

5.1 Resonant Interactions of Shear Alfvén Waves

For shear Alfvén waves only interactions with $n \neq 0$ are possible. These are referred to as *gyroresonances* because inserting the dispersion relation (10) in the resonance condition (12) yields for the resonant parallel wavenumber

$$k_{\parallel,A} = \frac{n\Omega_c}{\pm V_A - v\mu}, \tag{13}$$

which apart from very small values of $|\mu| \leq V_A/v$ equals the inverse of the cosmic ray particle's gyroradius $R_L = v/\Omega_c$,

$$k_{\parallel,A} \simeq n/R_L \tag{14}$$

and higher harmonics.

5.2 Resonant Interactions of Fast Magnetosonic Waves

In contrast, for fast magnetosonic waves the $n = 0$ resonance is possible for oblique propagation due its compressive magnetic field component. The $n = 0$ interactions are referred to as *transit-time damping*, hereafter TTD. Inserting the dispersion relation (11) into the resonance condition (12) in the case $n = 0$ yields

$$v\mu = \pm V_A/\cos\theta \tag{15}$$

as necessary condition which is independent of the wavenumber value k. Apparently all super-Alfvénic ($v \geq V_A$) cosmic ray particles are subject to TTD provided their parallel velocity $v\mu$ equals at least the wave speeds $\pm V_A$. Hence equation (15) is equivalent to the two conditions

$$|\mu| \geq V_A/v, \quad v \geq V_A. \tag{16}$$

In a low-beta plasma the requirement of super-Alfvénic velocities favors the preferential acceleration of electrons over hadrons if particles are accelerated out of the thermal population. Additionally, fast mode waves also allow gyroresonances ($n \neq 0$) at wavenumbers

$$k_F = \frac{n\Omega_c}{\pm V_A - v\mu\cos\theta}, \tag{17}$$

which is very similar to Eq. (13) and typically leads again to

$$k_F \simeq nR_L^{-1} \tag{18}$$

5.3 Implications for Stochastic Particle Acceleration

The simple considerations of the last two subsections allow us the following immediate conclusions:

(1) With TTD-interactions with undamped fast mode waves alone, it is not possible to scatter particles with $|\mu| \leq V_A/v$, i.e., particles with pitch angles near 90°. Obviously, these particles have basically no parallel velocity and cannot catch up with fast mode waves that propagate with small but finite speeds $\pm V_A$. In particular this implies that with TTD alone it is not possible to establish an isotropic cosmic ray distribution function. We always need gyroresonances to provide the crucial finite scattering at small values of μ.

(2) Conditions (15) and (16) reveal that TTD is no gyroradius effect. It involves fast mode waves at all wavenumbers provided the cosmic ray particles are super-Alfvénic and have large enough values of μ as required by Eq. (16). Because gyroresonances occur at single resonant wavenumbers only, see Eqs. (13) and (17), their contribution to the value of the Fokker–Planck coefficients in the interval $|\mu| \geq V_A/v$ is much smaller than the contribution from TTD.

Therefore for comparable intensities of fast mode and shear Alfvén waves, TTD will provide the overwhelming contribution to all Fokker–Planck coefficients $D_{\mu\mu}$, $D_{\mu p}$ and D_{pp} in the interval $|\mu| \geq V_A/v$. At small values of $|\mu| < V_A/v$ only gyroresonances contribute to the values of the Fokker–Planck coefficients involving according to Eqs. (14) and (18) wavenumbers at $k_{\|,A} = k_R \simeq \pm n\Omega_c/V_A$.

(3) We recall from Eq. (6) that the momentum diffusion coefficient A basically is given as the integral over D_{pp}. The value of D_{pp} then is determined by the dominant contribution from TTD, implying

$$A \simeq \int_{V_A/v}^{1} d\mu D_{pp}^{\mathrm{TTD}}(\mu). \tag{19}$$

(4) On the other hand, the spatial diffusion coefficient is given by the integral over the *inverse* of the Fokker–Planck coefficient $D_{\mu\mu}$, so that here the smallest values of $D_{\mu\mu}$ due to gyroresonant interactions in the interval $|\mu| < V_A/v$ determine the spatial diffusion coefficient

$$\kappa \simeq \frac{v^2}{8} \int_{-V_A/v}^{V_A/v} \frac{d\mu}{D_{\mu\mu}^{\mathrm{G}}(\mu)} \tag{20}$$

The gyroresonances can be due to shear Alfvén waves or fast magnetosonic waves. The analysis of Schlickeiser and Miller (1998) shows, that if the plasma turbulence is a mixture of shear Alfvén waves distributed slab-like and isotropically distributed fast magnetosonic waves, then the gyroresonances in Eq. (20) are due to shear Alfvén waves.

(5) Omitting more specific effects, we consequently conclude from these elementary considerations that fast magnetosonic waves are responsible for the

stochastic acceleration of cosmic ray particles whereas shear Alfvén waves are responsible for the spatial diffusion of cosmic ray particles. To arrive at this important result, it has been decisive to discard the magnetostatic approximation of the plasma wave turbulence and to consider in particular *finite* wave propagation effects of order V_A/v.

(6) The longest wavelength of an undamped standing wave in a physical system is limited by the size $\lambda_{max} \leq L$ of this system. This implies that the plasma wavenumbers are larger than $k \geq k_{min} = 2\pi/L$. The resonance conditions (14) and (18) then yield $R_L \leq nL/2\pi$ corresponding to a limit on the energy of the interacting cosmic ray particles:

$$W \simeq pc \leq W_{max}(n) \equiv \frac{|n|ZeB_0L}{2\pi} \qquad (21)$$

For parameter values typical for the galactic interstellar medium (1 pc= $3.086 \cdot 10^{16}$ m) Eq. (21) yields

$$W_{max}(n) = 2 \cdot 10^{16} \, |n|Z \, \text{ eV } \frac{B_0}{4\mu G} \frac{L}{30 \text{ pc}}$$

$$= 0.003 \, |n|Z \, \text{ J } \frac{B_0}{4 \cdot 10^{-10} \text{ T}} \frac{L}{30 \text{ pc}} , \qquad (22)$$

whereas for typical solar flare parameters we obtain

$$W_{max}(n) = 5 \cdot 10^{12} \, |n|Z \, \text{ eV } \frac{B_0}{100 \text{ G}} \frac{L}{10^4 \text{ km}} \qquad (23)$$

In case of parallel propagating waves, where $n = \pm 1$ are the only possible gyroresonance values, the maximum energy $W_{max}(n = 1)$ corresponds to the well-known Hillas-limit (Hillas 1984). However, as Eq. (22) demonstrates, for obliquely propagating waves there exists resonant acceleration through higher harmonics ($|n| \geq 2$) of cosmic ray particles with energies larger than the Hillas-limit $W_{max}(n = 1)$, although at a reduced level $\propto J_n^2(nW/W_{max}(n = 1)) \simeq \mathcal{O}(W_{max}(n = 1)/nW)$ as compared to energies below the Hillas-limit $W \leq W_{max}(n = 1)$.

This simplified picture becomes more involved if the different resonance conditions are combined with investigations on the dynamics of turbulence intensities, incorporating in particular effects due to wave cascading. We refer the reader to the review of Miller et al. (1997) for more details.

6 Diffusive Particle Acceleration near Shock Waves

The transport equation (5) has also been used to investigate the test-particle diffusive acceleration of cosmic ray particles in quasi-parallel shock waves. Here also finite wave speed effects are important. This concerns in particular the cosmic ray bulk speed V in Eq. (6) that can be different from the gas speed u for nonzero values of $D \neq 0$.

Following earlier work by McKenzie and Westphal (1969) and Scholer and Belcher (1971), Vainio and Schlickeiser (1998, 1999, 2001) calculated anew the transmission of small-amplitude parallel-moving Alfvén waves through a parallel super-Alfvénic shock. In their investigation Vainio and Schlickeiser combined the equations for

(1) the continuity of the transverse momentum

$$\left[\rho u_n \boldsymbol{u}_t - \frac{B_n \boldsymbol{B}_t}{4\pi}\right] = 0, \tag{24}$$

where the shock bracket $[X] \equiv X_1 - X_2$ denotes the difference of the upstream (index 1) and downstream (index 2) value of the physical quantity X, u_n (B_n) and $\boldsymbol{u}_t$ $(\boldsymbol{B}_t)$ are the normal (to the shock) and tangential gas flow velocity (magnetic field) components, respectively;

(2) continuity of the normal magnetic field

$$B_{n,1} = B_{n,2} = B_0; \tag{25}$$

(3) the continuity of the tangential electric field

$$\left[u_n \boldsymbol{B}_t - B_n \boldsymbol{u}_t\right] = 0; \tag{26}$$

(4) and the continuity of the mass flux

$$\left[\rho u_n\right] = 0 \tag{27}$$

with the different relation of velocity and magnetic field fluctuations for forward (f) and backward (b) moving Alfvén waves, i. e.

$$\delta \boldsymbol{u}^{\mathrm{f}} = -\frac{\delta \boldsymbol{B}^{\mathrm{f}}}{(4\pi\rho)^{1/2}}, \quad \delta \boldsymbol{u}^{\mathrm{b}} = \frac{\delta \boldsymbol{B}^{\mathrm{b}}}{(4\pi\rho)^{1/2}}. \tag{28}$$

To arrive at a complete set of equations for the downstream values, i.e., to be able to determine the gas compression ratio $r = \rho_2/\rho_1 = u_{n,1}/u_{n,2}$ of the shock, Eqs. (24) – (28) must be completed by yet two equations (e.g., Boyd and Sanderson 1969) describing the continuity of the normal momentum

$$\left[\rho u_n^2 + P + \frac{B_t^2}{8\pi}\right] = 0 \tag{29}$$

and energy flux (for an adiabatic equation of state, $P\rho^{-\gamma_g} = \mathrm{const.}$)

$$\left[\frac{1}{2}\rho u_n(u_n^2 + u_t^2) + \frac{\gamma_g P u_n}{\gamma_g - 1} + \frac{u_n B_t^2}{4\pi} - \frac{B_n(\boldsymbol{u}_t \cdot \boldsymbol{B}_t)}{4\pi}\right] = 0, \tag{30}$$

respectively. Two approaches with respect to this were taken:

(a) In their first paper, Vainio and Schlickeiser (1998) neglected the influence of the Alfvén waves in the normal momentum and energy flux equations to arrive at

$$r = \frac{\gamma_g + 1}{\gamma_g - 1 + \frac{2\beta}{M^2}}, \tag{31}$$

where $M = u_1/V_{A,1}$ and β are the upstream Alfvénic Mach number of the shock and the upstream plasma beta.

(b) In subsequent work (Vainio and Schlickeiser 1999) the Alfvén wave normal momentum and energy flux were taken into account in the respective conservation laws to arrive at a cubic equation for r that can be solved in a parametric form. For $\gamma_g = 5/3$,

$$M^2 = (1 + y)r(y) \tag{32}$$

$$r(y) = \frac{8y^2(y + 1) - 6\beta y^2 - q_{L,1}(y + 1)(5y - 3)}{2y^2(y + 1) + q_{L,1}(y + 1)(y + 3)}, \tag{33}$$

where $q_{L,1} = (\delta B_1/B_0)^2$, and the parameter y runs between

$$\frac{\beta - 1 + q_{L,1} + \sqrt{(\beta + 1 + q_{L,1})^2 - 4\beta}}{2} < y < \infty. \tag{34}$$

In both approaches *the downstream electromagnetic field properties can be calculated from the specified upstream electromagnetic field*. In particular, specifying the upstream Alfvén wave cross helicity state $H_{c,1}$, that indicates the relative fraction of forward and backward moving Alfvén waves, so that the upstream cosmic ray bulk speed is $V_1 = u_1 + H_{c,1}V_{A,1}$, Vainio and Schlickeiser calculated the resulting downstream cosmic ray bulk speed $V_2 = u_2 + H_{c,2}V_{A,2}$. This immediately yields the scattering center compression ratio

$$r_k \equiv \frac{V_1}{V_2} = r\frac{M + H_{c,1}}{M + r^{1/2}H_{c,2}}, \tag{35}$$

which in general is different from the gas compression ratio r. Because the shock wave is collisionless, *it is this scattering center compression ratio, and not the gas compression ratio, that determines the spectral index of the power law momentum spectrum of the accelerated cosmic ray particles*.

When downstream momentum diffusion is neglected, the particle differential energy spectrum at the shock is — up to a cut off determined by losses, particle escape, finite geometrical shock extent (see the discussion of Eq. (22) above), and finite acceleration time — a power law in momentum

$$dJ/dE \propto p^{-\Gamma}, \quad \Gamma = \frac{r_k + 2}{r_k - 1}, \tag{36}$$

whose spectral index Γ is solely determined by r_k. In Fig. 1 we show the calculated cosmic ray spectral index values as a function of the spectral index of

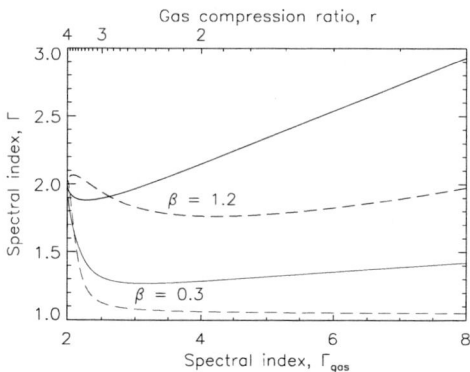

Fig. 1. Cosmic-ray spectral index produced by an adiabatic shock with a constant upstream plasma beta, neglecting stochastic acceleration in the downstream region. Dashed and solid lines give the results for $H_{c,1} = +1$ and -1, respectively. The magnetic amplitude of the upstream waves is $b = 0.1$ and their spectral index is $q = 1.5$. The Alfvén wave normal momentum and energy flux are includeded in deriving the shock's gas compression ratio. From Vainio and Schlickeiser (1999).

the conventional theory, $\Gamma_{\text{gas}} = (r + 2)/(r - 1)$ and as a function of the gas compression ratio for four different specified upstream states and an adiabatic index $\gamma_g = 5/3$. The results of Fig. 1 are based on approach (b), where the Alfvén wave normal momentum and energy fluxes have been included in the respective conservation laws. In all cases, the scattering center compression ratio r_k differs significantly from the gas compression ratio r. Practically never does the scattering center compression ratio r_K agree with the gas compression ratio r. In particular, for low upstream plasma beta small spectral index values $\Gamma \leq 2$ are possible, whereas the gas compression ratio value is limited to $r \leq 4$ for $M \to \infty$. Thus, the model, being able to generate particle power law spectra harder than the originally limiting value $\Gamma = 2$, avoids the discrepancy noted by, e.g., Lerche (1980), Drury (1983) and Dröge et al. (1987) that the original shock wave acceleration theory in its simplest test-particle form is not in accord with the observed flat particle spectra in shell-type supernova remnants and bright spiral galaxies. And it again is keeping track of different propagation speeds of forward and backward moving waves that leads to this significant result.

The principal difference between the gas compression ratio and the scattering center compression ratio, being equivalent to the difference between the effective plasma wave velocity and the gas velocity, and the possible consequences for the spectral index of the differential momentum spectrum of accelerated particles, has been already noted by Bell (1978), see his Eqs. (11) and (12); although he did no quantitative calculations of this effect. By calculating the correct transmission coefficients of Alfvén waves through the shock from the Rankine-Hugoniot continuity equations Vainio and Schlickeiser (1988, 1999) demonstrated that precisely this effect can account for the generation of particle spectral indices flatter than $\Gamma = 2$. We point out, however, that the studies of Vainio and Schlickeiser were restricted to circularly polarized Alfvén waves; to make the model fully self-consistent, one would need to add the fast mode waves into the upstream region, because they also are generated by the accelerated ion beam directed against the plasma inflow to the shock. As a final remark, we note that the studies of Vainio and Schlickeiser were concentrating on the effects at the step like gas shock, where the cosmic ray distribution function is continuous: thus the nonlinear particle pressure effects appearing at larger spatial scales (e.g. Pelletier 1999) were not considered, although in some shocks these effects may play a role in preventing the infinite accelerated particle energy densities from occurring.

7 Conversion of Bulk Motion to Relativistic Particles

As third particle acceleration process we consider the energisation of charged particles in bulk outflows by interactions with the surrounding medium. In case of active galactic nuclei and gamma-ray burst sources these outflows have relativistic velocities; however, the model also applies to non-relativistic outflow velocities as in solar and stellar coronal mass ejections and in the late phases of supernova explosions. Certainly, this process has to be regarded on a different footing than stochastic acceleration by resonant particle acceleration and diffusive shock acceleration because it does not generate high energy particles from a thermal pool. It rather involves some sort of kinetic friction between two particle populations with relative streaming operating in the relativistic regime, where one population has been already accelerated to a fast directed outflow.

In each case we model the bulk motion as a one-dimensional channeled outflow with, in general, relativistic bulk velocity V, consisting of cold electrons and protons of density n_b^*. For convenience we assume that the outflow is directed parallel to the uniform background magnetic field. This beam of protons and electrons propagates into the surrounding interstellar (or coronal) medium that consists of cold protons and electrons at rest of much smaller density n_i^*. Viewed from the coordinate system comoving with the outflow, the interstellar protons and electrons represent a proton-electron beam propagating with relativistic speed $-V$ antiparallel to the uniform magnetic field direction. Generalising the analysis of Pohl and Schlickeiser (2000) we demonstrate that very quickly the beam excites low-frequency magnetohydrodynamic Alfvén and whistler waves via a two-stream instability which isotropise the incoming protons and electrons in the outflow plasma. This pick-up of interstellar protons and electrons then is a source of isotropic, quasi-monoenergetic protons and electrons with Lorentz factor Γ in the blast wave frame.

If the surrounding interstellar medium is non-uniform, $n_i^* \neq$ const., the injection of relativistic particles in the outflow rest system by this pick-up process is highly variable and may explain the short time-variability of nonthermal radiation in active galactic nuclei and gamma-ray bursts.

7.1 Basic Equations of the Relativistic Pick-Up Model

We consider in the laboratory frame (all physical quantities in this system are indexed with $*$) the cold outflow electron-proton plasma of density n_b^* and thickness d^* in x-direction running into the cold interstellar medium of density n_i^*, consisting also of electrons and protons, parallel to the uniform magnetic field of strength B. In the comoving frame the total phase space distribution function of the plasma in the outflow region at the start thus is

$$f(p, t = 0) = \frac{1}{2\pi p^2} n_i \delta(\mu + 1)\delta(p - P) + \frac{1}{4\pi p^2} n_b \delta(p)$$
$$= n_b f_b(p, \mu, t = 0) + n_i f_i(p, \mu, t) , \tag{37}$$

where $\mu = p_\parallel/p$ the cosine of the pitch-angle of the particles in the magnetic field B, $P = \Gamma m V = mc\sqrt{\Gamma^2 - 1}$ where $\Gamma = 1/\sqrt{1 - (V^2/c^2)}$. The number densities transform as $n_i = \Gamma n_i^*$, $n_b = n_b^*/\Gamma$. In general, the particle density in the outflow region $n_b^* \gg n_i^*$ is much larger than the surrounding interstellar gas density n_i^*.

The beam distribution function f_i in Eq. (37) is unstable and excites low-frequency transverse plasma waves. We want to calculate the time t_f it takes these plasma waves to isotropise the incoming interstellar protons and electrons; if this relaxation time is much smaller than d/c an isotropic distribution of primary protons and electrons in the outflow region is effectively generated. We restrict our analysis here to parallel propagating low-frequency transverse plasma waves.

Because $n_i/n_b = \Gamma^2 n_i^*/n_b^* \ll 1$ the beam is weak. Therefore the contributions from the beam to the plasma wave dispersion relation at frequencies ω_R well below the non-relativistic electron gyrofrequency ($|\omega_R| \ll |\Omega_e|$) are perturbations to the real part of the dispersion relation for very subluminal (index of refraction $N = ck/|\omega_R| \gg 1$) transverse waves

$$\Re \Lambda_t = 1 - \frac{c^2 k^2}{\omega_R^2} + \frac{\omega_{p,e}^2}{\Omega_e^2} + \frac{\omega_{p,e}^2}{(\omega_R - \Omega_p)\Omega_e}$$

$$\simeq -\frac{c^2 k^2}{\omega_R^2} + \frac{\omega_{p,e}^2}{\Omega_e^2} + \frac{\omega_{p,e}^2}{(\omega_R - \Omega_p)\Omega_e} \tag{38}$$

in a single component electron-proton plasma ($n_i = 0$). In Eq. (38) $\omega_{p,e}$ and Ω_p denote the electron plasma frequency and the non-relativistic proton gyrofrequency, respectively.

The n possible plasma modes are given by the solution of the equation $\Re \Lambda_t = 0$ which for subluminal phase speeds yields

$$\omega_R^2 + \frac{V_A^2 k^2}{\Omega_p} \omega_R - V_A^2 k^2 = 0 \tag{39}$$

where we introduce the Alfvén velocity $V_A = (-\Omega_e \Omega_p / \omega_{p,e}^2)^{1/2} c = B_0 (4\pi m_p n_b)^{-1/2}$. The solutions of Eq. (39) can be written as

$$k^2 = \frac{\Omega_p \omega_R^2}{V_A^2 (\Omega_p - \omega_R)} \tag{40}$$

and

$$\omega_{R1,2} = \frac{V_A^2 k^2}{2\Omega_p} \left[\pm \sqrt{1 + \frac{4\Omega_p^2}{V_A^2 k^2}} - 1 \right], \tag{41}$$

which include

(1) forward ($\omega/k > 0$) and backward ($\omega/k < 0$) moving, right-handed ($\omega_R < 0$) and left-handed ($\omega_R < 0$) circularly polarized Alfvén waves,

$$\omega_{R1,2} = \pm V_A k \tag{42}$$

at small frequencies $|\omega_R| \ll \Omega_p$, corresponding to small wavenumbers $|k| \ll (2\Omega_p / V_A)$;

(2) forward (for $k < 0$) and backward (for $k > 0$) moving right-handed circularly polarized whistler waves

$$\omega_R = -\frac{V_A^2 k^2}{\Omega_p} = \Omega_e k^2 c^2 / \omega_{p,e}^2 \tag{43}$$

at frequencies between $\Omega_e < \omega_R < -\Omega_p$, corresponding to large wavenumbers $|k| \gg (2\Omega_p / V_A)$; and

(3) forward (for $k > 0$) and backward (for $k < 0$) moving left-handed circularly polarized ion-cyclotron waves

$$\omega_R = \Omega_p \left[1 - \frac{\Omega_p^2}{V_A^2 k^2} \right] \tag{44}$$

at large wavenumbers $|k| \gg (2\Omega_p / V_A)$.

Neglecting spatial dependencies, the time-dependent behaviour of the intensities of the excited waves of type n is independent of the polarisation state and given by

$$\frac{\partial I_n}{\partial t} = 2\psi_n I_n, \tag{45}$$

where the growth rate ψ_n is

$$\psi_n \simeq \frac{\pi \mathrm{sgn}(k)}{2\Gamma} [\omega_R^2 \frac{\partial \Re \Lambda_t}{\partial \omega_R}]^{-1} \sum_i \omega_{p,i}^2 H[|k| - R_i^{-1}](1 - \mu_i^2) \frac{\partial F_i}{\partial \mu} \delta(\mu - \mu_i) \tag{46}$$

in terms of the normalized phase space distribution function of the incoming interstellar particle beam

$$f_i = \delta(E - \Gamma)F_i(\mu, t)/[2\pi(m_i c)^3 \Gamma(\Gamma^2 - 1)^{1/2}] \tag{47}$$

H denotes the step function, $R_i = \Gamma V/|\Omega_i|$ is the gyroradius of the beam particles, $\omega_{p,i} = (4\pi n_i e_i^2/m_i)^{1/2} = \omega_{p,e}(m_e/m_i)^{1/2}(n_i/n_b)^{1/2}$ the beam plasma frequency, and

$$\mu_i = -\frac{\text{sgn}(q_i)}{R_i k} \tag{48}$$

According to Eq. (38)

$$\omega_R^2 \frac{\partial \Re \Lambda_t}{\partial \omega_R} = \frac{2c^2 k^2}{\omega_R} + \frac{\Omega_p c^2 \omega_R^2}{V_A^2(\omega_R - \Omega_p)^2} \tag{49}$$

Inserting Eq. (40) we obtain

$$\omega_R^2 \frac{\partial \Re \Lambda_t}{\partial \omega_R} = \frac{c^2 \Omega_p \omega_R}{V_A^2} \frac{2\Omega_p - \omega_R}{(\Omega_p - \omega_R)^2} \tag{50}$$

Especially for forward and backward moving Alfvén waves Eq.(50) becomes

$$[\omega_R^2 \frac{\partial \Re \Lambda_t}{\partial \omega_R}]_A \simeq \frac{2c^2 k^2}{\omega_R} = \pm \frac{2c^2 k}{V_A}, \tag{51}$$

for whistler waves

$$[\omega_R^2 \frac{\partial \Re \Lambda_t}{\partial \omega_R}]_W \simeq \frac{c^2 k^2}{\omega_R} = -\frac{\Omega_p^2 c^2}{V_A^2}, \tag{52}$$

and for ion-cyclotron waves

$$[\omega_R^2 \frac{\partial \Re \Lambda_t}{\partial \omega_R}]_{IC} \simeq \frac{V_A^2 c^2 k^4}{\Omega_p^3} \tag{53}$$

Consequently, for all three $(m = A,W,IC)$ modes we find from Eq. (46) for the growth rates of forward $(+)$ and backward $(-)$ moving waves

$$\psi_\pm = \pm \psi_m \tag{54}$$

where

$$\psi_m = \frac{\pi V_A^2}{2\Omega_p c^2 \Gamma} \frac{(\Omega_p - \omega_R)^2}{|\omega_R(2\Omega_p - \omega_R)|} \sum_i \omega_{p,i}^2 H[|k| - R_i^{-1}](1 - \mu_i^2)\frac{\partial F_i}{\partial \mu}\delta(\mu - \mu_i) \tag{55}$$

In particular,

$$\psi_A = \frac{\pi V_A}{4|k|c^2\Gamma} \sum_i \omega_{p,i}^2 H[|k| - R_i^{-1}](1 - \mu_i^2)\frac{\partial F_i}{\partial \mu}\delta(\mu - \mu_i), \tag{56}$$

$$\psi_W = \frac{\pi V_A^2}{2\Omega_p c^2 \Gamma} \sum_i \omega_{p,i}^2 H[|k| - R_i^{-1}](1 - \mu_i^2)\frac{\partial F_i}{\partial \mu}\delta(\mu - \mu_i), \tag{57}$$

and

$$\psi_{IC} = \frac{\pi \Omega_p^3}{2 V_A^2 c^2 \Gamma k^4} \sum_i \omega_{p,i}^2 H[|k| - R_i^{-1}](1 - \mu_i^2)\frac{\partial F_i}{\partial \mu}\delta(\mu - \mu_i), \tag{58}$$

respectively.

The two equations (45) then in each case yield the integrals

$$I_+(t)I_-(t) = I_+(t = 0)I_-(t = 0) \tag{59}$$

and

$$[I_+(t) - I_-(t)] - [I_+(t = 0) - I_-(t = 0)] = Z_m(k) \tag{60}$$

where

$$Z_m(k) = \frac{\pi V_A^2}{\Omega_p c^2 \Gamma} \frac{(\Omega_p - \omega_R)^2}{|\omega_R(2\Omega_p - \omega_R)|} \sum_i \omega_{p,i}^2 H[|k| - R_i^{-1}] \sum_\pm \int_0^t dt'(1 - \mu_i^2)$$

$$\frac{\partial F_i}{\partial \mu}\delta(\mu - \mu_i)I_\pm(k, t'), \tag{61}$$

The general solutions of Eqs. (59) and (60) at time t are

$$I_+(t) = \sqrt{Y + \frac{1}{4}(Z_m + I_+(0) - I_-(0))^2} + 0.5\ (Z_m + I_+(0) - I_-(0)) \tag{62}$$

and

$$I_-(t) = \sqrt{Y + \frac{1}{4}(Z_m + I_+(0) - I_-(0))^2} - 0.5\ (Z_m + I_+(0) - I_-(0)) \tag{63}$$

where

$$Y \equiv I_+(0)I_-(0) \tag{64}$$

For weak initial turbulence $I(k, 0) \ll |Z_m(k)|$ with a vanishing cross-helicity $I_+(k, 0) = I_-(k, 0) = I(k, 0)$ we obtain for Eqs. (62) − (63) approximately

$$I_\pm(k, t) \simeq \frac{1}{2}[|Z_m| \pm Z_m] + \frac{2I^2(k, 0)}{|Z_m|} \tag{65}$$

To proceed we have to evaluate $Z_m(k)$ for the three plasma modes (Alfvén waves, whistler waves and ion-cyclotron waves).

The influence of these excited waves on the beam particles is described by the quasilinear Fokker–Planck equation for the resonant wave-particle interaction.

Because the index of refraction for all waves is large compared to unity, the Lorentz force associated with the magnetic field of the waves is much stronger than the Lorentz force associated with the electric field, so that on the shortest time scale these waves scatter the particles in pitch angle μ but conserve their energy, i.e. the waves isotropise the beam particles. The Fokker–Planck equation for the phase space density then reads

$$\frac{\partial F_i}{\partial t} = \frac{\partial}{\partial \mu}\left[D_{\mu\mu}\frac{\partial F_i}{\partial \mu}\right], \tag{66}$$

where the pitch angle Fokker–Planck coefficient is determined by the two wave intensities

$$D_{\mu\mu} = \sum_{n=\pm}\frac{\pi\,\Omega_i^2\,(1-\mu^2)}{2\,B_0^2\Gamma^2}\int_{-\infty}^{\infty}dk\,I_n(k,t)\delta(\omega_R - kV\mu - \frac{\Omega_i}{\Gamma})$$

$$\simeq \sum_{n=\pm}\frac{\pi\,\Omega_i^2\,(1-\mu^2)}{2\,B_0^2\Gamma^2}\int_{-\infty}^{\infty}dk\,I_n(k,t)\delta(kV\mu + \frac{\Omega_i}{\Gamma}) \tag{67}$$

where we again use the limit $N \gg 1$. Integrating Eq. (66) over pitch angle and time and using Eq. (48) we find

$$\int_{-1}^{\mu}d\mu'\,[F_i(\mu',t) - F_i(\mu',t=0)] = \frac{\pi\Omega_i^2}{2B_0^2\Gamma_2}\sum_{\pm}(1-\mu^2)\int_{-\infty}^{\infty}dk\int_0^t dt'\,I_\pm(k,t')$$

$$\frac{\partial F_i}{\partial \mu}\delta(kV\mu + \frac{\Omega_i}{\Gamma}) = \frac{\pi\Omega_i^2}{2B_0^2V|\mu|}(1-\mu^2)\sum_{\pm}\int_0^t dt'\,I_\pm(\frac{\mu_i k}{\mu},t')\frac{\partial F_i}{\partial \mu} \tag{68}$$

Evaluating Eq. (68) at $\mu = \mu_i$ gives

$$\sum_{\pm}\int_0^t dt'\,I_\pm(k,t')\frac{\partial F_i}{\partial \mu}\delta(\mu - \mu_i) =$$

$$\frac{2B_0^2V\Gamma^2}{\pi\Omega_i^2}|\mu_i|\int_{-1}^{\mu_i}d\mu'\,[F_i(\mu',t) - F_i(\mu',t=0)] \tag{69}$$

which can be inserted into Eq. (61) to yield

$$Z_m(k) = \frac{2V_A^2B_0^2}{V\Omega_p c^2\Gamma}\frac{(\Omega_p - \omega_R)^2}{|\omega_R(2\Omega_p - \omega_R)|}\sum_{i=e,p}H[|k| - R_i^{-1}]\omega_{p,i}^2 R_i^2|\mu_i|$$

$$\int_{-1}^{\mu_i}d\mu'\,[F_i(\mu',t) - F_i(\mu',t=0)], \tag{70}$$

7.2 Self-excited Magnetohydrodynamic Turbulence

The solutions (62) – (65) relate the wave intensities and particle distribution functions F_i at any time t to their starting values at time $t = 0$. At the start of the wave and particle evolution ($t = 0$) there is the mono-energetic beam distribution (37), i.e. in terms of the normalised distribution

$$F_{p,e}(\mu, t = 0) = \delta(\mu + 1) \tag{71}$$

The final state of the isotropisation phase is reached at time t_f when both growth rate and temporal derivative of the distribution disappear, i.e. when $\partial F_i / \partial \mu = 0$. Consequently

$$F_{p,e}(\mu, t = t_f) = \frac{1}{2} \tag{72}$$

At this time the magnetohydrodynamic waves have completely isotropised the beam distribution.

Inserting Eqs. (71) and (72) in Eq. (70) allows us to calculate the final state of the self-excited magnetohydrodynamic turbulence spectra by integrating over μ' to obtain

$$\int_{-1}^{\mu_i} d\mu' [F_i(\mu', t_f) - F_i(\mu', t = 0)] = -\frac{1}{2}[1 - \mu_i] \tag{73}$$

so that

$$Z_m(k) = -\frac{V_A^2 B_0^2}{V \Omega_p c^2 \Gamma |k|} \frac{(\Omega_p - \omega_R)^2}{|\omega_R(2\Omega_p - \omega_R)|}$$

$$\sum_{i=e,p} H[|k| - R_i^{-1}] \omega_{p,i}^2 R_i [1 + \frac{\mathrm{sgn}(q_i)}{R_i k}] \tag{74}$$

Explicitly, for the three wave types Eq. (74) reads

$$Z_A(k) = -\frac{V_A B_0^2}{2V k^2 c^2 \Gamma} \sum_{i=e,p} H[|k| - R_i^{-1}] \omega_{p,i}^2 R_i [1 + \frac{\mathrm{sgn}(q_i)}{R_i k}] \tag{75}$$

$$Z_W(k) = -\frac{V_A^2 B_0^2}{V |k| \Omega_p c^2 \Gamma} \sum_{i=e,p} H[|k| - R_i^{-1}] \omega_{p,i}^2 R_i [1 + \frac{\mathrm{sgn}(q_i)}{R_i k}] \tag{76}$$

and

$$Z_{IC}(k) = -\frac{\Omega_p^3 B_0^2}{V V_A^2 c^2 |k|^5 \Gamma} \sum_{i=e,p} H[|k| - R_i^{-1}] \omega_{p,i}^2 R_i [1 + \frac{\mathrm{sgn}(q_i)}{R_i k}] \tag{77}$$

In all three cases $Z_m(k)$ is negative so that Eq. (65) reduces to

$$I_+(k, t_f) \simeq \frac{I^2(k, 0)}{|Z(k)|} \tag{78}$$

and

$$I_-(k, t_f) \simeq |Z(k)| \tag{79}$$

i.e. the beam generates mainly backward moving magnetohydodynamic waves. We obtain

$$I_{-,m}(k, t_f) = \frac{V_A^2 B_0^2}{V \Omega_p c^2 \Gamma |k|} \frac{(\Omega_p - \omega_R(k))^2}{|\omega_R(k)(2\Omega_p - \omega_R(k))|}$$

$$\sum_{i=e,p} H[|k| - R_i^{-1}] \omega_{p,i}^2 R_i |1 + \frac{\text{sgn}(q_i)}{R_i k}|, \tag{80}$$

Using

$$\omega_{p,i=e}^2 = \frac{n_i}{n_b} \omega_{p,e}^2, \quad \omega_{p,i=p}^2 = \frac{m_e n_i}{m_p n_b} \omega_{p,e}^2, \quad R_e = \frac{m_e}{m_p} R_p$$

Eq. (80) becomes

$$I_{-,m}(k, t_f) = \frac{V_A^2 B_0^2 \omega_{p,e}^2 R_p}{V \Omega_p c^2 |k|} \frac{m_e}{m_p} \frac{n_i}{n_b \Gamma} \frac{(\Omega_p - \omega_R(k))^2}{|\omega_R(k)(2\Omega_p - \omega_R(k))|}$$

$$\left[H[|k| - R_p^{-1}]|1 + \frac{1}{R_p k}| + H[|k| - \frac{m_p}{m_e} R_p^{-1}]|1 - \frac{m_p}{m_e R_p k}| \right] \tag{81}$$

In particular, for Alfvén waves

$$I_{-,A}(k, t_f) = \frac{B_0^2 R_p n_i}{n_b \Gamma} \frac{V_A}{V} \frac{m_e}{m_p} \frac{\omega_{p,e}^2}{2c^2 k^2}$$

$$\left[H[|k| - R_p^{-1}]|1 + \frac{1}{R_p k}| + H[|k| - \frac{m_p}{m_e} R_p^{-1}]|1 - \frac{m_p}{m_e R_p k}| \right], \tag{82}$$

for whistler waves

$$I_{-,W}(k, t_f) \simeq \frac{B_0^2 V_A^2 m_e n_i}{c^2 m_p n_b} \frac{\omega_{p,e}^2}{\Omega_p |k|} \tag{83}$$

and for ion-cyclotron waves

$$I_{-,IC}(k, t_f) \simeq \frac{B_0^2 m_e n_i}{V_A^2 c^2 m_p n_b} \frac{\omega_{p,e}^2 \Omega_p^2}{|k|^5} \tag{84}$$

7.3 Pick-Up Conditions

We can estimate the isotropisation length of the beam particles by using the fully-developed turbulence spectra for calculating the pitch angle Fokker–Planck coefficient. For ease of exposition we assume that the initial turbulence spectrum has the form $I(k,0) = I_0 k^{-2}$. As a consequence of pitch angle scattering the beam particles adjust to the isotropic distribution on a length scale given by the scattering length

$$\lambda = \frac{3v}{8} \int_{-1}^{1} d\mu \frac{(1 - \mu^2)^2}{D_{\mu\mu}(\mu)} \tag{85}$$

Eq. (85) is valid for scattering lengths λ larger than the gyroradius of the particles, $R_p = 3 \cdot 10^6 \Gamma_{100}/B_0(G)$ m and $R_e = 1.7 \cdot 10^3 \Gamma_{100}/B_0(G)$ m for protons and electrons, respectively. After straightforward integration and inserting typical parameter values for active galactic nuclei outflows we obtain in case of beam protons for the scattering length and the isotropisation time scale in the outflow plasma

$$\lambda_p \simeq 10^9 \frac{n_{b,8}^{1/2}}{\Gamma_{100}\, n_i^*} \text{ m} \tag{86}$$

and

$$t_{f,p} = \lambda/c \simeq 3.5 \frac{n_{b,8}^{1/2}}{\Gamma_{100}\, n_i^*} \text{s} \tag{87}$$

If the thickness d of the outflow region is larger than the scattering length, indeed an isotropic distribution of the inflowing interstellar protons and electrons with Lorentzfactor $\langle \Gamma \rangle = \Gamma(1 - \beta_A \beta) \simeq \Gamma$ in the blast wave frame is efficiently generated.

The diffusive escape of the relativistic particles from the outflow also is determined by the scattering length (86) through the spatial diffusion coefficient $\kappa = v\lambda/3$. The escape time scale is given by

$$T_E = \frac{d^2}{\kappa} = \frac{3d^2}{\lambda v} \simeq 10^6 \frac{d_{13}^2 \Gamma_{100}^{3/2} n_i^*}{(n_{b,8}^*)^{1/2}} \text{s} \tag{88}$$

The outflow region is a *thick target* for the pick-up protons and electrons if this escape time scale is longer than the relevant comoving radiation loss time scales.

In the blast wave frame the external density $n_i = \Gamma n_i^*$ and pick-up occurs at a rate

$$\dot{N}(\gamma) = \pi R^2 c n_i^* \sqrt{\Gamma^2 - 1}\, \delta(\gamma - \Gamma) . \tag{89}$$

The pick-up is a source of isotropic, quasi-monoenergetic protons and electrons with Lorentz factor Γ in the blast wave frame.

This shows that a relativistic outflow can pick-up ambient matter via a two-stream instability which provides relativistic particles in the outflow without requiring any acceleration process. By efficiently generating low-frequency magnetohydrodynamic waves, which isotropise the incoming interstellar beam distribution of protons and electrons, directed bulk motion is converted into accelerated relativistic particles within the outflow region, at the expense of decelerating the outflow. This relativistic pick-up model (Pohl and Schlickeiser 2000) and further variants (Pohl et al. 2002; Schuster et al. 2002; Schlickeiser et al. 2002) have been applied successfully to energetic active galactic nuclei and gamma-ray burst sources. Its success is based on the self-consistent calculation of the coupled time-evolution of particle distribution functions and magnetohydrodynamic wave intensities.

7.4 Application to Coronal Mass Ejections

This conversion of directed bulk motion into accelerated energetic particles also applies to non-relativistic outflows as solar coronal mass ejections (CMEs). CMEs are a major form of eruptive phenomena in the solar corona involving masses of the order $\simeq 10^{15} - 10^{16}$ g and kinetic energies of $\simeq 10^{23} - 10^{25}$ J. The leading edge speeds of CMEs within about 5 solar radii of the surface range from less than 50 to greater 2000 km s^{-1} (Howard et al. 1985; Sheeley, Hakala and Wang 2000). Recently, in a hybrid plasma simulation Wang et al. (2001) have investigated the time evolution of ion species in CMEs moving along open coronal magnetic field lines with super-Alfvénic velocities ($V = 7V_A$) with respect to the ambient solar wind, demonstrating that the solar wind ions are heated and accelerated via kinetic wave-particle interaction processes of similar type as discussed above. Without driving a shock wave CMEs would thus accelerate ions to kinetic energies of $E_{\mathrm{kin}} = 20(V/2000 \text{ km/s})^2$ keV. It will be most interesting to apply the analytical formalism developed above to this physical situation.

8 Kinetic Equation of Plasma Waves

In our previous discussion of the three particle energization mechanisms (stochastic acceleration, diffusive shock acceleration, conversion of bulk motion) it has proven essential to model adequately the dynamical evolution of the power spectra of the partially turbulent electromagnetic fields. A complete decription has to include the different coupling and interaction processes of cosmic plasma waves discussed in Sect. 2. Besides the various wave growth and wave damping mechanisms another relevant process that sets in at appreciable wave intensities is the cascading of waves. As a consequence of wave steepening spectral wave energy cascades to higher frequencies and wavenumbers (e.g. Marsch 1991).

The idea to describe the evolution of turbulence by a diffusion of energy in wavenumber space was pioneered by Leith (1967) in hydrodynamics, and subsequently introduced to magnetohydrodynamics by Zhou and Matthaeus (1990).

This approach provides a simple framework to take into account at least approximately turbulence evolution in space physics applications. Zhou and Matthaeus (1990) present a general transport equation for the wave spectral density in case of isotropic turbulence, which includes terms for spatial convection and propagation, non-linear transfer of energy across the wavenumber spectrum, and a source and sink of wave energy. The kinetic equation for a 3-dimensional spectral density $\bar{W}_i(\mathbf{k})$ of mode i, which denotes the wave energy density per unit volume of wavenumber space of the plasma mode i, is given by the usual conservation equation

$$\frac{\partial \bar{W}_i(\mathbf{k})}{\partial t} = -\frac{\partial}{\partial \mathbf{k}} \cdot \mathbf{F}(\mathbf{k}), \tag{90}$$

where the flux

$$\mathbf{F}(\mathbf{k}) = -D \frac{\partial \bar{W}_i(\mathbf{k})}{\partial \mathbf{k}} \tag{91}$$

is expressed as a diffusive term with the diffusion coefficient in wavenumber space

$$D = k^2/\tau_s(k) \tag{92}$$

and the spectral energy transfer time scale $\tau_s(k)$. For isotropic turbulence one obtains for the associated one-dimensional spectral density $W_i(k) = 4\pi k^2 \bar{W}_i(\mathbf{k})$ the simplified diffusion equation

$$\frac{\partial W_i}{\partial t} = \frac{\partial}{\partial k}\Big[\frac{k^4}{\tau_s(k)}\frac{\partial}{\partial k}(k^{-2}W_i)\Big] + \Gamma_i W_i + S_i(k), \tag{93}$$

where we include a term for the damping or growth of waves and a wave energy injection and/or sink term $S_i(k)$. Tsap (2000) has discussed the relevant damping processes of fast magnetosonic waves.

The spectral energy transfer time scale (or the wavenumber diffusion coefficient (92)) depends upon the cascade phenomenology. In the Kolmogorov treatment the spectral energy transfer time at a particular wavelength λ is the eddy turnover time $\lambda/\delta v$, where δv is the velocity fluctuation of the wave. In the Kraichnan treatment the transfer time is longer by a factor $V_A/\delta v$. Both phenomenologies are further discussed in Zhou and Matthaeus (1990) and yield

$$\tau_s(k) \simeq \frac{1}{V_A k^{3/2}} \begin{cases} \sqrt{\frac{2U_B}{W_i}} & \text{(Kolmogorov)} \\ \frac{2U_B}{k^{1/2}W_i} & \text{(Kraichnan)} \end{cases} \tag{94}$$

where $U_B = B_0^2/8\pi$ denotes the energy density of the ordered magnetic field. Substituting these transfer time scales into Eq. (93), and assuming a steady state with no damping, we obtain $W_i = W_0 k^{-s}$, where $s = 5/3$ for the Kolmogorov case and $s = 3/2$ for the Kraichnan phenomenology. The diffusion equation (93) in either case is non-linear.

It is obvious that for small enough turbulence intensities the spectral energy transfer times τ_s are much longer than the wave growth or wave damping time scales, so that the wavenumber diffusion term in Eq. (93) can be neglected with respect to the term representing wave damping and/or wave growth. In this limit Eq. (93) reduces to the simplified Eq. (45) used as starting point for the wave evolution in Sect. 7.

Besides the noted generation of turbulence by kinetic instabilities (see Sect. 7), very often wave cascading from low to high wavenumbers is an important way of producing broadband wave spectra. One possibility, discussed e.g. in the context of solar flares (Miller and Roberts 1995), is that long-wavelength turbulence results from the rearrangement of large-scale magnetic fields and/or a shear flow instability, so that it is reasonable to assume the deposition of wave energy peaked at long wavelength, probably comparable to the physical size of the system, as the primary energy release. Cascading as described by the non-linear diffusion term in Eq. (93) will then transfer this spectral energy to higher wavenumbers, where the waves will be able to resonate with progressively lower energy charged particles, until they eventually interact with the charged particles in the tail of the backgroung thermal distribution, which is described by the various wave damping rates.

9 Summary and Conclusions

We have reviewed three acceleration processes of charged particles in cosmic plasmas: resonant stochastic acceleration, diffusive shock acceleration and conversion of bulk motion to individual charged particle energies by relativistic pick-up. All three processes rely on the interactions of charged particles with partially random electromagnetic fields which are theoretically described within quasilinear theory. We demonstrate that in all three cases the modeling of the dynamics of the electromagnetic plasma turbulence is most crucial, and we discuss the relevant wave-particle and wave-wave interaction processes that control the turbulence dynamics.

In case of resonant stochastic acceleration it is essential to discard the magnetostatic approximation of the plasma wave turbulence and to consider finite wave propagation effects. It then appears that transit-time damping of fast magnetosonic waves provides the dominant contribution to the acceleration of charged particles below the Hillas limit which is reached when the gyroradius of the particle equals the longest wavenumber of the plasma waves in the system. We pointed out that acceleration to even higher energies is possible by gyroresonant contributions from higher harmonics.

In case of diffusive shock acceleration the downstream electromagnetic field properties can be calculated from the specified upstream electromagnetic fields from the Rankine-Hugoniot shock relations modified by the finite plasma wave normal momentum and energy flux. Again it is essential to keep track of the different propagation speeds and the different transmission coefficients through the quasi-parallel shock of forward and backward moving plasma waves. This results in a difference between the gas compression ratio of the shock wave and the scattering center compression ratio which accounts for the generation of spectral indices flatter than 2 for the momentum power law distribution function of accelerated particles.

In case of the relativistic pick-up process we show that a directed relativistic outflow can pick-up ambient matter via a two-stream instability which provides relativistic particles in the outflow without requiring any acceleration process. By efficiently generating low-frequency magnetohydrodynamic waves, which isotropise the incoming interstellar beam distribution of protons and electrons, directed bulk motion is converted into accelerated relativistic particles within the outflow region, at the expense of decelerating the outflow. Here it is essential to calculate self-consistently from the coupled wave and particle kinetic equations the time-evolution of the magnetohydrodynamic wave intensities.

I thank the organisers for inviting me to this workshop. I am very grateful to Dr. K.-L. Klein and the anonymous referee for helpful and constructive comments to this manuscript.

10 Appendix A: Derivation of the Diffusion–Convection Transport Equation

Starting point is the Fokker–Planck equation for the gyrotropic particle phase space density $f(z, p, \mu, t)$ where z denotes the spatial variable along the ordered uniform magnetic field $\boldsymbol{B_0} = B_0(e)_z$ and $\mu = p_\parallel/p$. The Fokker–Planck equation reads

$$\frac{\partial f}{\partial t} + v\mu\frac{\partial f}{\partial z} - S_0(z, p, t) = \frac{\partial}{\partial \mu}\left[D_{\mu\mu}\frac{\partial f}{\partial \mu} + D_{\mu p}\frac{\partial f}{\partial p}\right] +$$

$$\frac{1}{p^2}\frac{\partial}{\partial p}p^2\left[D_{\mu p}\frac{\partial f}{\partial \mu} + D_{pp}\frac{\partial f}{\partial p}\right] \tag{95}$$

where the Fokker–Planck coefficients $D_{\mu\mu}, D_{\mu p}, D_{pp}$ are given in Eqs. (7). We restrict our analysis to isotropic source terms $S(z, p, t)$.

We now make the basic assumption of diffusion theory that the particle distribution function $f(z, p, \mu, t)$ under the action of low-frequency magnetohydrodynamic waves adjusts very quickly to a quasi-equilibrium through pitch-angle diffusion which is close to the isotropic equilibrium distribution. Defining the isotropic part of the phase space density $F(z, p, t)$ as the μ-averaged phase space density

$$F(z, p, t) \equiv \frac{1}{2}\int_{-1}^{1} d\mu\ f(z, p, \mu, t) \tag{96}$$

we follow the analysis of Jokipii (1971) and Hasselmann and Wibberenz (1968) to split the total density f into the isotropic part F and an anisotropic part g,

$$f(z, p, \mu, t) = F(z, p, t) + g(z, p, \mu, t) \tag{97}$$

where because of Eq. (96)

$$\int_{-1}^{1} d\mu \, g(z,p,\mu,t) = 0 \tag{98}$$

Substituting Eq. (97) into Eq. (95) yields

$$\frac{\partial F}{\partial t} + \frac{\partial g}{\partial t} + v\mu\frac{\partial F}{\partial z} + v\mu\frac{\partial g}{\partial z} - S_0(z,p,t) = \frac{\partial}{\partial \mu}\left[D_{\mu\mu}\frac{\partial g}{\partial \mu} + D_{\mu p}\frac{\partial F}{\partial p} + D_{\mu p}\frac{\partial g}{\partial p}\right] +$$

$$\frac{1}{p^2}\frac{\partial}{\partial p}p^2\left[D_{\mu p}\frac{\partial g}{\partial \mu} + D_{pp}\frac{\partial F}{\partial p} + D_{pp}\frac{\partial g}{\partial p}\right] \tag{99}$$

Averaging Eq. (99) over μ gives

$$\frac{\partial F}{\partial t} + \frac{v}{2}\frac{\partial}{\partial z}\int_{-1}^{1} d\mu \, \mu g - S_0(z,p,t) = \frac{1}{2}\left[D_{\mu\mu}\frac{\partial g}{\partial \mu} + D_{\mu p}\frac{\partial F}{\partial p} + D_{\mu p}\frac{\partial g}{\partial p}\right]_{\mu=-1}^{\mu=1} +$$

$$\frac{1}{2p^2}\frac{\partial}{\partial p}p^2\left[\int_{-1}^{1} d\mu D_{\mu p}\frac{\partial g}{\partial \mu} + \int_{-1}^{1} d\mu D_{pp}\frac{\partial F}{\partial p} + \int_{-1}^{1} d\mu D_{pp}\frac{\partial g}{\partial p}\right] \tag{100}$$

The first paranthesis on the right-hand side of Eq. (100) vanishes because $D_{\mu\mu}, D_{\mu p}$ and D_{pp} all include the factor $(1-\mu^2)$ which becomes zero for $\mu \to \pm 1$. We then obtain

$$\frac{\partial F}{\partial t} + \frac{v}{2}\frac{\partial}{\partial z}\int_{-1}^{1} d\mu \, \mu g - S_0(z,p,t) = \frac{1}{2p^2}\frac{\partial}{\partial p}p^2\left[\int_{-1}^{1} d\mu D_{\mu p}\frac{\partial g}{\partial \mu}\right.$$

$$\left. + \int_{-1}^{1} d\mu D_{pp}\frac{\partial F}{\partial p} + \int_{-1}^{1} d\mu D_{pp}\frac{\partial g}{\partial p}\right] \tag{101}$$

Subtracting Eq. (101) from Eq. (99) and transposing results in

$$v\mu\frac{\partial F}{\partial z} - \frac{\partial}{\partial \mu}\left[D_{\mu\mu}\frac{\partial g}{\partial \mu} + D_{\mu p}\frac{\partial F}{\partial p} + D_{\mu p}\frac{\partial g}{\partial p}\right] =$$

$$-\frac{\partial g}{\partial t} - v\mu\frac{\partial g}{\partial z} + \frac{v}{2}\frac{\partial}{\partial z}\int_{-1}^{1} d\mu \, \mu g + \frac{1}{p^2}\frac{\partial}{\partial p}p^2\left[\left(D_{\mu p}\frac{\partial g}{\partial \mu} - \frac{1}{2}\int_{-1}^{1} d\mu D_{\mu p}\frac{\partial g}{\partial \mu}\right)\right.$$

$$\left. + \left(D_{pp} - \frac{1}{2}\int_{-1}^{1} d\mu D_{pp}\right)\frac{\partial F}{\partial p} + \left(D_{pp}\frac{\partial g}{\partial p} - \frac{1}{2}\int_{-1}^{1} d\mu D_{pp}\frac{\partial g}{\partial p}\right)\right] \tag{102}$$

which together with Eq. (101) is still exact.

The diffusion approximation applies if the isotropic particle density is slowly evolving, i.e.

$$\frac{\partial F}{\partial t} = \mathcal{O}(\frac{F}{T}), \quad \frac{\partial F}{\partial z} = \mathcal{O}(\frac{F}{L}) \tag{103}$$

with typical length scales $L \gg \lambda$ and time scales $T \gg \tau$ much larger than the mean free path $\lambda = v\tau$ and the pitch angle scattering relaxation time $\tau \simeq \mathcal{O}(1/D_{\mu\mu})$, respectively. In this case the particles have enough time to adjust locally to a near-equilibrium, so that the anisotropy is small i.e. $g \ll F$. If we then regard g as of order τ, when F is of order 1, and recall that $D_{\mu p}$ and D_{pp} are of order $\epsilon = V_{\rm ph}/v = V_A/v \ll 1$ and ϵ^2, respectively, smaller than $D_{\mu\mu} = \mathcal{O}(1/\tau)$, we may characterize the differential operators in Eq. (102) by different time scales. The first three terms on the left-hand side of Eq. (102) are of order 1, whereas the fourth term can be neglected because it is of order $(\epsilon\tau)$ smaller. The first three terms on the right-hand side of Eq. (102) are of order τT^{-1}, $\tau v T^{-1} = \lambda T^{-1}$, and λT^{-1}, respectively, and therefore small as compared to the left-hand side of Eq. (102). The last six terms are at least second-order as $\epsilon\tau$, $\epsilon\tau$, ϵ^2, ϵ^2, $\epsilon^2\tau$, $\epsilon^2\tau$ in small quantities and therefore negligible, too. So to lowest order we approximate Eq. (102) by

$$\frac{\partial}{\partial\mu}\left[D_{\mu\mu}\frac{\partial g}{\partial\mu} + D_{\mu p}\frac{\partial F}{\partial p}\right] \simeq v\mu\frac{\partial F}{\partial z} \tag{104}$$

Integrating over μ we obtain

$$D_{\mu\mu}\frac{\partial g}{\partial\mu} + D_{\mu p}\frac{\partial F}{\partial p} = c_1 + \frac{v\mu^2}{2}\frac{\partial F}{\partial z} \tag{105}$$

where the integration constant c_1 is determined from the requirement that the left-hand side of Eq. (104) vanishes for $\mu = \pm 1$, yielding

$$c_1 = -\frac{v}{2}\frac{\partial F}{\partial z}$$

so that Eq. (105) becomes

$$\frac{\partial g}{\partial\mu} = -\frac{(1-\mu^2)v}{2D_{\mu\mu}}\frac{\partial F}{\partial z} - \frac{D_{\mu p}}{D_{\mu\mu}}\frac{\partial F}{\partial p} \tag{106}$$

Integrating Eq. (106) over μ results in

$$g(z,p,\mu,t) = c_2 - \frac{v}{2}\frac{\partial F}{\partial z}\int_{-1}^{\mu}dx\frac{1-x^2}{D_{\mu\mu}(x)} - \frac{\partial F}{\partial p}\int_{-1}^{\mu}dx\frac{D_{\mu p}(x)}{D_{\mu\mu}(x)} \tag{107}$$

The integration constant c_2 is determined by the condition (98) yielding

$$c_2 = \frac{v}{4}\frac{\partial F}{\partial z}\int_{-1}^{1}d\mu\frac{(1-\mu)(1-\mu^2)}{D_{\mu\mu}(\mu)} + \frac{1}{2}\frac{\partial F}{\partial p}\int_{-1}^{1}d\mu\frac{(1-\mu)D_{\mu p}(\mu)}{D_{\mu\mu}(\mu)} \tag{108}$$

The anisotropy (107) consists of two components. The first is related to pitch angle scattering and the spatial gradient of F. The second contribution to the anisotropy is from the momentum gradient of F and is related to the Compton-Getting effect (Compton and Getting 1935).

Neglecting the last term on the right-hand side of Eq. (101) because $g \ll F$ gives

$$\frac{\partial F}{\partial t} = -\frac{v}{2}\frac{\partial}{\partial z}\int_{-1}^{1} d\mu\mu g + S_0(z,p,t)+$$

$$\frac{1}{2p^2}\frac{\partial}{\partial p}p^2\left[\int_{-1}^{1} d\mu D_{\mu p}\frac{\partial g}{\partial \mu} + \int_{-1}^{1} d\mu D_{pp}\frac{\partial F}{\partial p}\right] \tag{109}$$

From Eq. (106) we obtain

$$\int_{-1}^{1} d\mu D_{\mu p}\frac{\partial g}{\partial \mu} = -\frac{v}{2}\frac{\partial F}{\partial z}\int_{-1}^{1} d\mu\frac{D_{\mu p}}{D_{\mu\mu}} - \frac{\partial F}{\partial p}\int_{-1}^{1} d\mu\frac{D_{\mu p}^2}{D_{\mu\mu}} \tag{110}$$

while Eq. (107) yields

$$\int_{-1}^{1} d\mu\mu g = -\frac{v}{2}\frac{\partial F}{\partial z}\int_{-1}^{1} d\mu\mu\int_{-1}^{\mu} dx\frac{1-x^2}{D_{\mu\mu}(x)} - \frac{\partial F}{\partial p}\int_{-1}^{1} d\mu\mu\int_{-1}^{\mu} dx\frac{D_{\mu p}(x)}{D_{\mu\mu}(x)} =$$

$$-\frac{v}{4}\frac{\partial F}{\partial z}\int_{-1}^{1} d\mu\frac{(1-\mu^2)^2}{D_{\mu\mu}(\mu)} - \frac{1}{2}\frac{\partial F}{\partial p}\int_{-1}^{1} d\mu\frac{(1-\mu^2)D_{\mu p}(\mu)}{D_{\mu\mu}(\mu)} \tag{111}$$

where we partially integrated the right-hand side.

Inserting Eqs. (110)-(111) into Eq. (109) we obtain

$$\frac{\partial F}{\partial t} - S_0(z,p,t) = \frac{\partial}{\partial z}(\kappa\frac{\partial F}{\partial z}) + \frac{1}{p^2}\frac{\partial}{\partial p}(p^2 A\frac{\partial F}{\partial p})$$

$$+\frac{v}{4}\frac{\partial}{\partial z}(a_1\frac{\partial F}{\partial p}) - \frac{1}{4p^2}\frac{\partial}{\partial p}p^2 va_1\frac{\partial F}{\partial z} \tag{112}$$

where

$$\kappa = \frac{v^2}{8}\int_{-1}^{1} d\mu\frac{(1-\mu^2)^2}{D_{\mu\mu}(\mu)} \tag{113}$$

$$A = \frac{1}{2}\int_{-1}^{1} d\mu\left[D_{pp}(\mu) - \frac{D_{\mu p}^2(\mu)}{D_{\mu\mu}(\mu)}\right] \tag{114}$$

and

$$a_1 = \int_{-1}^{1} d\mu\frac{(1-\mu^2)D_{\mu p}(\mu)}{D_{\mu\mu}(\mu)} \tag{115}$$

Noting that

$$\frac{v}{4}\frac{\partial}{\partial z}(a_1\frac{\partial F}{\partial p}) - \frac{1}{4p^2}\frac{\partial}{\partial p}p^2 va_1\frac{\partial F}{\partial z} = \frac{v}{4}\frac{\partial a_1}{\partial z}\frac{\partial F}{\partial p} - \frac{1}{4p^2}\frac{\partial(p^2 a_1 v)}{\partial p}\frac{\partial F}{\partial z} \quad (116)$$

yields for Eq. (112) the diffusion-convection equation

$$\frac{\partial F}{\partial t} - S_0(z,p,t) = \frac{\partial}{\partial z}(\kappa\frac{\partial F}{\partial z}) + \frac{1}{p^2}\frac{\partial}{\partial p}(p^2 A\frac{\partial F}{\partial p})$$

$$+\frac{v}{4}\frac{\partial F}{\partial p}\frac{\partial a_1}{\partial p} - \frac{1}{4p^2}\frac{\partial(p^2 va_1)}{\partial p}\frac{\partial F}{\partial z} \quad (117)$$

which agrees with Eq. (5) if the Fokker–Planck coefficients are calculated in the rest frame of the wave-carrying background medium.

References

1. U. Achatz, J. Steinacker, R. Schlickeiser, Astr. Astrophys. 250, 266 (1991)
2. A. R. Bell, Monthly Not. Roy. Astr. Soc. 182, 147 (1978)
3. T. J. M. Boyd, J. J. Sanderson, Plasma Dynamics, Nelson and Son, London (1969)
4. A. H. Compton, I. A. Getting, Physical Rev. 47, 817 (1935)
5. W. Dröge, I. Lerche, R. Schlickeiser, Astr. Astrophys. 178, 252 (1987)
6. L. O'C. Drury, Space Science Rev. 36, 57 (1983)
7. J. A. Earl, Astrophys. J. 180, 227 (1973)
8. E. Fermi, Physical Rev. 75, 1169 (1949)
9. E. Fermi, Astrophys. J. 119, 1 (1954)
10. D. E. Hall, P. A. Sturrock, Physics of Fluids 10, 2620 (1967)
11. K. Hasselmann, G. Wibberenz, G., Zeitschrift für Geophysik 34, 353 (1968)
12. A. M. Hillas, Ann. Rev. Astron. Astrophys. 22, 425 (1984)
13. R. A. Howard, N. R. Sheeley, M. J. Koomen, D. J. Michels, J. Geophys. Res. 90, 8173 (1985)
14. J. R. Jokipii, Astrophys. J. 146, 480 (1966)
15. J. R. Jokipii, Rev. Geophys. Space Phys. 9, 27 (1971)
16. J. R. Jokipii, Space Science Rev. 36, 27 (1983)
17. C. F. Kennel, F. Engelmann, Physics of Fluids 9, 2377 (1966)
18. J. G. Kirk, R. Schlickeiser, P. Schneider, Astrophys. J. 328, 269 (1988)
19. J. A. Krommes, in: A. A. Galeev & R. N. Sudan (eds.), Basic Plasma Physics II, North-Holland, Amsterdam, p. 183 (1984)
20. C. E. Leith, Physics of Fluids 10, 1409 (1967)
21. D. Le Quéau, A. Roux, Solar Phys. 111, 19 (1987a)
22. D. Le Quéau, A. Roux, Solar Phys. 111, 59 (1987b)
23. I. Lerche, Physics of Fluids 11, 1720 (1968)
24. I. Lerche, Astr. Astrophys. 85, 141 (1980)
25. I. Lerche, R. Schlickeiser, Astr. Astrophys. 383, 319 (2002)
26. Y. E. Litvinenko, this volume (2002)
27. E. Marsch, Rev. Mod. Astron. 4, 145 (1991)
28. J. K. McKenzie, K. O. Westphal, Planetary and Space Sci. 17, 1029 (1969)
29. J. A. Miller, D. A. Roberts, Astrophys. J. 452, 912 (1995)
30. J. A. Miller, P. J. Cargill, A. G. Emslie, et al., J. Geophys. Res. 102, 14631 (1997)

31. B. T. Park, V. Petrosian, R. A. Schwartz, Astrophys. J. 489, 358 (1997)
32. E. N. Parker, in: G. Setti (ed.), The Physics of Non-Thermal Radio Sources, Reidel, Dordrecht, p. 137 (1976)
33. G. Pelletier, Astr. Astrophys. 350, 705 (1999)
34. G. Pelletier, H. Sol, E. Asséo, Physical Rev. A 38, 2552 (1988)
35. M. Pohl, R. Schlickeiser, Astr. Astrophys. 354, 395 (2000)
36. M. Pohl, I. Lerche, R. Schlickeiser, Astr. Astrophys. 383, 309 (2002)
37. R. Schlickeiser, Astrophys. J. , 336, 243 (1989)
38. R. Schlickeiser, Cosmic Ray Astrophysics, Springer, Berlin (2002)
39. R. Schlickeiser, U. Achatz, J. Plasma Phys. 49, 63 (1993)
40. R. Schlickeiser, J. A. Miller, Astrophys. J. 492, 352 (1998)
41. R. Schlickeiser, J. Steinacker, Solar Phys. 122, 95 (1989)
42. R. Schlickeiser, R. Vainio, M. Böttcher, I. Lerche, M. Pohl, C. Schuster, Astr. Astrophys. , in press (2002) (astro-ph/0207066)
43. M. Scholer, this volume (2002)
44. M. Scholer, J. W. Belcher, Solar Phys. 16, 472 (1971)
45. C. Schuster, M. Pohl, R. Schlickeiser, Astr. Astrophys. 382, 829 (2002)
46. N. R. Sheeley, W.N. Hakala, Y.-M. Wang, J. Geophys. Res. 105, 5081 (2000)
47. J. Steinacker, W. Dröge, R. Schlickeiser, Solar Phys. 115, 313 (1988)
48. D. A. Swanson, Plasma Waves, Academic Press, Boston (1989)
49. R. A. Treumann, N. Dubouloz, R. Pottelette, Adv. Space Res. 18(8), 291 (1996)
50. R. A. Treumann, R. Pottelette, Adv. Space Res. 23, 1705 (1999)
51. Y. T. Tsap, Solar Phys. 194, 131 (2000)
52. R. Vainio, R. Schlickeiser, Astr. Astrophys. 331, 793 (1998)
53. R. Vainio, R. Schlickeiser, Astr. Astrophys. 343, 303 (1999)
54. R. Vainio, R. Schlickeiser, Astr. Astrophys. 378, 309 (2001)
55. A. A. Vedenov, E. P. Velikhov, R. Z. Sagdeev, Nuclear Fusion Suppl. 2, 465 (1962)
56. S. Wang, X. Y. Wang, C. S. Wu, Y. Li, J. K. Chao, T. Yeh, Solar Phys. 202, 385 (2001)
57. Y. Zhou, W. H. Matthaeus, J. Geophys. Res. 95, 14881 (1990)

Part III

Aspects of Current Research

Recent Progress in Understanding
Energy Conversion and Particle Acceleration
in the Solar Corona

Bernhard Kliem[1], Alec MacKinnon[2], Gérard Trottet[3], and Tim Bastian[4]

[1] Astrophysikalisches Institut Potsdam, 14482 Potsdam, Germany
[2] Department of Adult and Continuing Education, University of Glasgow, Glasgow, G3 6LP, UK
[3] LESIA, Observatoire de Paris, Section d'Astrophysique de Meudon, 92195 Meudon-Cedex, France
[4] National Radio Astronomy Observatory, Charlottesville, VA 22903, USA

Abstract. We report on results of the working group sessions at the CESRA 2001 workshop on "Energy Conversion and Particle Acceleration in the Solar Corona" which was focused on radio observations and related modeling. Progress reached in the following areas is summarized: (1) diagnostics of coronal magnetic fields and the morphology of the field in flares and filament eruptions; (2) evidence of magnetic reconnection and MHD turbulence in radio emissions; (3) acceleration site, propagation, and trapping of radio-emitting energetic particles in flares; (4) the sites of particle acceleration in long duration events, as evidenced by the 2000 July 14 ("Bastille Day") flare; (5) radio imaging of CMEs and filament eruptions; (6) the relationship of coronal and interplanetary shock waves to flares, CMEs, and other coronal waves; and (7) the origin of solar energetic particles.

1 Introduction

In this contribution, we present a summary of the main issues raised and the new observations presented in the working group sessions of the present workshop. Two groups were concerned with radio observations and their diagnostic possibilities, although recent progress on the use of multi-wavelength observations and particle data were also emphasized. These observations served to renew discussion of: the mechanisms of the energy release in solar flares and coronal mass ejections (CMEs) (Sect. 2); the magnetic configurations in which these phenomena occur (Sect. 3); signatures of magnetic reconnection and MHD turbulence (Sect. 4); acceleration and transport of electrons in flares (Sect. 5); characteristics of large-scale eruptive events and production of solar energetic particles (SEPs) (Sect. 6). Our report is organized along the main topics of the working group sessions. We endeavor to place results and their discussion into a broader context instead of reporting each individual contribution in a detailed manner. We acknowledge that we were unable to cite all of the many interesting and illuminating presentations made during the course of the meeting but nevertheless hope that we capture important current trends in solar radio astronomy. The remainder of our report is structured as above, with Sect. 7 giving brief Conclusions.

2 Energy Release in the Solar Corona: A Brief Overview

Models of the basic instability in flares may be categorized according to whether the flare is eruptive or confined. CMEs are typically associated with eruptive flares, so eruptive models are thought to apply to both phenomena. We will discuss the "standard model" of eruptive flares and its more recent extensions, known as the "tether cutting model" and the "magnetic breakout model". We will also sketch the "loop-loop interaction model," which is often applied to non-eruptive events, despite recent claims that tether cutting can be applied here as well. Three-dimensional aspects of these models are discussed in greater detail in [1].

The Standard (CSHKP) Flare Model

The model for eruptive flares advanced primarily by Carmichael [2], Sturrock [3], Hirayama [4], and Kopp & Pneuman [5] has become the "standard model" of solar flares. It consists of two main phases: (1) the opening of a closed magnetic configuration, originally supposed to be closely related to the eruption of a filament/prominence , which creates an inverted Y-shaped magnetic configuration with a current sheet extending to greater heights above a closed magnetic configuration, and (2) long-lasting magnetic reconnection in this current sheet leading to the energy release in the main flare phase (e.g., Fig. 3c in [1]). The latter includes the partial reclosing of the configuration by reconnected field lines in the downward reconnection outflow . The released energy is dumped at the magnetic footpoints in the chromosphere by energetic particle precipitation and heat conduction . This results in the formation of flare ribbons and of hot and dense flare loops through chromospheric evaporation ; these loops turn later into cooling postflare loops.

Phase 2 is well-supported by a variety of observations of eruptive flares (e.g., [6, 7, 8, 9, 10, 11, 12]) and by MHD simulations (e.g., [13]). However, some questions remain. Foremost are questions concerning the spatial and temporal scales on which the reconnection occurs (i.e., whether it is stationary (Petschek-like) or highly dynamic and fragmented in space), whether MHD turbulence is excited and fills substantial volumes, and whether the downward reconnection outflow jet indeed forms a standing fast-mode shock upon hitting the newly-formed flare and postflare loops. See Sect. 4 for recent observations pertaining to these questions.

The processes which open magnetic fields in flares and CMEs (phase 1) are poorly understood. The observations indicate that eruptive events originate in highly sheared magnetic flux systems oriented along a line of magnetic polarity inversion (the neutral line) in the photosphere, lying underneath a less sheared, stabilizing magnetic arcade. The erupting sheared core flux does not always contain a filament , which suggests that magnetic, not the thermodynamic effects are fundamental for its loss of balance. Here we briefly describe two current models for the opening of the core flux; some relevant new observations are discussed in Sects. 3 and 6.

The Tether Cutting Model

Moore & Roumeliotis [14] suggested that eruptions result from a catastrophic loss of balance between the upward-directed magnetic pressure force and the downward-directed magnetic tension force *within* a sheared core flux system. They proposed that slow magnetic reconnection at the footpoints of the core flux system replaces short arched field lines by longer ones for which the stabilizing influence of the photospheric anchoring of the footpoints is reduced. Regarding the short arched field lines as tethers of the core flux, this reconnection can be viewed as "tether cutting". Moore & Roumeliotis suggested that the core flux then starts to slowly rise, dragging in material from the sides and forming a current sheet in which magnetic reconnection occurs. As soon as fast reconnection becomes operative, field lines of both the core flux and the overlying stabilizing flux system are efficiently cut on either side of the neutral line, and moreover, the upward reconnection outflow accelerates the further rise of the core flux system. A catastrophic loss of balance – an eruptive flare or a CME – can result. *Yohkoh* Soft X-ray Telescope (SXT) observations motivated a recent refinement of the model: loss of balance due to reconnection between two sheared core flux systems [15]. Noting that some non-eruptive flares start in much the same way as eruptive events, these authors suggest that the basic mechanism applies to all flares.

The tether cutting model finds some support in the observations (see [14, 15]). It predicts activity within the sheared core field, i.e., *close* to the neutral line, shortly before and at the beginning of eruptive events. This can be tested, e.g., by imaging microwave observations, which are sensitive to emissions by energetic particles near the footpoints of magnetic field lines.

The Magnetic Breakout Model

The magnetic breakout model (Antiochos, DeVore, & Klimchuk [16]) assumes that the eruption results from the loss of balance *between* sheared core flux and overlying arcade-like flux and that the overlying flux is composed of two oppositely directed flux systems (which *requires* a quadrupolar field configuration, topologically similar to the early model by Sweet [17]). Magnetic reconnection between these flux systems, possibly triggered by a swelling of the core flux, cuts the "tethers" formed by the overlying flux so that the core flux can "break out". This reconnection transports part or all of the overlying flux to the sides, which is one way to circumvent the consequences of the Aly–Sturrock theorem ([18, 19] - the energy of the magnetic field associated with a given boundary condition is bounded above by the open field line configuration). The breakout model is also appealing because it can be generalized to the three-dimensional magnetic field configuration of delta sunspots [20], which are known to be the most prolific producers of big, eruptive flares (e.g., [21]). On the other hand, if the reconnection *above* the sheared core field is regarded to be the main effect, then the flare ribbons forming at the footpoints of the field lines that emerge from the reconnection region are expected to move towards the photospheric neutral line, which is opposite to the observations. One can expect, however, that reconnection is triggered also below the rising core flux, similarly to the tether cutting model. Furthermore, it is not clear whether the supposed quadrupolar configuration is a characteristic of all eruptive events.

Some observational support was given in [22, 23], but further observations are clearly needed. Imaging at microwaves is a sensitive tool to check for the implied presence of particles, accelerated at or near the reconnection region above the core flux, at magnetic footpoints *remote* from the neutral line.

The Loop-Loop Interaction Model

Magnetic reconnection between two loops leads to two new loops – a transition between two closed configurations releasing energy. It is thus a viable model for confined (non-eruptive) flares . The process need not be restricted to a pair of single loops; groups of loops or flux bundles or the interaction between newly emerged flux and preexisting closed flux are conceivable as well. Melrose [24] has investigated the process and identified the conditions required for large energy release for two favorable loop configurations: (1) interaction of two loops at a large angle to each other, with one of the resulting loops being very short and carrying the larger current, and (2) interaction of two coaligned loops to form a longer loop and a nested shorter loop. Imaging observations in support of loop interactions are summarized in the next section.

3 Diagnostics of Coronal Magnetic Fields

It is widely accepted that stressed magnetic fields contain the reservoir of free energy required to drive energetic phenomena on the Sun. Yet there remain many outstanding questions regarding the coronal magnetic field: How are magnetic fields generated? How do they emerge into the corona? By what means are magnetic fields stressed? What physical processes trigger energy release? In what magnetic configurations? Which physical processes convert magnetic energy to hot plasma, energetic particles, and mass motions?

To answer these questions detailed, quantitative knowledge of the coronal magnetic field is needed. This knowledge has been slow to accumulate due to the difficulty in making the required measurements. Work has proceeded along three lines:

1. The coronal magnetic field has been inferred through extrapolation of longitudinal or vector magnetic field measurements in the photosphere.
2. The coronal magnetic field morphology has been observed in a non-quantitative way by high-resolution imaging observations of coronal emissions (EUV, soft X-rays, radio).
3. Quantitative measurements or constraints of the coronal magnetic field have been made using a wide variety of radio techniques. More recently, techniques in the infrared have been explored.

The technique of magnetic field extrapolation is illustrated elsewhere in this volume [25]. Morphological studies and quantitative measurements of coronal magnetic fields are discussed in the next two subsections, respectively. While we emphasize radio studies we note that recent progress has been made in exploiting IR spectral lines to measure magnetic fields in prominences [26] and in the solar corona (e.g., [27]) .

3.1 Morphological Studies of Coronal Magnetic Fields in Flares

Morphological studies of coronal magnetic fields in flares date back to the *Skylab* era [28], and have continued with studies with the *Solar Maximum Mission* satellite [29], *Yohkoh*/SXT (e.g., [30, 31, 32]) and with *TRACE* ([33, 7], and references therein). The *Yohkoh* Hard X-ray Telescope (HXT) has also contributed to studies of flaring magnetic fields by allowing unambiguous identification of conjugate magnetic footpoints and their evolution in time [34]. More recently, interest in 3D magnetic reconnection has motivated careful studies of the 3D topology of active region magnetic fields and elsewhere (e.g., working group presentation by Chertok et al. [35]) in order to identify separatrices or "quasi-separatrix layers" (e.g., [36, 37]), believed to be the sites of energy release through magnetic reconnection in complex topologies [38]. These and similar studies have sought to establish the structure and evolution of magnetic loops and loop systems by studying the 3D magnetic topology, changing magnetic connectivities (reconnection), and/or topological relaxation following energy release. Morphological studies at radio wavelengths have also played a prominent role in guiding ideas regarding energy release in active phenomena, which we now briefly discuss.

Confined Flares. Beginning with some of the first Very Large Array (VLA) maps in the early 1980s [39, 40, 41] certain radio observations have been interpreted in terms of energy release in interacting loop systems , or of emerging magnetic flux into pre-existing flux systems. More recently, joint imaging observations of impulsive flares in microwaves, soft X rays (SXR), and hard X rays (HXR) have been used to demonstrate such interaction [42, 43]. Typically two or three cospatial HXR and microwave sources were identified, which were interpreted as footpoint regions of two loops, one of them small (extent $< 20''$) and unresolved, the other significantly larger (footpoint distance 30–80$''$). The loops were oriented in different directions, suggesting the possibility of magnetic reconnection between them (but they were not antiparallel in general). The different sizes indicate that interaction of a newly emerging (small) loop with a preexisting larger loop is the most common type of loop-loop-interaction in flares. Additional radio-imaging and spectroscopic data, as well as X-ray data, in support of loop models for impulsive flares is reviewed in [44].

Eruptive Phenomena. Several workshop contributions were relevant to coronal magnetic fields as they relate to eruptive flares. Altyntsev et al. [45] presented a morphological study of radio, EUV and magnetic observations of the powerful, two-ribbon long duration flare of 1998 September 23. They made comparisons between the morphology of the radio brightness distribution in maps obtained by the Siberian Solar Radio Telescope (SSRT) at a frequency of 5.7 GHz and the magnetic loops seen in EUV images obtained by *TRACE* and magnetic field extrapolations . Prior to the impulsive phase, three series of microwave sub-second brightenings (SSBs) were seen, the first two of which correlated poorly with the HXR emission. The third series was well correlated with HXR emission. The first two series of microwave SSBs were associated with a neutral line source (NLS) while the third series originated from a source remote from the neutral line. The authors attribute the first two series of SSBs to plasma radiation from a dense ($n_e \sim 10^{17}$ m^{-3}) source while the third is attributed to bursty gyrosynchrotron emission . They also suggest that the flare progresses from low-lying, dense, compact magnetic loops in which trapping is negligible, to larger scale loops in which trapping becomes effective, leading to a large increase in the (gyrosynchrotron) radio flux during the impulsive phase of the flare and a disappearance of temporal fine structure. The authors also compare potential field extrapolations of the magnetic field of the active region in which the flare occurred to successive *TRACE* EUV images. They show that the magnetic field relaxes to a more nearly potential configuration as the flare progresses and argue that, when taken jointly, the preflare evolution of the active region, the low-to-high evolution of microwave emission from the flaring magnetic loops, and the relaxation of the magnetic field to a more potential configuration suggest the flare might be viewed as a "reversed movie" of the pre-flare evolution of the magnetic field.

We note in passing that, more generally, radio NLS have been discussed previously as a feature of active regions (e.g., [46]) and as a precursor activity of flares (e.g., [47]). Precursor activity in close proximity to the neutral line is expected within the framework of the tether cutting model. Bogod et al. presented multi-wavelength observations from the RATAN 600 showing fascinating spatial and spectral fine structure in the polarization of active regions prior to flare activity. Further studies of NLS and additional polarization and spectroscopic signatures prior to eruptive phenomena are needed to further develop radio diagnostics of the relevant physical processes operating in eruptive flares.

Results by Uralov and co-workers were also presented, a detailed analysis of SSRT (5.7 GHz), Nobeyama Radio Heliograph (NoRH; 17 GHz), and *SOHO*/EIT and LASCO observations of a complex filament eruption and CME associated with a two-ribbon flare that occurred on 2000 September 4. Their analysis leads to the suggestion that, in this case at least, the filament eruption was initiated by a modified tether-cutting scenario, wherein the initial rise of the filament was in fact caused by the interaction and merging of two filaments which caused the net flux overlying both filaments to rise. Tether cutting then commenced, resulting in a helical field around the dual filament system. The combined action of the filament merging and tether cutting allowed the filament to rise and the standard model to come into play. Uralov et al. suggested in addition that elements of the magnetic breakout model contribute as the CME erupts out into the high corona although it is unclear whether the filament system was embedded in the requisite quadrupolar magnetic field environment.

3.2 Radio Diagnostics of Coronal Magnetic Fields

Radio imaging, polarization, and spectroscopic measurements offer a variety of techniques for measuring coronal magnetic fields. Some of these techniques date back several decades (see [48] for an early compilation) and have been used to establish the average coronal magnetic field or to measure the field in specific instances. However, since comprehensive context observations have rarely been available, measurements of magnetic fields using radio techniques have not been systematically exploited in practice. As new radio and optical/IR instruments and instrument upgrades are completed, and as multiwavelength context observations become more effectively exploited, the future for quantitative magnetic field measurements using radio/IR techniques looks very promising.

In this section we draw on workshop contributions relevant to diagnosis of coronal magnetic fields. Additional techniques not touched on at the workshop include: coronal free-free emission (circular polarization [49]; tomography [50]); gyroresonance emission ([51]; depolarization and Faraday rotation [52, 53]). This general area has also been reviewed in [51].

Gyrosynchrotron Emission from Flares.   Gyrosynchrotron emission from extremely hot thermal or energetic nonthermal electrons is the dominant radio emission mechanism from flares at centimeter and shorter wavelengths. Radio observations at frequencies above 100 GHz have become possible only recently (e.g., [54]). Lüthi et al. and Raulin et al. presented the first submillimeter observations of three flares, one of them being detected up to 405 GHz. This newly opened spectral domain will provide diagnostics of relativistic electrons (e.g., spectral hardness, upper energy cut-off, synchrotron losses) that are much more sensitive than those obtained from > 10 MeV gamma-ray observations and thus offer new constraints on acceleration models. Moreover, at these high frequencies, the smooth, long duration emission detected after the synchrotron burst traces, at least in some cases, the response of the chromosphere to the flare energy deposition [55]. The radiative transfer being much simpler than for optical lines, this kind of observation will provide new and valuable constraints on dynamical models of the low atmosphere [56].

Gyrosynchrotron emission is of considerable importance because it offers both morphological and quantitative insight into the details of flaring magnetic loops. In particular, images of gyrosynchrotron emission in flares trace out those magnetic loops to which energetic electrons have access at any given time. The details of the spectrum and polarization are sensitive to the electron distribution function and the magnetic field strength and orientation in the source. However, sparse frequency coverage, poor image quality, and/or poor angular and temporal resolution can all greatly limit the success with which gyrosynchrotron emission can be exploited for either type of study.

To date, gyrosynchrotron radiation has been underexploited for lack of imaging over a broad spectral range with the requisite spatial and temporal resolution, although many efforts have attempted to place meaningful constraints on the magnetic field and other physical parameters in flaring volumes by fitting models to available data in a self-consistent fashion. Recognizing that HXR data can be used to constrain the electron distribution function, numerous studies have analyzed joint HXR/radio observations of flares. These include (i) analyses of spatially unresolved fixed-frequency radio observations and spatially unresolved HXR spectroscopic data (e.g., [57, 58]); (ii) analyses of spatially resolved radio data in a small number of frequency bands and spatially unresolved HXR spectroscopic data (e.g., [59]; also the presentation by Melnikov (Sect. 5)); or (iii) analyses of imaging data in both the radio and HXR bands (e.g., as also presented by Nindos [60]). This latter work involved joint 17/34 GHz imaging observations of three impulsive flares by the NoRH, supplemented by *Yohkoh*/HXT and SXT imaging. The authors attempted to model the flaring radio sources as gyrosynchrotron emission from a dipolar field configuration, but failed to find self-consistent solutions. Remarkably, they find the data are most easily reconciled if the sources emit in magnetic loops which vary little in field strength or in cross section along their length. On the other hand, the marginal angular resolution of the NoRH for the flares in question, for which the loop lengths were $\sim 20''$, suggests that careful follow-up is needed.

Fine Structures in Burst Emissions.   An intriguing paper by Fleishman et al. [61] presented periodic, millisecond, narrowband pulsations . An analysis of the delay between the right and left-hand circularly polarized radiation showed that the observations could be understood in terms of a group delay between the two polarization channels induced by propagation of radiation from a compact, unpolarized source through the magnetized corona. The authors were able to deduce a number of source parameters in addition to the density and magnetic field strength of the background plasma.

Observations were presented of "zebra-pattern" structure and fiber bursts in spectral records at frequencies of a few GHz [62]. Interpreting the bursts as a manifestation of the whistler instability, these authors were able to deduce the magnetic field in the source for reasonable values of the source temperature and density. Zebra patterns superimposed on fiber bursts in the 1–2 GHz range were recently discovered [63]; these pose a challenge for the unified model of both phenomena proposed in [62]. An analysis of zebra pattern bursts was also presented by Zlotnik et al. Assuming upper-hybrid wave instability at double plasma resonance for the zebras (see Sect. 4.4) and hydrostatic density stratification, they derived the height dependence of the magnetic field strength in the source.

4 Signatures of Magnetic Reconnection and MHD Turbulence

Dynamic radio spectra in the decimetric and metric wavelength range are particularly suited to reveal details of energy release and particle acceleration processes. This is due to their sensitivity to even small energy releases and to the high spectral and temporal resolution typically achieved. Radio (and HXR) data show that the energy release generally occurs in a highly time-variable manner and, consequently, at small spatial scales. The following subsections focus on recently detected fine structures that are signatures of magnetic reconnection or MHD turbulence (possibly resulting from reconnection) in solar flares.

4.1 Drifting Pulsation Structures

Broadband pulsations are a common phenomenon in type IV continuum bursts (see, e.g., [64, 65, 66]). Recently it has been shown that the various drifting pulsation structures (DPS) in the decimetric range are associated with the impulsive phase of at least some flares [67, 68]. In one case, the pulsating radio flux was (negatively) correlated with the HXR time profile [67]. Another case of correlation between the pulses in a DPS and peaks in the HXR time profile, a positive correlation in this case, was reported at the workshop by Saint-Hilaire et al. for the flare on 1999 September 8. These associations and their typical frequency range make the DPS a prime candidate for a close relationship with the primary flare energy release.

DPS models can be grouped into three categories: (1) MHD oscillations of the loop which modulate the radio emissivity; (2) intrinsic oscillations of the flux created by an oscillatory nonlinear regime of the kinetic plasma instabilities that emit the radio waves; and (3) intrinsic oscillations of the flux due to an oscillatory particle source (e.g. [64]). A new approach to DPS models in category (3) was taken by Kliem et al. [67]. They proposed that a dynamical regime of magnetic reconnection (so-called impulsive bursty reconnection) in the coronal current sheet of the standard flare model simultaneously leads to the formation of a growing plasmoid , which becomes strongly accelerated along the sheet, and to a pulsating particle source at the magnetic X line adjacent to the plasmoid. The radio source is then formed in or near the plasmoid. This model represents an improvement over previous DPS models in that it provides a direct link between the pulsating particle source and the radio source. If the model can be substantiated, the DPS phenomenon provides a diagnostic of the plasma densities very near the acceleration (and energy release) region and a diagnostic of the temporal characteristics of particle acceleration by magnetic reconnection.

Several new cases of decimetric DPS observations were reported at the workshop by Karlický and coworkers, and by Yan and coworkers (see also [69]). The DPS were generally found to be associated with the impulsive phase of flares (although they can occur before or after the HXR burst peak) and with plasma ejections (SXR plasmoids or ejecta seen in the EUV or in Hα), which supports the model by Kliem et al. [67]. The statistical study of burst spectra from the 0.8–2.0 GHz Ondřejov spectrograph presented by Jiřička et al. [63] shows that one third of all pulsations in this frequency range are DPS.

Khan et al. [70] presented the first spatially resolved observation of a DPS source, which strongly supports the association of DPS sources with plasmoid ejections suggested by Kliem et al. [67]. The near-limb M1.4 flare on 2000 August 25, 14:23 UT showed a DPS during the impulsive phase, which drifted slowly from > 550 MHz to < 220 MHz, and an ejection imaged by *Yohkoh*/SXT, which ascended with a projected velocity of $290 \pm 60\,\mathrm{km\,s^{-1}}$. The Nançay Radioheliograph (NRH) images at 327 MHz show the rather large DPS source initially located slightly above but overlapping with the SXR plasmoid, then moving outward jointly with the plasmoid. This source location and motion clearly excludes previous DPS models, which place the pulsating radio source in a flare loop below the reconnection region, for this event.

These new observations provide mounting evidence that the DPS are associated with plasmoid/flux rope ejections and linked to a pulsating particle source in a current sheet below the plasmoid, which supports the standard flare model in general. The DPS presumably provide a diagnostic of the density in or near flare ejecta and of the intrinsically time-variable reconnection and acceleration processes. Further combined observations of DPS spectra, DPS source positions, and hot flare ejecta (at SXR or in the EUV) are required to substantiate the new model and to finally exploit the diagnostic potential of the DPS.

4.2 Sawtooth Fine Structure

Sawtooth oscillations in the time profiles of the SXR emission are known to be a typical signature of minor disruptions in tokamaks. A "classical" sawtooth consists of a nearly linear rise of the SXR flux (reflecting the rise of temperature and density in the tokamak core plasma due to the applied heating), which shows a transition to rapidly growing "precursor" oscillations at a certain level of the current concentration in the core. These oscillations terminate abruptly in a rapid crash of the SXR emission, core temperature, and core density to values only slightly above the initial values of the cycle, which can repeat many times. Several types of sawtooth oscillations exist in different devices. The physics of these oscillations is not yet fully understood, but it appears to involve the kink displacement of the core, magnetic reconnection, turbulence, and rapid (anomalous) transport (e.g., [71]).

Klassen et al. [72] reported the discovery of a morphologically similar spectral fine structure in the meter-decimeter wave emission during the impulsive phase of a number of flares. Each sawtooth of a series is formed by a very narrow-band emission stripe that drifts toward lower frequencies for a few seconds, followed by a sudden jump to a value near the start frequency. This pattern is always associated with a Type II burst, whose onset coincides on at least some occasions with termination of the sawtooth. The jumps are often correlated with a fast-drift burst. The bandwidth of the sawtooth emission stripe is only $\sim 1\,\%$, which suggests it is due to plasma emission from a small source. The morphological resemblance to sawteeth in laboratory plasmas is striking and has led to the suggestion that they are a signature of magnetic reconnection in the energy release site of a flare. One should bear in mind, however, that there is a significant difference to laboratory sawtooth emission: the frequency drift implies that the source density is decreasing during the ramp phase – opposite to the tokamak phenomenon. Also, the relatively low emission frequency of meter-wavelength sawteeth undermines the idea that they originate in the region of the primary energy release.

Nevertheless, not only the timing and the spectral shape but also the correlation with fast-drift bursts suggest a relationship between the sawtooth fine structure and energy release and particle acceleration in flares. Once a convincing model is developed, the sawtooth phenomenon may provide a useful diagnostic.

4.3 Type II-Like Bursts Without Frequency Drift

It is often assumed that the downward reconnection outflow in the standard flare model forms a fast-mode termination shock upon hitting the underlying newly reconnected, hot, dense magnetic loops. A low value of the plasma parameter β , a condition widely met in the solar corona, supports the formation of a shock at the head of the reconnection outflow (even in the absence of an obstacle). Such a shock is an essential element in some variants of the standard flare model, e.g., it was supposed to be the site of the efficient particle acceleration in [73]. There are hints pointing to the existence of this shock, primarily the loop-top HXR sources seen in some limb flares [74], and also numerical simulations of the standard flare model have supported the hypothesis that a termination shock is formed for $\beta \ll 1$ [75, 13]. These simulations must be taken as suggestive, since they cannot include the full inhomogeneity of density and field strength in the corona. More observations are required to confirm the picture.

Radio observations of a non-drifting type II burst-like feature in the dynamic spectrum of a flare, presented at the workshop by Aurass et al. [76], may be the first direct signature of the supposed termination shock. The feature was observed in a C6.9 flare on 1997 April 7. The impulsive phase of this flare led to a broadband, complex type IV continuum spanning the whole observed spectrum (40–800 MHz). A classical type II burst emerged from the continuum and drifted beyond the instrumental cutoff at 40 MHz in ≈ 20 min. About an hour past the impulsive phase, a type II-like emission appeared in the 300-400 MHz range and consisted of irregular patches forming two main bands at about 340 and 380 MHz that stayed nearly stationary in frequency. It was interpreted as the radio signature of a shock wave staying at nearly constant plasma density – the fast-mode termination shock standing in the downward reconnection outflow. The extreme rarity of such a signature, the timing, and the unusual appearance of the fast drifting (herringbone-like) fine structures render this interpretation still tentative, however.

Noting the delay between the onset of the non-drifting type II-like burst and the impulsive phase of the flare, Aurass et al. [76] suggest that unambiguous

detection of such features near the time of the impulsive flare phase is very difficult due to the disturbed and strongly non-stationary state of the flaring volume.
Subsequently, however, Aurass et al. [77] presented another case of type II-like
(herringbone-like) emission, only slowly drifting at frequencies clearly higher
than the associated regular type II burst, and this time occurring during the
impulsive phase of the flare (2001 March 29); it was interpreted as a signature
of the termination shock as well. These investigations are of high relevance for
progress in understanding the dynamics of the reconnection outflow jets in flares.

4.4 Narrow-Band Decimetric Spikes and Lace Bursts

A new phenomenological model of narrow-band decimetric spike bursts was presented by Bárta and Karlický [78]. It is based on the assumptions that the radio
emission originates from places in the corona where the double plasma resonance
condition is fulfilled (analogous to the well-known models for zebra patterns
[79, 80]). The double plasma resonance is given by $\omega_{uh} = (\omega_p^2 + \Omega_B^2)^{1/2} = s\,\Omega_B$,
where ω_{uh}, ω_p, and Ω_B, are the upper hybrid, plasma, and cyclotron frequencies
of the thermal electron component, respectively, and s is the cyclotron harmonic
number. The growth rate of electrostatic upper hybrid waves in plasmas composed of a thermal background and an energetic electron component with excess
perpendicular energy (a loss-cone, bi-Maxwellian, or oblique beam distribution)
peaks strongly if the double resonance is satisfied, thus permitting narrow bandwidth of the emission. Locations where this condition is satisfied and the resulting emission frequency are shifted stochastically if MHD turbulence introduces
local variations of the density or magnetic field in the inhomogeneous corona.

Synthetic dynamic radio spectra were constructed which displayed spikes
arranged in irregular chains when emission from a single model flux tube was
considered. Clouds of spikes were obtained by adding the emissions from a number of flux tubes which resemble the appearance of observed ones rather well. A
rather high level of turbulent density fluctuations (having an r.m.s. of 10 % of
the average background density) was required to obtain these spectra, however.

The same model was applied to lace bursts reported by Jiřička, Karlický,
Bárta, and coworkers [63, 81]. Lace bursts are a rare, new pattern of fine structure in the dynamic radio spectrum discovered in the ~1–2 GHz range during
the impulsive or main phase of solar flares. They are composed of many, overlapping, narrow-band (~50 MHz) emission lanes that can appear on a background
of continuum emission. Individual lanes are characterized by a rapid, irregular
frequency drift. These often last only a few seconds and sometimes last less than
a second. Again, synthetic spectra resemble some of the lace burst observations
remarkably well.

A significant level of background plasma turbulence is required for both decimetric spike bursts and lace bursts. It has been suggested that the reconnection
outflow jets are the source of the turbulence for both burst types. This is in line
with some models of the flare energy release [82] and with models of stochastic
particle acceleration [83, 84], which are based on the same hypothesis. Numerical

investigations of reconnection have in fact provided some evidence that the outflows become turbulent (but only in the non-resistive description [85]) and that
wave excitations in the vicinity of the magnetic X line lead to stochastic electric
field enhancements [86]. Alternatively, the particle precipitation that leads to
chromospheric evaporation creates turbulence [87]. If the basic assumptions of
the new model for decimetric spikes and lace bursts can be substantiated by
further quantitative work, then these bursts may become an important diagnostic, revealing the timing, typical densities and fluctuation levels of turbulent
plasmas in flares. See Sect. 5.2 for new results pertaining to metric spikes, and a
discussion of decimetric and metric spikes in [66].

5 Particle Acceleration and Transport

The means by which large numbers of particles are accelerated in flares, the
extent to which they transport energy throughout the flaring volume and the
factors which act on them as they do this have all been outstanding questions in
flare research for some decades now. The evidence for flare site particle acceleration and its implications for energy distribution, numbers etc. of particles are
described in detail elsewhere in this volume [88]. Several contributions to this
workshop bore in a timely way on aspects of these questions.

Ideally we would know all details of the evolution of electromagnetic fields
in a flare. The production of particles of various energies would follow in a natural way from these, and the various flare radiations would be produced either
directly by these particles [88], or as a result of their propagation in a greater
volume (see [56]). In practice, of course, we start from the observations and we
have to work back to try to deduce what the electromagnetic fields have been
doing. In order to make any progress at all, certain assumptions have to be
made, some of which are so prevalent that we may usefully repeat them here.
The problems of "transport" and "acceleration" are decoupled: it is supposed
that a spatially localized accelerator acts briefly on particles at some location
in the corona, playing no further role in their evolution. The hope is that the
resulting, separate transport problem may be solved sufficiently decisively as
to allow deduction of the properties of the acceleration region from observed
flare radiations. That said, there are plenty of examples of models where this
strict separation between transport and acceleration is not made (e.g., [89, 90]).
A further, fairly standard assumption is that the magnetic geometry in which
particles subsequently move is that of a single loop, possibly with magnetic field
strengthening towards the chromosphere - although there is lots of evidence for
frequent involvement of multiple, more elaborate structures in flares (examples
in [91, 92, 7]). Coronal trapping may then play a key role in particles' evolution,
as suggested by temporal evolution of high-energy radiations (e.g., [93, 94]; see
also [88]), with its overall importance depending on whether the pitch-angle scattering time is much longer than the loop transit time or comparable to it (the
"weak" and "moderate" scattering regimes, respectively - [95]). There is also the
possibility of "strong" scattering, in which the scattering time is much less than

the loop transit time, and the continual randomization of velocity actually improves the effectiveness of coronal trapping [96]. Most of this discussion assumes "pure" scattering in wave-particle interactions [97], unaccompanied by significant energy change; if scattering occurs only via binary collisions, evolution in energy needs to be simultaneously treated (e.g., [98, 99]). It is also commonly supposed that background plasma properties are not significantly affected so that, e.g. wave frequencies (though not growth rates!), test particle drift and diffusion coefficients, etc. need not be recalculated because of the presence of the accelerated particles.

While loop geometry is often an ingredient of these discussions, more complex geometries may of course be found (e.g., [100]); if field strength varies along individual field lines, particles may nonetheless be trapped in restricted regions, and the ideas above continue to apply. Also it is clear that particles may equally well be accelerated on open field lines, or in much more extended structures sampling a wide range of heights and conditions.

Although the accelerator is treated as a black box in many such discussions, there are of course well-established candidates: stochastic acceleration by plasma turbulence (e.g., [101]), in the presence of shocks (also discussed in [101]), or in the electric fields that must accompany reconnection [102]. A recent review of all of these, in the flare context, is that of [84]. There are obvious aims for particle acceleration theories: production of enough fast particles, with the correct distribution of energies, fast enough (as reviewed in [88]). In view of the evident importance of fast particles in the overall flaring process, we must also aim for theories which fit nicely into an overall picture of flare energy release and radiations, and whose details have been elaborated to a significant degree of realism e.g. we must understand how they will operate in the "noisy", turbulent conditions of a flare.

The preceding comments apply equally to accelerated ions and electrons. Most contributions to the workshop under this heading concentrated on electrons, however, and this is reflected in the remainder of this section.

5.1 Moving, Coronal Hard X-ray Source

In [103] a coronal HXR source was described, of great interest as a kind of source for which many or all of the above assumptions may not hold (see also Sect. 6.2). Evidently associated with a large flare more than 20° behind the limb, this source was detected in the 33 - 53 keV energy bands of the *Yohkoh*/HXT. Source complexity and/or temporal development prevented reconstruction of a straightforward set of images, but *Yohkoh*'s fan beam collimators were used to construct a sequence of one-dimensional images which clearly showed upward motion. Observations with the NoRH at 17 and 34 GHz yielded more detailed information, in particular revealing a bright, compact moving source apparently producing free-free radio radiation . The plasma density in the radio source was deduced, and combined with the *Yohkoh* observations to give the number and energy content of the X-ray emitting electrons. The simplest, "thin target" interpretation deduces the number and energy distribution of electrons that must be instantaneously present to account for the observed X-ray flux. On this assumption the X-ray emitting electrons above 20 keV in energy constitute 0.2 % of all electrons present, and have a total energy content of 6×10^{21} J [103]. Most significant of all, the pressure in nonthermal electrons appears to be comparable to that of the ambient plasma. Any further consideration of escape or slowing-down of the emitting electrons, will tend to increase their numbers and energy content. So in this instance, accelerated electrons are sufficient in number and energy to exert a non-negligible influence on their surroundings.

It is not clear if magnetic field convergence alone can account for the observed, high altitude, compact source, or if some extra physical effects are needed. Although there is at present no particular reason to discard simple, expanding loop pictures, one is tempted to think of the detached plasmoid which forms in the reconnection process [104], a very different sort of magnetic geometry which could conceivably result in an isolated source with a large population of fast particles. A situation in which fast particles contribute non-negligible pressure also leads us to recall the so-called "thermal" X-ray source models (e.g. [105, 106]), in which X-rays are produced much more efficiently because most particles have comparable individual energies and binary collisions serve only to redistribute energy among all particles, not to transfer energy from a privileged minority of particles to the "cold majority". Existing estimates of particle number and energy content yield a source not in this regime, but nevertheless closer to a thermal source than anything else yet observed. We must hope for further, detailed observations of this sort of source from the *Ramaty High Energy Solar Spectroscopic Imager (RHESSI)*.

5.2 Metric Spike Bursts and Acceleration Region Site

Among the huge variety of solar radio bursts, the narrow-band metric and decimetric bursts potentially play a special role. While radiation mechanisms, and even the relationship between the metric and decimetric variants, remain unclear, they nevertheless show a suggestive relationship to other manifestations of solar particle acceleration, as reviewed in [66]. In particular, a recent study [107] has produced convincing evidence that metric spikes mark the acceleration sites of type III bursts. Spatial information from the NRH allows the coronal trajectories of type III electron streams to be reconstructed. Extrapolation of these to lower altitudes finds starting positions in close coincidence with the locations of metric spikes. The precise position of the type III acceleration region along this extrapolated trajectory remains uncertain: the acceleration region could in principle lie below the spike burst source. In two cases, however, separate type III bursts in the same group follow divergent paths, both consistent with the same spike burst location (see Fig. 3 of [66]). Unless the diverging field lines bend together in an unlikely way below the spike burst location, these particular observations argue for near identity of acceleration region and spike burst source.

The cases with divergent trajectories are particularly interesting when overlaid with *Yohkoh*/SXT images ([107]; Fig. 3 of [66]). These together reveal starting positions for the type III bursts, and spike burst sources, near a location where the connectivity of field lines changes, i.e. a key location in the occurrence of reconnection (e.g. [108]). Moreover, sequentially occurring, divergent Type III bursts suggest a varying location for the accelerator.

It seems clear that metric spike bursts mark the location of the type III acceleration process rather accurately. Some further questions come to mind, however (see also [66]). Precisely what spike bursts tell us about particle acceleration will depend on their degree of intimacy with the acceleration process. On the one hand, some interpretations of radio spikes regard them as a very direct reflection of the acceleration process, with their narrow bandwidths and short durations indicating a highly fragmented primary energy release [109]. Other proposed mechanisms, involving maser action or wave-wave interaction (see [66]), assign the bandwidths and durations to the radiation process, and studies of such mechanisms often display the qualitative feature of rapidly fluctuating emissions in spite of a steady driver [110, 111]. Whether the spike bursts come directly from the acceleration region, or from its immediate vicinity is clearly an important question for sorting out exactly what they tell us, in particular whether they point to an *intrinsically* fragmented acceleration process (e.g. as discussed in [88]).

5.3 Propagation of Electron Beams and the Type III Radio Bursts

The streams of energetic electrons giving rise to the type III radio bursts occur both during and away from flares. How they manage to both generate Langmuir waves leading to radio emission and propagate at least to 1 AU is an old problem, "Sturrock's dilemma" [112], solved by the consistent treatment of the interaction of particles and waves. Waves emitted at the head of the electron stream are re-absorbed at the back, so that waves and electrons are able to travel *en masse* as a self-consistent structure - [113, 114]. Modern computers allow simulations, in the weak turbulence regime, of increasingly realistic systems, in particular including gradually varying background plasma parameters, for instance with declining density as in the corona [115], and in a medium with small scale density fluctuations [116]. Interestingly, the propagation of the electron stream in the latter case is not greatly affected, even although Langmuir wave energy density becomes quite spatially clumpy, growing most rapidly in regions of positive density gradient. Although detailed calculations of consequent electromagnetic radiation were not presented at the workshop, this situation would presumably lead to a concentration of observed radiation at certain frequencies. Thus it was clear that this beam-plasma system, studied in the first instance for its centrality to the understanding of type III bursts, may yield models for some other sorts of bursts if density inhomogeneities in the background plasma are allowed to play an important role.

Electron streams produce type III bursts when they are accelerated on open field lines, allowing them to propagate freely in an (at least comparatively) unmagnetized medium of decreasing density. In flares, it is clear that electrons also propagate in lower atmosphere, denser, closed field line structures, producing related sorts of radio burst (U bursts and reverse drift bursts) and HXR emission. At the same time they transport energy throughout the flare volume and give rise to at least some substantial component of the thermal flare phenomena. Thus electron (beam) propagation is a serious issue for the totality of flare phenomena. Methods like those applied to the type III problem have also been applied to understanding HXR bursts, although the greater influence of binary collisions on beam electrons is one complication [117, 118], while the multiplicity of plasma modes in a more strongly magnetized plasma is a more serious one e.g. [119, 120].

5.4 Particle Propagation in Coronal Magnetic Traps

As mentioned at the start of this section, the competition between trapping by magnetic convergence and pitch angle scattering has been a staple in interpretation of radio and X-ray observations for the last three decades. Still uncertain, even controversial, however, are the degree of magnetic convergence in flaring loops, the dominant mechanisms of pitch angle scattering and the regime in which scattering operates. Substantial variations from one flare to the next, even across spatial structures within a single flare, are also quite possible.

First we note that several observations of flare hard X-rays may apparently be interpreted in terms of simple magnetic trapping plus scattering only via binary collisions (see [88]). On this assumption it was possible to explain temporal evolution of spatially resolved microwave data from Owens Valley, and even to deduce the injected pitch angle distribution [121]. Centimeter wavelength burst time profiles were presented which appear to be consistent with the evolution, under the influence only of binary collisions, of an isotropic population of electrons produced at the top of a loop [122]. At this workshop, Melnikov presented further observations of four events, spatially resolved at 17 and 34 GHz with the NoRH. The maximum of the radio brightness distribution appeared to be found at the tops of loop structures, in contradiction with simple expectations for an optically thin source uniformly filled with fast electrons. Again, an interpretation in terms of an isotropic population of electrons, produced at the loop top and allowed to evolve collisionally, appears at least qualitatively consistent with these observations. Because the collisional pitch-angle scattering rate falls off more rapidly with energy than the energy loss rate (e.g. [123]), different behavior may be expected in radiation signatures reflecting deka-keV and MeV electron energies.

We must expect at least some electrons to be scattered more rapidly by collisionless effects, likely of their own making. There is a clear, qualitative relationship to the problems associated with streaming electrons, described in the previous section. In this case anisotropy results not from small pitch-angle electrons leaving larger pitch-angle electrons behind, but from the preferential loss to the denser atmosphere of small pitch-angle electrons and the resulting development of a loss-cone type anisotropy. Depending on the details of the ("fast" plus "ambient") electron distribution and the ratio $\omega_{\mathrm{p}}/\Omega_B$, fastest growing waves may be in whistler , upper hybrid or direct electromagnetic (cyclotron maser) modes [124].

The possibility of electron scattering by whistlers consequent on the development of a loss-cone distribution has been considered in detail [125]. This scattering mechanism may apply *a priori* to electrons above $\simeq 30$ keV in energy. Although the growth rate is large, the whistlers are also strongly Landau damped throughout much of the loop as a result of the continuously changing magnetic field direction. Scattering will take place consequently in the weak or moderate regimes. Spatially resolved X-ray observations have previously been interpreted in terms of single loops , concentrating on the relationship between the intensities and spectra of "coronal" and "footpoint" portions of the loop (e.g., [126, 99]). Some observations previously interpreted in terms of binary collisional scattering in the corona may also be consistent with scattering by whistlers, but a much wider range of behaviors is possible. For instance, footpoint X-ray spectra may be determined entirely by the scattering rate and may yield no direct information on the original injected electron distribution (although spatially unresolved, flare-integrated observations should still yield the injected distribution via the thick-target results of [127], as long as scattering is indeed unaccompanied by significant energy change).

The large event of 1999 August 28 , 00:55 UT, observed at 17 and 34 GHz with the NoRH, was much discussed at the workshop (e.g.,[128]; contributions by Melnikov et al.). Most notable was a source which appeared to spread along a loop structure of about 60000 km extent, at a velocity of ~ 12000 km s^{-1}. At this high frequency, the radiation mechanism would undoubtedly be (gyro)synchrotron emission of electrons with energies in the MeV range, i.e. with speeds not much less than that of light. Clearly the observed source expansion does not reflect rectilinear motion of the emitting electrons. An exciting interpretation of this event suggests itself in the context of the three regimes of trapping and propagation suggested above (Melnikov et al., workshop presentation). Specifically, as the radiating electrons attempt to stream along the loop they cause waves to grow (in this instance low-frequency whistlers); in turn the growing waves scatter the electrons strongly in pitch angle. Straightforward streaming along the field is impeded and the electrons instead travel *en masse*, in a diffusive way, behind a front which travels at the whistler group velocity. Propagation is in the strong regime of [96].

The range of possible line-of-sight effects and the lack of good context observations from any other wavelength render this interpretation provisional. A simple alternative interpretation, for instance, involves emitting electrons gaining access to neighboring field lines sequentially, so that the evolution of the source structure would reflect evolving connectivity of field lines sampling the acceleration region, rather than the propagation of accelerated particles. A subsequent injection of electrons (in the same flare) apparently propagated rectilinearly, requiring a radical change in loop plasma parameters. Nevertheless this is a suggestive observation, among the first that might clearly be interpreted in terms of strong scattering, and a powerful demonstration of the insights to particle acceleration and propagation that, at least until *RHESSI*, are unique to the radio domain.

5.5 Developments in Particle Acceleration Theory

Elsewhere in this volume we have detailed reviews of particle acceleration by plasma waves and in shocks [101], and in reconnection electric fields [102, 104]. Here we focus on two workshop contributions that addressed particular issues of current interest.

Isotropisation. Most acceleration mechanisms in which particles repeatedly encounter the accelerator need to either involve or invoke a scattering agent for encounters to continue to be effective. The original, converging shock version of Fermi acceleration gives the prototypical example, but in fact this is a requirement for all shock or turbulent acceleration scenarios (e.g., [129]). Above we saw possible examples of the role that such scattering can play in particle propagation, impeding rectilinear propagation or enhancing scattering into the loss cone. In a sense this issue blurs the distinction, mentioned at the beginning of this section, between "acceleration" and "transport" phases of a particle's history: factors which will certainly be important in transport are also essential ingredients of the acceleration process.

The possible role of the Electron Firehose instability in isotropising accelerated electrons was considered by Paesold and Benz [130]. This instability can operate when electron anisotropy favors directions parallel to the magnetic field. For plausible flare conditions, waves are excited at frequencies above the proton gyrofrequency. Since it is a non-resonant instability it can involve the bulk of the accelerated electrons without needing satisfaction of a possibly restrictive resonance condition. Particle-in-cell simulations have confirmed the occurrence of this instability and its isotropising action, also demonstrating that ions may be simultaneously heated [131].

Turbulent Dreicer Field. Arzner [132] presented a study of particle acceleration by a DC electric field, E, in the presence of static magnetic inhomogeneities, thought to represent a snapshot of turbulence. Such magnetic inhomogeneities trap the particles for small values of E, randomizing their acquired directed momentum. For sufficiently large values of E, the particles, while traversing one length scale of the inhomogeneities, gain a sufficient amount of parallel momentum to overcome the trapping effect and are thus essentially freely accelerated. The field strength required was determined and termed "turbulent Dreicer field" in analogy to the classical Dreicer field required to overcome randomization by binary collisions. Stationary magnetic inhomogeneities may thus inhibit acceleration by a DC field, however, a fully turbulent plasma *contributes* to particle acceleration due to the induced electric fields connected with moving magnetic mirrors [84, 101].

6 Large-Scale Eruptive Phenomena and Solar Energetic Particles

Large scale eruptive phenomena, discussed in this section, include: long duration flares, filament eruptions, CMEs, and large scale coronal and interplanetary (IP) shocks. These large-scale coronal disturbances, which usually occur simultaneously, constitute potential accelerators of both interacting and escaping particles. The main topics discussed at the workshop were: radiative signatures from non-thermal particles during long duration flares, the relationship between flares, ejecta and shocks, and the origin, shock versus coronal acceleration, of gradual Solar Energetic Particle (SEP) events.

6.1 Multi-wavelength Observations of Long Duration Flares

Radiative signatures of interacting electrons observed during long duration flares have provided evidence for distinct episodes of acceleration: acceleration in the low corona, as traced by the hard X-rays and the microwaves, followed by time-extended acceleration (tens of minutes to hours) within much larger volumes of the magnetically stressed low and middle corona (at heights below, say, 1 $R_\odot$ above the photosphere) during the decay of the X-ray signatures which usually last much longer than the hard X-rays, e.g., [133, 134] and references therein. Gamma-ray (GR) observations of a few flares have revealed that relativistic ions may also be accelerated over several hours at the Sun, e.g., [135, 136]. During the workshop, the characteristics of these different episodes of acceleration have been extensively discussed for the 2000 July 14 ("Bastille Day") flare (see also [25]).

The 2000 July 14 event at $\sim$ 10 UT was associated with a class X (3B) two-ribbon flare that occurred in active region AR 9077 (N17, E01). It comprised the eruption of a filament followed by the formation of the flare ribbons, a fast halo-type CME, and coronal and IP shocks. It was furthermore accompanied by a large SEP event including a rapid rise of relativistic proton fluxes measured by neutron monitors. This data pool reveals that electrons were accelerated over durations of several hours at different sites in the corona and that signatures of particle acceleration had different timing in different spectral ranges:

- Two pieces of evidence show that the bulk of the HXR emission was produced by electrons accelerated in the low corona probably within loops connecting the two flare ribbons: (i) Yohkoh/HXT images above 33 keV indicate that the HXR are produced at loop foot points along the flare ribbons initially in the western (leading) part of AR 9077 [10], and later on the strongest HXR sources occurred at the same locations as the strongest EUV sources in the eastern (trailing) part of AR 9077 [137] (see Fig.9 in [7]) and (ii) type III (upward moving electron beams), reverse drift (downward moving electron beams) and U (propagation of electron beams in loops) radio bursts have also been detected in the 1–7.6 GHz frequency range, see [69], suggesting regions of acceleration with electron densities in the range of $(0.1–7) \times 10^{17}$ m^{-3} for fundamental plasma emission. Ions were also accelerated in the low corona as revealed by the detection of GR lines till the end of the HXR emission [138].

- Below 1 GHz the dynamic spectrum shows a complex pattern which includes type IIIs, wide band type IV like continua, type II emission from coronal shocks (mostly below 170 MHz) and numerous fine structures [139, 140]. The NRH images show a time varying and complex distribution of radio emitting sources at long decimetric–short metric wavelengths [139, 141]. The strongest radio emission was a bright structureless continuum with a slow drift towards lower frequencies which began during the decay and lasted till the end of the X-ray emission. At metric wavelengths the emission came first from sources located in the north-west quadrant. Later on radio sources brightened in the north-east quadrant and persisted for several hours. These radio sources were most likely produced by electrons accelerated in the middle corona at much higher altitudes than HXR producing electrons . The comparison of EIT and NRH images suggested that such an acceleration is the consequence of large scale magnetic restructuring, triggered far away from the active region by the interaction of the magnetic configuration of the eruptive filament with pre-existing large scale structures [141]. In the north-western quadrant magnetic restructuring was also outlined by moving type IV sources above the regions traversed by the filament.

The observations obtained during the 2000 July 14 flare and previous multi-wavelength studies of complex CME-flare events suggest that long lasting acceleration of electrons is due to reconnection of coronal structures of various spatial scales previously opened by an eruptive prominence or by a CME, e.g. [136, 142]. The acceleration sites revealed by the radio sources in the middle corona may spread over an area comparable to that spanned by a CME, e.g. [143]. Such an environment allows particles rapid access to the outer corona and to space. Well connected regions of coronal activity at lower heights than the CME may thus significantly contribute to SEP events , e.g. [144].

6.2 CMEs and Shock Waves

CMEs are eruptions of magnetized plasma from the Sun which move through the corona with speed from a few tens up to ~ 2000 km s^{-1}. CMEs are closely related to flares, viewed by their SXR emission, and filament eruptions, although not on a one to one basis. It seems now well accepted that flares, filament eruptions and CMEs most likely reflect different manifestations, in different environments, of the same magnetic energy release triggered by e.g. shear or reconnection in the magnetic structure, see [145, 146, 147]. However, the cause-and-effect relationships between these three phenomena are not yet firmly established. In the following we outline some radio observations related to the early stage of ejecta, imaging of CMEs and filament eruptions and we briefly discuss large-scale wave-like phenomena associated with flare-CME events.

Early Stage of CMEs. Joint SXT, EIT and LASCO observations of a reasonable sample of events indicate that CMEs are associated with flares exhibiting ejections of hot plasma around their onsets, but the CME onsets occurred at earlier times [148]. For a few events it was found that the CME acceleration takes place during the rise of the associated SXR flare from $\sim$ 0.3 to 3.6 $R_\odot$ [147]. At 212 and 405 GHz fast radio spikes have been detected close to CME onsets a few tens of minutes prior to or close to the associated flare onset [54]. Sheiner reported the occurrence of narrow band, weak bursts at 15.5 GHz several hours before one CME. On the other hand, Nitta has presented a few examples where CME-related changes in the X-ray and EUV images are probably seen after the flare onset. These findings are broadly consistent with those of earlier statistical studies indicating that CME and associated flare onsets can occur at any time within several tens of minutes of one another, e.g., [149]. This points to the difficulties/ambiguities to precisely define onset times associated to the release of energy which most probably has been stored during an extended period of time.

Radio imaging observations in the decimeter–meter domain have provided unique information which may help to discriminate between models of filament eruptions and CMEs. For example, Marqué et al. have presented detailed studies of decimeter–meter radio bursts associated with the early stage of two filament eruptions [150, 151]. These radio bursts are most likely emitted by electrons accelerated at sites of magnetic reconnection. For one event, no CME was detected, and the locations of the radio bursts, which are detected at the beginning of the acceleration phase, are far from the eruptive filament as well as from the flare loops seen by EIT. This broadly agrees with breakout class models which predict weak reconnection signatures after an initial evolution at sites far from the magnetic neutral line [16]; see also Sect. 2. On the other hand, for the second event which is associated with a halo CME, the bursts are detected at the beginning of the slow ascending phase. They are located on one side of the filament near a parasitic magnetic polarity. This suggests a reconnection process between the parasitic polarity and the overlying arcades as modeled in, e.g., [152].

Radio Imaging of CMEs and Filament Eruptions. The advantages of imaging CMEs and filament eruptions at radio wavelengths are that the radio emission can be detected on the disk over a wide range of altitudes (up to a few $R_\odot$) and that radio images are obtained with higher time resolution than EUV and X-ray full-disk images. However, radio imaging of the magnetic structure associated with eruptive filaments and with CME expanding loops has only been obtained for a few events.

Radio free-free emission from a few CMEs has been imaged at meter–decameter wavelengths by the Culgoora and the Clark Lake radioheliographs [153, 154, 155]. For one filament eruption associated with a halo CME, regions of brightness temperatures below that of the quiet Sun have been detected by the NRH in the 150-450 MHz domain [151]. Whereas some of these depressions were signatures of static dimmings associated to large scale coronal restructuring, one of them was unambiguously identified as the coronal cavity associated to the filament. Such cavities, which correspond to regions at coronal temperatures with lower density than the surrounding corona, were recognized long ago in white-light during eclipses and in X-rays, e.g., [156, 157]. The combination of NRH and LASCO observations provided the evidence that the projected motions of the CME above ~ 2 $R_\odot$ and of the radio cavity can be fitted to a common trajectory. This suggests that the cavity is linked to the magnetic structure that supports the filament against gravity and that this magnetic structure, expanding as the filament moves up, is part of the CME.

Nonthermal radio emission from an expanding ensemble of loops, which closely resembled a part of a white-light CME observed by LASCO, was detected by the NRH in the 164–421 MHz range [158]. At the four observing frequencies, the source configuration and the radio flux spectrum were found to be consistent with synchrotron radiation from ~ 0.5–5 MeV electrons interacting with magnetic fields of $\sim 10^{-5}$ to a few 10^{-4} T. Further evidence for the presence of energetic electrons in expanding loops forming a part of a flare-associated CME has been provided by the observation of a high-speed coronal ejection in 33–53 keV HXR, detectable to an altitude of some 0.3 $R_\odot$ above the photosphere [103]. Additional examples of radio and HXR CMEs have to be identified and studied in order to establish whether or not the presence of energetic electrons within some loops of flare-associated CMEs is a usual feature and to infer where and by which mechanism these electrons have been accelerated.

Flare and CME Related Shock Waves. It is well documented that radio type II bursts are radiative signatures of shock waves in both the low and middle corona (meter–decameter waves) and the IP medium (hectometer–kilometer waves). The theory of type II usually assumes that they are radiated by electrons accelerated by the shock . These electrons excite Langmuir waves at each level in the corona and IP medium which are partially converted into electromagnetic waves at the fundamental and the harmonics of the local plasma frequency [159]. In dynamic radio spectra type IIs usually appear as fundamental–harmonic bands of enhanced emission slowly drifting from high to low frequencies. Some type IIs appear as highly fragmented and highly polarized, i.e. a typical fragment shows a single band with a constant drift, disappears and reappears at a frequency higher or lower than expected. Such bursts have been observed in the IP medium [160]. Zlobec & Thejappa have shown one example of such bursts in the low corona. These morphological differences mostly reflect changes from event to event in both the shock parameters and the physical conditions of the medium through which the shock propagates. This can be best studied in the IP medium where radio spectral observations can be combined with in situ measurements of shocks. For example it was proposed that the presence of sharp density gradients associated with structures like shocks, CMEs or co-rotating interaction regions induces large fluctuations in the efficiency of Langmuir wave conversion leading to fragmented type II emission at the fundamental [160].

It seems now well established that: (i) all IP shocks which generate kilometric type IIs are associated with Interplanetary Coronal Mass Ejection (ICMEs)[161] which are believed to be the IP counterparts of CMEs and (ii) IP type IIs are generated upstream of ICME-driven shocks [162, 163]. In the corona the situation is more puzzling. Because meter–decameter type IIs generally do not extend into hectometer type IIs, coronal and IP shocks do not seem to be related (see the review [164] and references therein). This led to the conclusion that coronal shocks are mainly blast waves triggered by flare-related energy release, e.g., [165]. However, the results of a recent statistical study, based on temporal coincidence and velocity data derived from height-time plots of CMEs and type II drift rates, suggest that 30% of the analyzed coronal type II shocks might be driven by CME leading edges [166]. Nonetheless some caution is required because the type II shock velocities, derived from density models, are very approximate so that the comparison with CME projected velocities can be somewhat misleading. Indeed, imaging observations show that generally the type II source does not have the appropriate position or speed consistent with the CME bow-shock [167, 168]. Evidence for radio bursts which are cospatial with the CME leading edges was given in [169]. However, it could not be firmly established that these very weak bursts which, like type IIs, drift slowly towards lower frequencies, are signatures of shocks. In fact the nature of a coronal shock may change along its path. One clear example of a coronal type II shock which is initially driven by a fast short-lived X-ray ejecta issued from the flare region and which continues later on as a blast wave was reported in [170].

Type II burst emission is not the only signature of large scale wave-like phenomena in the corona. Moreton waves, e.g. [171], which consist of rapidly expanding fronts (800–2500 km s^{-1}) resulting in Hα signatures, were interpreted many years ago as weak fast-mode MHD shock waves [172, 173]. Observations from EIT [174, 175] and *TRACE* [176] revealed that large-scale disturbances (so called EIT waves) occur in close association with flares, CMEs and type II meter–decameter bursts [177]. The mean velocity of EIT waves is $\sim$ 270 km s^{-1} with a small dispersion [177]. It was argued that this is significantly above the sound and the Alfvén speeds, estimated to be, respectively, $\sim$ 180 km s^{-1} and 203 km s^{-1}, so that EIT waves can be regarded as fast magnetosonic waves ([177] and references therein). Because of the uncertainties on the sound and Alfvén speeds, this interpretation is not unique. As an alternative, EIT waves may also be regarded as signatures of large-scale structural changes in the medium itself [178]. More recently flare–associated large-scale wave-like phenomena have been detected in X-rays by SXT ([179] and references therein). Multi-wavelength studies of two events that occurred on 1997 November 3 [180] and 1998 May 6 [179], have provided some evidence that the X-ray disturbances are signatures of coronal shocks, i.e.: rapid motion ($\sim$ 600 km s^{-1}) from the flare core, close temporal and even spatial relationship with type IIs, and association with a Moreton wave. It should be noted that in both events an EIT wave, propagating in the same direction as the X-ray wave, was also detected. This suggests that for these events the EIT feature is a wave. For the 1998 May 6 event a Mach number close to unity was inferred from emission measure and temperature variations measured as the wave passes through the medium and compresses it [179]. It is unlikely that this slow shock was driven by the X-ray ejecta detected during the 1998 May 6 event because the ejecta move at a lower speed and in a different direction than the X-ray wave. It was thus concluded that the slow shock seen in X-rays was a blast wave generated by plasma motions close to the flare core.

In summary, it seems rather firmly established that IP type II radio bursts are driven waves by ICMEs in which the moving medium continues to energize the wave. The close relationship between flares, CMEs and other kinds of ejecta makes difficult a clear identification of the physical origin of coronal type IIs. Studies of events for which X-ray, EUV, coronagraph and radio imaging observations were available seem to indicate that meter type IIs are blast waves arising either from pressure pulses resulting from the flare energization or from waves initially driven by fast short-lived X-ray ejecta from the flare region. Although there have been some suggestions, the detection of the radiative signatures of the CME bow-shock in the corona remains an open issue. This may be the consequence of two opposite effects which influence the evolution of a shock wave in the corona: the increase of the Alfvén velocity with height between ~ 0.2–0.8 $R_\odot$ and ~ 1.8 $R_\odot$ suggested in [181] and the acceleration of the CME which occurs within a similar range of altitudes (e.g., [147]).

6.3 The Origin of SEP Events

The current view is that there are two classes of SEP events: impulsive and gradual events resulting, respectively, from particles accelerated by the HXR/GR flare itself and by the CME bow shock [182]. In situ measurements of the characteristics of both SEP populations are outlined in [183, 184]. However, this classification, which is mainly based on in situ measurements of particles with a few MeV/n and on statistical studies of the association between SEPs, CMEs, HXR/GR bursts and related phenomena, is questionable:

- Even at energies of a few MeV/n many events defy such a simple picture ([183, 184] and references therein).
- The distinction between impulsive and gradual SEP events, in terms of injection cones, charge states and composition, was suggested at low energies but is not borne out at higher energies (above, say, a few tens of MeV/n).
- It is not yet established that the CME driven shock is able to accelerate particles up to relativistic energies (see discussions in [144, 183]).
- Evidence has been given that for some events neither the HXR flare nor the CME bow shock are related to acceleration of high energy (above a few tens of MeV/n) particles, instead acceleration occurs minutes to tens of minutes after the HXR flare in coronal regions with heights in the range 0.1–1 $R_\odot$ (see [141] and references therein, [185] and Sect. 6.1). However, which acceleration process is able to produce high energy particles from these coronal sources remains to be investigated. For example, proton acceleration to GeV energies in reconnecting current sheets , as discussed in [102], seems to require unrealistically high magnetic field strengths.

The above remarks pointed out the difficulty of establishing a link between solar processes and particles at 1 AU. This essentially stems from the simultaneous presence of several potential accelerators (flares, CMEs, shocks) as discussed in the previous sections and the destruction of acceleration signatures by particle transport in the turbulent IP magnetic field. Although the origin of SEPs remains still an open question, it is clear that detailed studies of a large number of events, combining particle data with imaging and spectral observations of related phenomena in the corona and in the IP medium, are needed in order to avoid misleading associations and to bring new input to theory and modeling. Such an approach will benefit from new X-ray imaging observations by *RHESSI* and from *STEREO* measurements which will provide particle data and CME images at two locations in the IP medium.

7 Concluding Remarks

The working group sessions at the CESRA workshop in 2001 were focused on problems of energy release in flares and CMEs, magnetic reconnection, particle acceleration, particle transport, SEPs, and coronal and IP shocks, but have touched many further aspects of solar radio physics. Since the previous workshops in this series, significant progress was reached primarily by multi-wavelength studies including radio observations, by regular (routine) imaging observations with the NRH and the NoRH, by systematic exploration of spectral fine structures in the decimetric and metric wavelength ranges, and by opening the 100–400 GHz range for solar observations. New types of radio sources, e.g., DPS sources in SXR ejecta, the free-free emission from erupting filaments, or synchrotron emission from CMEs, and new types of spectral fine structures, e.g., lace bursts, sawtooth bursts, and zebra patterns in fiber bursts, were discovered. New models were proposed for some of these emissions. Further progress will be achieved by continuing along these lines, but a necessity to achieve regularly available imaging capability in combination with sensitive spectrographs in the ~ 0.5–10 GHz range was felt in order to fully exploit the diagnostic potential of solar radio emissions in this range. Required advances in theory include the extension of reconnection modeling to collisionless effects (to bridge the gap between energy release and particle acceleration), the formation and propagation of shocks in inhomogeneous media (to clarify the nature of coronal shocks and their detailed relation to flare blast waves and CMEs), more quantitative modeling of the various regimes of particle scattering, and development of modeling techniques encompassing the huge range of scales evidently involved in solar energy release.

References

1. B. Vršnak: *this volume* (2002)
2. H. Carmichael: in *AAS-NASA Symposium on the Physics of Solar Flares (NASA SP-50)* (1964), p. 451
3. P.A. Sturrock: Nature **211**, 695 (1966)
4. T. Hirayama: Solar Phys. **34**, 323 (1974)
5. R.A. Kopp, G.W. Pneuman: Solar Phys. **50**, 85 (1976)
6. Z. Svestka: *Solar Flares* (Reidel, Dordrecht, 1976)
7. L. Fletcher, H.P. Warren: *this volume* (2002)
8. S. Tsuneta: Astrophys. J. **456**, 840 (1996)
9. A. Czaykowska, B. de Pontieu, D. Alexander, G. Rank: Astrophys. J. **521**, L75 (1999)
10. S. Masuda, T. Kosugi, H.S. Hudson: Solar Phys. **204**, 55 (2001)
11. M.J. Aschwanden, D. Alexander: Solar Phys. **204**, 91 (2001)
12. T. Yokoyama, K. Akita, T. Morimoto, et al.: Astrophys. J. **546**, L69 (2001)
13. T. Yokoyama, K. Shibata: Astrophys. J. **549**, 1160 (2001)
14. R.L. Moore, G. Roumeliotis: "Triggering of Eruptive Flares - Destabilization of the Preflare Magnetic Field Configuration", in *IAU Colloq. 133: Eruptive Solar Flares* (1992), p. 69
15. R.L. Moore, A.C. Sterling, H.S. Hudson, J.R. Lemen: Astrophys. J. **552**, 833 (2001)
16. S.K. Antiochos, C.R. Devore, J.A. Klimchuk: Astrophys. J. **510**, 485 (1999)
17. P.A. Sweet: "The Neutral Point Theory of Solar Flares", in *IAU Symp. 6: Electromagnetic Phenomena in Cosmical Physics* (1958), p. 123
18. J.J. Aly: Astrophys. J. **375**, L61 (1991)
19. P.A. Sturrock: Astrophys. J. **380**, 655 (1991)
20. S.K. Antiochos: Astrophys. J. **502**, L181 (1998)
21. I. Sammis, F. Tang, H. Zirin: Astrophys. J. **540**, 583 (2000)
22. G. Aulanier, E.E. DeLuca, S.K. Antiochos, et al.: Astrophys. J. **540**, 1126 (2000)
23. A.C. Sterling, R.L. Moore, J. Qiu, H. Wang: Astrophys. J. **561**, 1116 (2001)
24. D.B. Melrose: Astrophys. J. **486**, 521 (1997)
25. Y. Yan: *this volume* (2002)
26. H. Lin, M.J. Penn, J.R. Kuhn: Astrophys. J. **493**, 978 (1998)
27. H. Lin, M.J. Penn, S. Tomczyk: Astrophys. J. **541**, L83 (2000)
28. S. Kahler: Astrophys. J. **214**, 891 (1977)
29. L.G. Bagalá, C.H. Mandrini, M.G. Rovira, et al.: Solar Phys. **161**, 103 (1995)
30. T. Sakurai, K. Shibata, K. Ichimoto, et al.: Proc. Astr. Soc. Japan **44**, L123 (1992)
31. S. Tsuneta, H. Hara, T. Shimizu, et al.: Proc. Astr. Soc. Japan **44**, L63 (1992)
32. S. Tsuneta: Astrophys. J. **456**, 840 (1996)
33. M.J. Aschwanden, A.I. Poland, D.M. Rabin: Ann. Rev. Astron. Astrophys. **39**, 175 (2001)
34. T. Sakao: Ph.D. Thesis (1994)
35. I.M. Chertok, E.I. Obridko, V.N. Mogilevsky, et al.: Astrophys. J. **567**, 1225 (2002)
36. P. Démoulin, L.G. Bagalá, C.H. Mandrini, et al.: Astron. Astrophys. **325**, 305 (1997)
37. L.G. Bagalá, C.H. Mandrini, M.G. Rovira, P. Démoulin: Astron. Astrophys. **363**, 779 (2000)

38. E.R. Priest, P. Démoulin: J. Geophys. Res. **100**, 23443 (1995)
39. M.R. Kundu, E.J. Schmahl, T. Velusamy: Astrophys. J. **253**, 963 (1982)
40. R.F. Willson: Solar Phys. **83**, 285 (1983)
41. R.K. Shevgaonkar, M.R. Kundu: Astrophys. J. **292**, 733 (1985)
42. M. Nishio, K. Yaji, T. Kosugi, et al.: Astrophys. J. **489**, 976 (1997)
43. Y. Hanaoka: Solar Phys. **173**, 319 (1997)
44. M. Aschwanden: "Radio and Hard X-ray Observations of Flares and their Physical Interpretation", in *Proceedings of the Nobeyama Symposium, held in Kiyosato, Japan, Oct. 27-30, 1998, Eds.: T. S. Bastian, N. Gopalswamy and K. Shibasaki, NRO Report No. 479., p.307-319* (1999), p. 307
45. A.T. Altyntsev, R.A. Sych, V.V. Grechnev, et al.: Solar Phys. **206**, 155 (2002)
46. C.E. Alissandrakis, G.B. Gelfreikh, V.N. Borovik, et al.: Astron. Astrophys. **270**, 509 (1993)
47. R.A. Sych, A.M. Uralov, A.N. Korzhavin: Solar Phys. **144**, 59 (1993)
48. G.A. Dulk, D.J. McLean: Solar Phys. **57**, 279 (1978)
49. K. Golap, C.V. Sastry: Solar Phys. **150**, 295 (1994)
50. A. Grebinskij, V. Bogod, G. Gelfreikh, et al.: Astron. Astrophys. Suppl. **144**, 169 (2000)
51. C.E. Alissandrakis: "The Magnetic Field of the Solar Corona", in *Advances in Solar Physics*, ed. by G. Belvedere, M. Rodonò, G.M. Simnett (Springer, 1994), no. 432 in LNP, p. 109
52. C.E. Alissandrakis, F. Chiuderi-Drago: Astrophys. J. Lett. **428**, L73 (1994)
53. C.E. Alissandrakis, F. Borgioli, F.C. Drago, et al.: Solar Phys. **167**, 167 (1996)
54. P. Kaufmann: *this volume* (2002)
55. G. Trottet, J.P. Raulin, P. Kaufmann, et al.: Astron. Astrophys. **381**, 694 (2002)
56. P. Heinzel, M. Karlický: *this volume* (2002)
57. K.L. Klein, G. Trottet, A. Magun: Solar Phys. **104**, 243 (1986)
58. G. Bruggmann, N. Vilmer, K.L. Klein, S.R. Kane: Solar Phys. **149**, 171 (1994)
59. R.F. Willson, K.R. Lang, K.L. Klein, et al.: Astrophys. J. **357**, 662 (1990)
60. M.R. Kundu, A. Nindos, S.M. White, V.V. Grechnev: Astrophys. J. **557**, 880 (2001)
61. G.D. Fleishman, Q.J. Fu, G.L. Huang, et al.: Astron. Astrophys. **385**, 671 (2002)
62. G.P. Chernov, L.V. Yasnov: Chinese Journal of Astronony and Astrophysics **1**, 525 (2001)
63. K. Jiřička, M. Karlický, H. Mészárosová, V. Snížek: Astron. Astrophys. **375**, 243 (2001)
64. M.J. Aschwanden: Solar Phys. **111**, 113 (1987)
65. J. Kuijpers: in *Radio Physics of the Sun, Kundu M.R., Gergely T.E. (eds.)*Vol. 86 (1980), p. 341
66. A.O. Benz: *this volume* (2002)
67. B. Kliem, M. Karlický, A.O. Benz: Astron. Astrophys. **360**, 715 (2000)
68. M. Karlický, Y. Yan, Q. Fu, et al.: Astron. Astrophys. **369**, 1104 (2001)
69. S. Wang, Y. Yan, R. Zhao, et al.: Solar Phys. **204**, 155 (2001)
70. J.I. Khan, N. Vilmer, P. Saint-Hilaire, A.O. Benz: Astron. Astrophys. **388**, 363 (2002)
71. R.J. Hastie: Astrophys. Space Sci. **256**, 177 (1998)
72. A. Klassen, H. Aurass, G. Mann: Astron. Astrophys. **370**, L41 (2001)
73. S. Tsuneta, T. Naito: Astrophys. J. **495**, L67 (1998)
74. S. Masuda, T. Kosugi, H. Hara, et al.: Nature **371**, 495 (1994)
75. T.G. Forbes, J.M. Malherbe: Astrophys. J. **302**, L67 (1986)

76. H. Aurass, B. Vršnak, G. Mann: Astron. Astrophys. **384**, 273 (2002)
77. H. Aurass, M. Karlický, B.J. Thompson, B. Vršnak: "Radio shocks from reconnection outflow jet? – new observations", in *Multi-Wavelength Observations of Coronal Structure and Dynamics – Yohkoh 10th Anniversary Meeting* (2002)
78. M. Bárta, M. Karlický: Astron. Astrophys. **379**, 1045 (2001)
79. V.V. Zheleznyakov, E.Y. Zlotnik: Solar Phys. **44**, 461 (1975)
80. J. Kuijpers: Astron. Astrophys. **40**, 405 (1975)
81. M. Karlický, M. Bárta, K. Jiřička, et al.: Astron. Astrophys. **375**, 638 (2001)
82. J. Jakimiec: "Turbulent Model of the Flare Energy Release", in *Ninth European Meeting on Solar Physics: Magnetic Fields and Solar Processes, ESA SP-448*, ed. A. Wilson. (1999), p. 729
83. S. Tsuneta: Publ. Astr. Soc. Japan **47**, 691 (1995)
84. J.A. Miller, P.J. Cargill, A.G. Emslie, et al.: J. Geophys. Res. **102**, 14 631 (1997)
85. J.F. Drake, D. Biskamp, A. Zeiler: Geophys. Res. Lett. **24**, 2921 (1997)
86. M. Hesse, K. Schindler, J. Birn, M. Kuznetsova: Phys. Plasmas **6**, 1781 (1999)
87. E. Antonucci, A.H. Gabriel, L.W. Acton, et al.: Solar Phys. **78**, 107 (1982)
88. N. Vilmer, A.L. MacKinnon: *this volume* (2002)
89. A. Anastasiadis, L. Vlahos, M.K. Georgoulis: Astrophys. J. **489**, 367 (1997)
90. L. Fletcher, P.C.H. Martens: Astrophys. J. **505**, 418 (1998)
91. M.J. Aschwanden, T. Kosugi, Y. Hanaoka, et al.: Astrophys. J. **526**, 1026 (1999)
92. L. Fletcher, T.R. Metcalf, D. Alexander, et al.: Astrophys. J. **554**, 451 (2001)
93. T. Takakura, K. Kai: Publ. Astr. Soc. Japan **18**, 57 (1966)
94. N. Vilmer, G. Trottet, S.R. Kane: Astron. Astrophys. **108**, 306 (1982)
95. C.F. Kennel, H.E. Petschek: J. Geophys. Res. **71**, 1 (1966)
96. P.A. Bespalov, V.V. Zaitsev, A.V. Stepanov: Solar Phys. **114**, 127 (1987)
97. C.F. Kennel: Rev. Geophys. **7**, 379 (1969)
98. M. Walt, W.M. MacDonald: Rev.Geophys. **4**, 543 (1964)
99. A.L. MacKinnon: Astron. Astrophys. **242**, 256 (1991)
100. Y. Lau, T.G. Northrop, J.M. Finn: Astrophys. J. **414**, 908 (1993)
101. R. Schlickeiser: *this volume* (2002)
102. Y. Litvinenko: *this volume* (2002)
103. H.S. Hudson, T. Kosugi, N.V. Nitta, M. Shimojo: Astrophys. J. **561**, L211 (2001)
104. M. Scholer: *this volume* (2002)
105. J.C. Brown, D.S. Spicer, D.B. Melrose: Astrophys. J. **228**, 592 (1979)
106. D.F. Smith, C.G. Lilliequist: Astrophys. J. **232**, 582 (1979)
107. G. Paesold, A.O. Benz, K.L. Klein, N. Vilmer: Astron. Astrophys. **371**, 333 (2001)
108. J.M. Greene: J. Geophys. Res. **93**, 8583 (1988)
109. J. Kuijpers, P. van der Post, C. Slottje: Astron. Astrophys. **103**, 331 (1981)
110. R.M. Winglee, G.A. Dulk, P.L. Pritchett: Astrophys. J. **328**, 809 (1988)
111. L. Vlahos, R.R. Sharma, K. Papadopoulos: Astrophys. J. **275**, 374 (1983)
112. P.A. Sturrock: in *AAS-NASA Symposium on Physics of Solar Flares*, ed. by W.N. Hess (1964), p. 357
113. R.J.M. Grognard: "Propagation of electron streams", in *Solar Radiophysics: Studies of Emission from the Sun at Metre Wavelengths* (1985), p. 253
114. V.N. Mel'nik, V. Lapshin, E. Kontar: Solar Phys. **184**, 353 (1999)
115. E.P. Kontar: Solar Phys. **202**, 131 (2001)
116. E.P. Kontar: Astron. Astrophys. **375**, 629 (2001)
117. K.G. McClements: Astron. Astrophys. **175**, 255 (1987)
118. R.J. Hamilton, V. Petrosian: Astrophys. J. **321**, 721 (1987)
119. G.D. Holman, M.R. Kundu, K. Papadopoulos: Astrophys. J. **257**, 354 (1982)

120. S.M. White, D.B. Melrose, G.A. Dulk: Astrophys. J. **308**, 424 (1986)
121. J. Lee, D.E. Gary: Astrophys. J. **543**, 457 (2000)
122. V.F. Melnikov, A. Magun: Solar Phys. **178**, 591 (1998)
123. V. Petrosian: Astrophys. J. **299**, 987 (1985)
124. R.R. Sharma, L. Vlahos: Astrophys. J. **280**, 405 (1984)
125. A.V. Stepanov, Y. Tsap: Solar Phys., submitted (2002)
126. A.L. MacKinnon, J.C. Brown, J. Hayward: Solar Phys. **99**, 231 (1985)
127. J.C. Brown: Solar Phys. **18**, 489 (1971)
128. K. Shibasaki: *this volume* (2002)
129. D.B. Melrose: *Plasma Astrophysics: Nonthermal Processes in Diffuse Magnetized Plasmas. Volume 2 - Astrophysical Applications* (Gordon and Breach Science Publishers, New York, 1980)
130. G. Paesold, A.O. Benz: Astron. Astrophys. **351**, 741 (1999)
131. P. Messmer: Astron. Astrophys. **382**, 301 (2002)
132. K. Arzner: Journal of Physics A Mathematical General **35**, 3145 (2002)
133. M. Pick: Solar Phys. **104**, 19 (1986)
134. G. Trottet: Solar Phys. **104**, 145 (1986)
135. G. Kanbach, D.L. Bertsch, C.E. Fichtel, et al.: Astron. Astrophys. **97**, 349 (1993)
136. V.V. Akimov, P. Ambrož, A.V. Belov, et al.: Solar Phys. **166**, 107 (1996)
137. L. Fletcher, H. Hudson: Solar Phys. **204**, 69 (2001)
138. G.H. Share, R.J. Murphy, A.J. Tylka, et al.: Solar Phys. **204**, 41 (2001)
139. C. Caroubalos, C.E. Alissandrakis, A. Hillaris, et al.: Solar Phys. **204**, 165 (2001)
140. I.M. Chertok, V.V. Fomichev, A.A. Gnezdilov, et al.: Solar Phys. **204**, 139 (2001)
141. K.L. Klein, G. Trottet, P. Lantos, J.P. Delaboudinière: Astron. Astrophys. **373**, 1073 (2001)
142. Z. Švestka, E.W. Cliver: "History and Basic Characteristics of Eruptive Flares", in *IAU Colloq. 133: Eruptive Solar Flares* (1992), p. 1
143. D. Maia, A. Vourlidas, M. Pick, et al.: J. Geophys. Res. **104**, 12 507 (1999)
144. K.L. Klein, G. Trottet: Space Science Reviews **95**, 215 (2001)
145. R.A. Harrison: Astron. Astrophys. **162**, 283 (1986)
146. N. Srivastava, R. Schwenn, B. Inhester, et al.: Space Science Reviews **87**, 303 (1999)
147. J. Zhang, K.P. Dere, R.A. Howard, et al.: Astrophys. J. **559**, 452 (2001)
148. N. Nitta, S. Akiyama: Astrophys. J. **525**, L57 (1999)
149. R.A. Harrison: Astron. Astrophys. **304**, 585 (1995)
150. C. Marqué, P. Lantos, K.L. Klein, J.M. Delouis: Astron. Astrophys. **374**, 316 (2001)
151. C. Marqué, P. Lantos, J.L. Delaboudinière: Astron. Astrophys. **387**, 317 (2002)
152. P.F. Chen, K. Shibata: Astrophys. J. **545**, 524 (2000)
153. K.V. Sheridan, B.V. Jackson, D.J. McLean, G.A. Dulk: Proceedings of the Astronomical Society of Australia **3**, 249 (1978)
154. N. Gopalswamy, M.R. Kundu: Astrophys. J. **390**, L37 (1992)
155. N. Gopalswamy, M.R. Kundu: Solar Phys. **143**, 327 (1993)
156. K. Saito, E. Tandberg-Hanssen: Solar Phys. **31**, 105 (1973)
157. G.S. Vaiana, J.M. Davis, R. Giacconi, et al.: Astrophys. J. **185**, L47 (1973)
158. T.S. Bastian, M. Pick, A. Kerdraon, et al.: Astrophys. J. **558**, L65 (2001)
159. G.J. Nelson, D.B. Melrose: "Type II bursts", in *Solar Radiophysics: Studies of Emission from the Sun at Metre Wavelengths* (1985), p. 333
160. G. Thejappa, R.J. MacDowall: J. Geophys. Res. **106**, 25 313 (2001)
161. H.V. Cane, N.R. Sheeley, R.A. Howard: J. Geophys. Res. **92**, 9869 (1987)

162. M.J. Reiner, M.L. Kaiser: J. Geophys. Res. **104**, 16 979 (1999)
163. G. Thejappa, R.J. MacDowall: Astrophys. J. **544**, L163 (2000)
164. M. Pick: "Radio and Coronagraph Observations: Shocks, Coronal Mass Ejections and Particle Acceleration", in *Proceedings of the Nobeyama Symposium, held in Kiyosato, Japan, Oct. 27-30, 1998, Eds.: T. S. Bastian, N. Gopalswamy and K. Shibasaki, NRO Report No. 479* (1999), p. 187
165. N. Gopalswamy, M.L. Kaiser, R.P. Lepping, et al.: J. Geophys. Res. **103**, 307 (1998)
166. H.T. Classen, H. Aurass: Astron. Astrophys. **384**, 1098 (2002)
167. W.J. Wagner, R.M. MacQueen: Astron. Astrophys. **120**, 136 (1983)
168. N. Gopalswamy, M.R. Kundu: "Surprises in the Radio Signatures of CMEs", in *Coronal Magnetic Energy Releases, Proceedings of the CESRA Workshop Held in Caputh/Potsdam, Germany, 16-20 May 1994* (Springer-Verlag, Berlin Heidelberg New York, 1995), p. 223
169. D. Maia, M. Pick, A. Vourlidas, R. Howard: Astrophys. J. **528**, L49 (2000)
170. K.L. Klein, J.I. Khan, N. Vilmer, et al.: Astron. Astrophys. **346**, L53 (1999)
171. G.E. Moreton: Astronom. J. **65**, 494 (1960)
172. Y. Uchida: Solar Phys. **4**, 30 (1968)
173. Y. Uchida: Solar Phys. **39**, 431 (1974)
174. B.J. Thompson, S.P. Plunkett, J.B. Gurman, et al.: Geophys. Res. Lett. **25**, 2465 (1998)
175. B.J. Thompson, J.B. Gurman, W.M. Neupert, et al.: Astrophys. J. **517**, L151 (1999)
176. M.J. Wills-Davey, B.J. Thompson: Solar Phys. **190**, 467 (1999)
177. A. Klassen, H. Aurass, G. Mann, B.J. Thompson: Astron. Astrophys. **141**, 357 (2000)
178. C. Delannée, G. Aulanier: Solar Phys. **190**, 107 (1999)
179. H.S. Hudson, J.I. Khan, J.R. Lemen, et al.: Solar Phys., submitted (2002)
180. J.I. Khan, H. Aurass: Astron. Astrophys. **383**, 1018 (2002)
181. G. Mann, A. Klassen, C. Estel, B.J. Thompson: ESA SP **446**, 477 (1999)
182. D.V. Reames: Space Science Reviews **90**, 413 (1999)
183. W. Dröge: *this volume* (2002)
184. S. Krucker: *this volume* (2002)
185. H.V. Cane, W.C. Erickson, N.P. Prestage: J. Geophys. Res. in press (2002)

Solar Observations
at Submillimeter Wavelengths

Pierre Kaufmann[1,2]

[1] CRAAM, Universidade Presbiteriana Mackenzie, São Paulo, SP, Brazil
[2] part-time researcher at CSS, UNICAMP, Campinas, SP, Brazil

Abstract. We review earlier to recent observational evidences and theoretical motivations leading to a renewed interest to observe flares in the submillimeter (submm) - infrared (IR) range of wavelengths. We describe the new solar dedicated submillimeter wave telescope which began operations at El Leoncito in the Argentina Andes: the SST project. It consists of focal plane arrays of two 405 GHz and four 212 GHz radiometers placed in a 1.5-m radome-enclosed Cassegrain antenna, operating simultaneously with one millisecond time resolution. The first solar events analyzed exhibited the onset of rapid submm-wave spikes (100-300 ms), well associated to other flare manifestations, especially at X-rays. The spikes positions were found scattered over the flaring source by tens of arcseconds. For one event an excellent association was found between the gamma-ray emission time profile and the rate of occurrence of submm-wave rapid spikes. The preliminary results favour the idea that bulk burst emissions are a response to numerous fast energetic injections, discrete in time, produced at different spatial positions over the flaring region. Coronal mass ejections were associated to the events studied. Their trajectories extrapolated to the solar surface appear to correspond to the onset time of the submm-wave spikes, which might represent an early signature of the CME's initial acceleration process.

1 Introduction

1.1 Why Explore the Unobserved Submm-IR Spectral Region?

The observations of solar flares at submillimeter waves are now receiving considerable attention, although their importance has been stressed for more than forty years, both on observational grounds and on theoretical indications. The first observations of radio bursts with fluxes up to 30 GHz were obtained by Hachenberg and Wallis (1961). One example is shown in Fig. 1, together with model fittings, which will be further discussed later. It shows examples of radio burst spectra varying in shape with time during the non-thermal impulsive phase of the event, with flux increasing up to 30 GHz, followed by the characteristic thermal post-burst-increase. Hachenberg and Wallis (1961) have suggested that these spectra were the composition of multiple synchrotron spectra with different turnover frequencies. Other models are shown in Fig. 1 which also fit these observations. Synchrotron emissions by ultrarelativistic electrons were suggested, extending the turnover frequency to the IR and visible ranges, in order to explain white-light flare emissions (Stein and Ney, 1963; Shklovsky, 1964) or, alternatively,

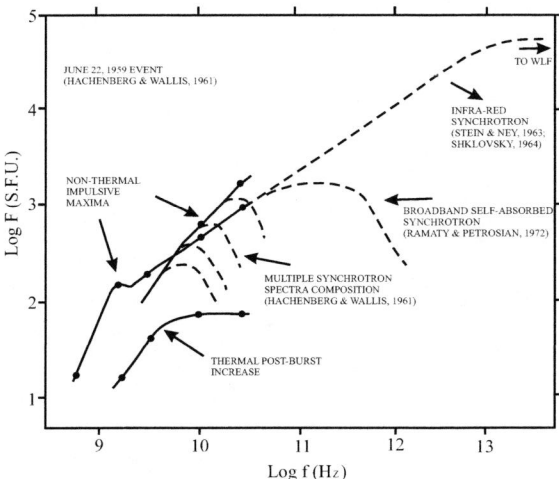

Fig. 1. Solar burst spectra in the radio-far IR range as suggested by early short microwave observations and model predictions.

with a turnover in the THz region caused by the effect of free-free absorption combined with gyrosynchrotron self-absorption (Ramaty and Petrosian, 1972). The first radio observations up to 90 GHz were summarized by Croom (1973), indicating the existence of several spectral classes, shown in Fig. 2. There are events with radio emission in the decimeter-meter range only, bursts with fluxes increasing up to the shortest mm-waves, and events exhibiting several spectral maxima in that range of wavelengths.

The first attempt to observe flares in the submm-wavelength range (meant here to address frequencies higher than 100 GHz) was made by Clark and Park (1968) using a cooled bolometer at 250 GHz, at the 1.5-m Queen Mary College optical telescope, UK. They have found brightenings of about 100 K on active regions, not necessarily well related to optical flares, on time scales of 1 min. limited by the raster mode of observation, shown in Fig. 3. The theoretical impacts of this discovery were discussed by Beckman (1968). Hudson (1975), using cooled bolometers at the 1.5-m optical telescope at Mount Graham, Arizona, USA, reported 10-50 K fluctuations at 850 GHz (1 min. time resolution) and at 12 THz (1 s time resolution). However there were no other observations since then, possibly because of the experimental difficulties existing for this range of wavelengths, the technology required not easily available, and for the severe limitations caused by atmospheric opacity on ground level observations.

There were many solar bursts exhibiting complex spectra at shorter microwaves, many of which with fluxes increasing with frequency up to the highest frequencies where these events have been measured, i.e. about 100 GHz. The most well known spectra of this class are shown in Fig. 4. They are found both for large and for small bursts. These results provide observational suggestion

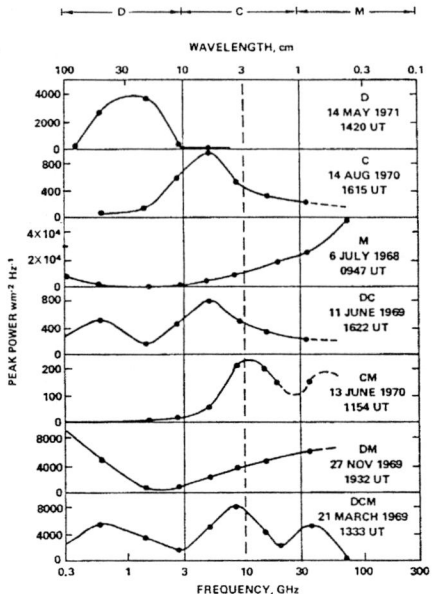

Fig. 2. Classes of solar burst spectra in the metric-millimeter range of wavelengths (after Croom, 1973).

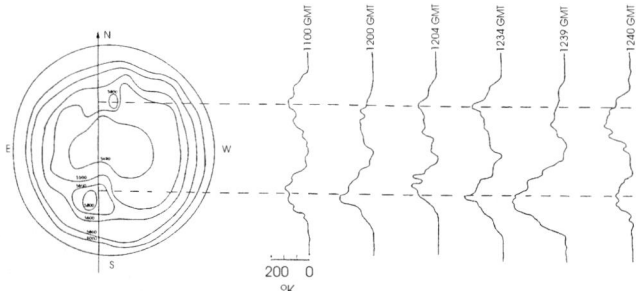

Fig. 3. First 250 GHz measurements made by successive N-S raster scans across the solar disc on April 4, 1968, have shown brightenings of about 100 K on active regions (after Clark and Park, 1968).

that important emissions and/or spectral components are to be expected in the submillimeter-infrared range.

Approaching the IR region from the higher energy gamma-, X-ray, UV and visible gives a similar impression of large unobserved submm-IR fluxes. The luminosity distribution for a large flare with data points due to different emission mechanisms (lines, thermal, non-thermal, etc.), show that luminosity increases from the highest energies toward the EUV and visible, giving the strong sug-

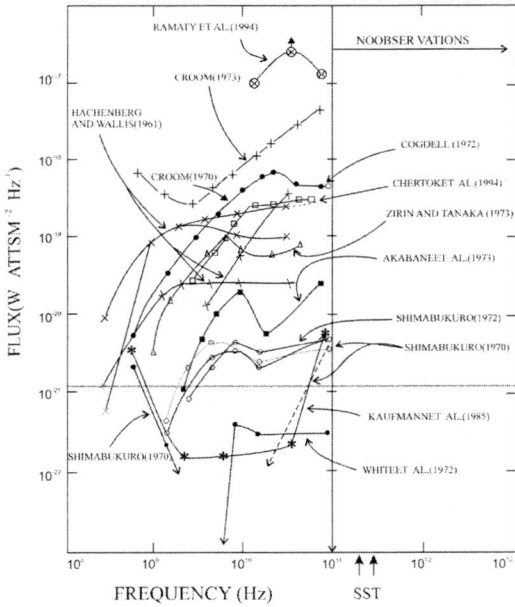

Fig. 4. Examples of observed solar burst spectra, for small and large events exhibiting fluxes increasing for shorter microwave-mm waves, suggesting an extension into the submm-IR region (after Kaufmann, 1995, 1996).

gestion that the maximum luminosity for this flare might well have been in the unobserved submm-IR spectral region (Ramaty and Mandzhavidze, 1994).

1.2 Emission mechanisms

In summary, Fig. 5 gives one simplified diagram of the complete electromagnetic spectrum for a flare with moderate importance. The submm-IR range occupies a frequency range of almost three decades, where a number of thermal and non-thermal models of emission may fit in to explain the spectra observed at longer (radio) and at shorter wavelengths (visible, UV, X- and gamma-ray).

Although emissions are due to different mechanisms, all can be thought as direct or indirect response to the primary flare energy production process. It is widely accepted that the first particles accelerated in flare site(s) produce radio emission in the cm-mm range by synchrotron losses in the active centers' magnetic field as they precipitate into the denser regions of the solar atmosphere, where they produce hard X-rays and gamma-rays by bremsstrahlung (Ginzburg and Syrovatskii, 1965; Takakura, 1967; Brown, 1976; Trottet et al., 1993; Chupp, 1996; Vilmer and MacKinnon, *this volume*). However, the physical parameters of the magnetoactive plasma where the bursts are produced must be taken into account as for their influence on the net escaping radio emissions, resulting in a complex composition of the above emissions with a long lasting non-thermal

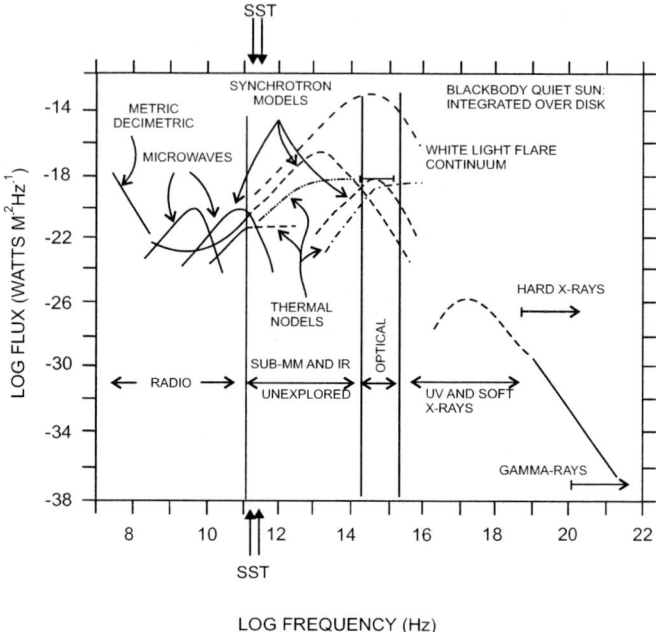

Fig. 5. A simplified description of the complete electromagnetic spectrum for a moderately intense solar flare. Important tests on emission mechanisms are expected to be possible with observations in the submm-IR region, unexplored for flare emissions. The two arrows labelled 'SST' indicate the observing frequencies of the Solar Submillimeter-wavelength Telescope (cf. Sect. 2).

component from trapped electrons, emissions from thermalized particles, all undergoing absorptions in the burst source and the surrounding medium (Ramaty, 1973; Ohki and Hudson, 1975; Klein, 1987; Dulk and Marsh, 1982; de Jager, 1986; Bastian, Benz and Gary, 1998).

The synchrotron radiation flux from a compact optically thick source of energetic electrons increases with frequency to the power 5/2 up to a "turnover" frequency close to the values calculated using Ginzburg and Syrovatskii's (1965) equation for a single electron, which is applicable to an ensemble of monoenergetic electrons and is a qualitatively good approximation to other populations dominated by higher energy electrons:

$$\nu_m = 10^{-5.3} B [E(eV)]^2 \text{ Hz}, \tag{1}$$

where B is the magnetic field (Gauss) and E the energy of the electrons accelerated. For $\nu > \nu_m$ the source becomes optically thin, and fluxes reduce with increasing frequency exhibiting different spectral indices, which are a function of the energies and spectral indices of the accelerated electron populations. Therefore, the observed radio burst spectra with turnover frequencies larger than 10^{11} Hz (100 GHz), as suggested by several examples given in Fig. 4, require dense sources of electrons accelerated to ultrarelativistic energies (> 5–15 MeV), for

B=1000-100 G. Electrons in the same range of energies are needed to explain gamma-rays observed at the onset of impulsive bursts (Forrest and Chupp, 1983; Chupp, 1996).

1.3 Signatures of Burst Energy Release Time Scales

The burst energy released with time at the origin of the flaring process, dE/dt, where E is the energy and t the time, is usually assumed to be proportional to the observed flux time profiles, at shorter microwaves to submm-waves, and at hard X- and gamma-rays. The discrete or fragmented nature of energy release in the impulsive burst phase became more evident as more sensitive detectors were used together with better time resolution. A pulsating X-ray burst with a 16 s period was detected in a balloon experiment (Parks and Winckler, 1969). Repetitive energy production was suggested from X-ray rapid burst structures (tens of seconds) observed by the OSO-V satellite (Frost, 1969). Well defined X-ray rapid time structures (seconds) in solar bursts, detected by the ESRO TD-1A satellite, led to the proposal of the concept of Elementary Flare Bursts, EFBs (van Beek, de Feiter and de Jager, 1974; de Jager and de Jonge, 1978). Shorter time structures (<1 s) were found in hard X-ray bursts by a number of different experiments (Kiplinger et al., 1983; Hurley et al., 1983; Machado et al., 1993).

In the radio band we must first be aware that there are two frequency regions reflecting entirely distinct mechanisms of burst emission (Kundu, 1965 and references therein; see Fig. 5). The well known rapid subsecond time structures at longer wavelengths (metric-decametric) are due to coronal plasma excited by electron beams and waves. A variety of spectral shapes are found, depending on the nature and drift of the exciter. The narrowband spikes have fluxes observed only in the respective bandwidths. In general there is an overall flux increase toward smaller frequencies (Wild, Smerd and Weiss, 1963, and references therein). Fast time structures found in the intermediate decimetric-centimetric radio region (Dröge, 1977; Slottje, 1978; Allaart et al., 1990; Benz et al., 1992 and Benz, *this volume*) are difficult to analyse because they are produced by non-linear processes (cf. Benz, *this volume*).

However, there is no ambiguity for the identification of the subsecond time structures found at shorter cm-mm wavelengths (Kaufmann, Piazza and Raffaelli, 1977; Gaizauskas and Tapping, 1980; Kaufmann et al. 1980; Nakajima, 2000; Altyntsev et al., 2000), which are linked to the main synchrotron burst emission. It has been found that the main burst emission flux at mm-wavelengths was proportional to the repetition rate of the fast superimposed time structures suggesting a quasi-quantization of the burst energetic content (Kaufmann et al. 1980; Raulin et al, 1998).

One remarkable solar event observed on May 21, 1984 exhibited the main flux increasing for shorter mm-wavelengths, with fast repetitive superimposed spikes, at 30 and 94 GHz, well correlated to $\geq$24 keV hard X-rays to within 128 ms, as shown in Fig. 6 (Kaufmann et al. 1985). Other very suggestive correlations have been found between fast burst time structures at mm-wavelengths and hard

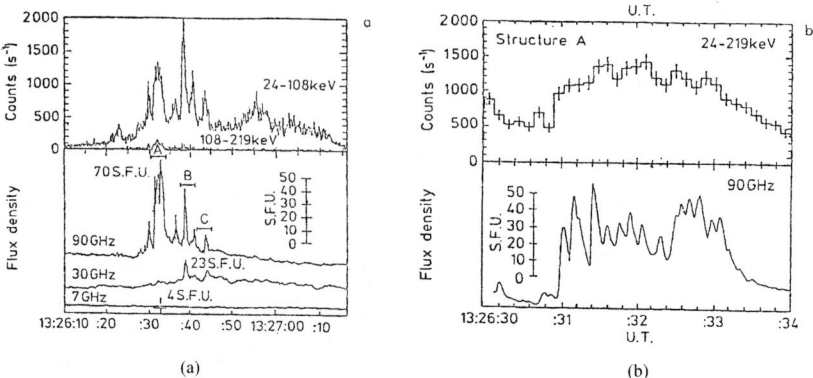

Fig. 6. The solar burst of May 21, 1984 as observed at 7 GHz, 30 GHz, 90 GHz, at Itapetinga Radio Observatory (Brazil), and at hard X-rays by the HXRBS experiment onboard the SMM satellite. (b) displays on an expanded scale a 4 s interval of the major structure A, showing 14 spikes in 2 s at 90 GHz, with the hard X-ray structure correlated to that major time structure. Finer details at hard X-rays cannot be identified because of the HXRBS limiting time resolution of 128 ms (after Kaufmann et al., 1985)

X-rays (Zodi Vaz et al., 1985; Takakura et al., 1983; Kaufmann et al., 2000), which favor the idea that they might represent signatures of the same energetic injections at the origin of the flaring process.

These observations bring some insights into the energy production at the origin of flares. One interesting suggestion is that small or large flares might be conceived as the result of a smaller or larger number of primary energetic injections, each one with comparable energy release rate and total content (Kaufmann, 1985). Qualitatively, the mean fluxes can be represented as being proportional to the rate of pulse production, R(t), quantized with a mean energy $\langle \epsilon \rangle$. Fluxes might then be described in this simplified approximation:

$$I(t) \propto R(t) \qquad (2)$$

And the total flare energy content over a duration T:

$$W \approx \langle \epsilon \rangle \int_0^T R(t)\, dt \approx n \langle \epsilon \rangle \qquad (3)$$

where n is the total number of energetic injections.

1.4 Constraints Imposed by the Short Time Scales

It is difficult to reconcile the short subsecond time scales observed to purely synchrotron and collision losses. Some possibilities have been considered assuming that the lifetime of the high energy electrons might be reduced by other mechanisms, such as by inverse Compton quenching in multiple compact synchrotron sources, which would require an emission spectrum with a turnover in the far IR region (Kaufmann et al., 1986). On the other hand, the short time scales might simply represent the scale sizes where the energy release process occurs (Emslie, Mehta and Machado, 1994). These results bring a number of constraints for interpretation difficult to explain without adopting too many assumptions (Beckman, 1968; McClements and Brown, 1986; Klein, 1987).

Other physical scenarios might be conceived to account for the short time scales. For example the energy release sites in flares might be placed very deep in the solar atmosphere where large magnetic fields would account for high turnover frequencies of synchrotron emission. Dense optically thick thermal sources might have turnover frequencies in the submm-IR range (Ohki and Hudson, 1975). Collision losses may become very effective to reduce lifetimes for both thermal and non-thermal particles due to the high densities found in the chromosphere-photosphere region. On the other hand, mechanisms of energy release in flares, in repetitive and short time scales, are predicted by various models of plasma instabilities such as reconnections in twisted magnetic fields (Sturrock and Uchida, 1981); networks of unstable magnetic fluxules (Sturrock et al., 1984); formation and explosive disruptions of magnetic islands by coalescence instabilities (Sakai et al., 1986; Sturrock, 1986).

2 The Solar Submillimeter-Wavelength Telescope

A new dedicated Solar Submillimeter-wavelength Telescope (SST) is now filling this observational gap (Kaufmann et al. 1994; 2002a). It was installed in 1999 at the El Leoncito Astronomical Complex, CASLEO, at 2550-m altitude, in the Argentinean Andes (Fig. 7a). Important adjustments were necessary after the SST subsystems were assembled at the site. The most critical was the successful mechanical recovery of the reflector, which had been damaged prior to the final installation in El Leoncito. The SST has a 1.5-m radome-enclosed Cassegrain reflector. The front-end has a focal plane arrangement of four 212 GHz and two 405 GHz room temperature, heterodyne, total power radiometers operated simultaneously with time resolution of 1 ms, receiving single linear polarizations, orthogonal for the two frequencies (Fig. 7b). Three 212 GHz beams partially overlap each other to allow burst angular position determinations (Giménez de Castro et al., 1999, and references therein). The six beams are shown projected on the solar disk in Fig. 8(a) for March 22, 2000, when beams were still large, before optimization was made, and Fig. 8(b) for April 6, 2001, with nominal angular sizes of 2 and 4 arcminutes at 405 and 212 GHz, respectively.

Atmospheric opacity was systematically measured at El Leoncito and found to be excellent, comparable to a number of other submm-wavelength sites at higher altitude. Zenith attenuation was most commonly found to be of 0.18 nepers at 212 GHz and of 0.8 nepers at 405 GHz. Nearly 85% of measurements taken during five months in 2001 indicated opacities less than 0.4 and 1.5 nepers, for the two frequencies, respectively (Melo et al. 2002).

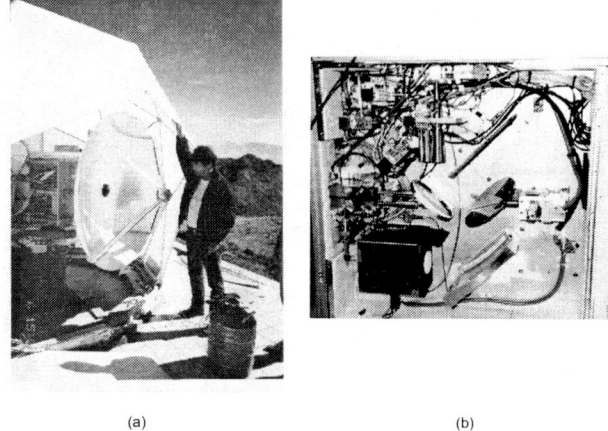

(a)   (b)

Fig. 7. (a) The SST 1.5-m antenna installed at the El Loncito Astronomical Complex, CASLEO, near San Juan, in the Argentina Andes. (b) View of the SST front-end box, as seen from the subreflector. The flat mirror (at right) reflects incoming signals (from the subreflector, or from the calibration loads at room - above - and hot - below, left - temperatures) to the two 405 GHz horns, left side, through the polarizing grid, in one plane of polarization. The orthogonal plane is reflected by the grid to the cluster of four 212 GHz horns, at the top. The resulting SST beams are shown projected on the solar disk in Fig. 8(a), before main reflector repairs, and Fig. 8(b) the nominal beamsizes, after the main repairs and adjustments were done.

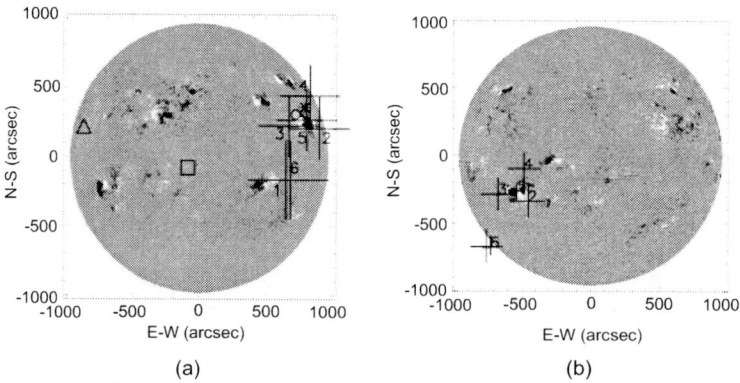

(a)   (b)

Fig. 8. (a) The SST beams (before the main reflector was repaired) on March 22, 2000, projected on a Kitt Peak Observatory magnetogram. Symbols correspond to positions tracked near the solar center (square) and at the limb (triangle), with 5 seconds sampled outputs as displayed in Fig. 10. (b) The six nominal beams (after repairs) on April 6, 2001, projected on a SOHO MDI magnetogram.

3 The First Solar Flare Observations at Submillimeter Waves

Short solar observing campaigns were carried out at El Leoncito in 1999 and 2000 while SST was still undergoing repairs, integration, tests and alignment. Longer runs began after April 2001. Several events were recorded and results for two flares studied in more detail are shown in the following sections.

3.1 The Solar Flare of March 22, 2000

SST observations were obtained for a GOES class X1.1 flare which occurred in NOAA AR 8910 at about 1830-1930 UT (Kaufmann et al., 2001a). The SST six beams are shown projected on the solar KPNO magnetogram in Fig. 8(a). The main burst emission at submillimeter waves was very weak for this event, as shown in the compressed time scale time profiles in Fig. 9: (a) at 212 GHz. obtained with large time integration, together with GOES X-rays, Itapetinga (Brazil) microwaves and HASTA Hα light curves; (b) the main synchrotron component identified after minimizing the small fluctuations by "beamswitching" subtraction of signals of beam 2 (on source) and beam 1 (off source), with

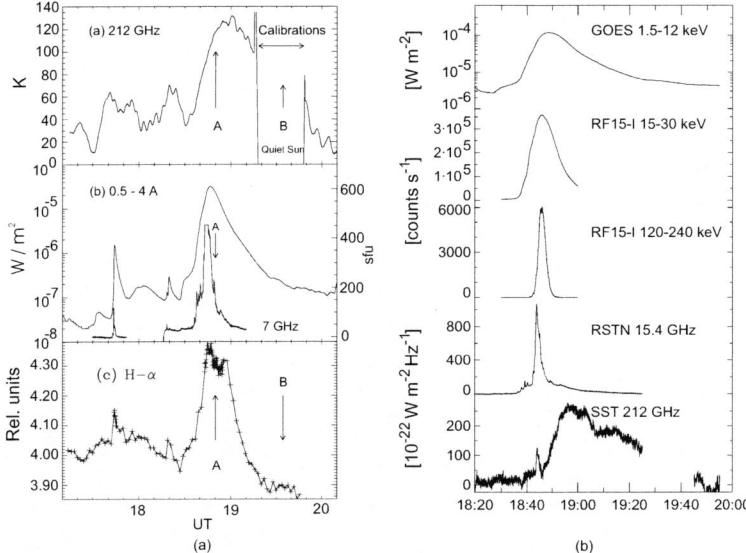

Fig. 9. . The March 22, 2000 flare compressed time profiles: (a) highly time-integrated 212 GHz flux density at the top, soft X-rays from GOES, 7 GHz from Itapetinga, Hα from HASTA (Max-Planck Institute Hα Solar Telescope for Argentina) (after Kaufmann et al., 2001b).(b) the non-thermal 212 GHz component was isolated using the beam-switching method between beams 2 and 1, shown with RSTN (Radio Solar Telescopes Network) microwaves, Interball and GOES X-rays time profiles (Trottet et al., 2002).

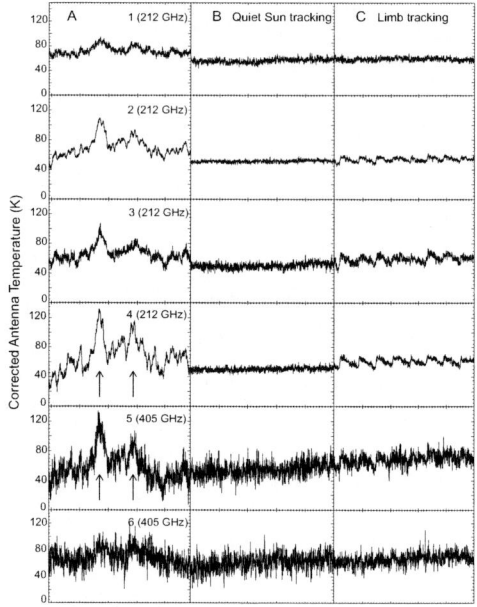

Fig. 10. 5 seconds sections, sampled with 5 ms time resolution, examples of rapid submillimeter spikes (left) as detected in the SST channels at the time labeled A in Fig. 9(a), in antenna corrected temperature scale, compared to the radiometers' output tracking the solar center (B) and limb (C) (after Kaufmann et al., 2001a).

intensity consistent with the optically thin emission of ≥ 12 MeV electrons (Trottet et al., 2002).

Numerous fast submillimeter spikes (100-300 ms) were detected. Examples of larger pulses are shown in Fig. 10, in a 5 seconds sample, 5 ms time resolution. Their flux densities are larger at 405 GHz (about 500 sfu, 1 sfu = 10^{-22} Wm^{-2}Hz^{-1}) than at 212 GHz (about 220 sfu). The occurrence rate with time exhibits substantial increases well correlated to the slower burst emission components at GOES X-rays, as it can be seen in the plots for March 21, 22 and 23, 2000 (Fig. 11). The first large enhancement in pulse occurrence in March 22 coincided with the sudden emergence of a large magnetic loop structure in AR 8910 sometime between 1715:37-1741:45 UT, as seen in the series of TRACE UV frames (Freeland, 2000, pers. comm.).

3.2 The Solar Flare of April 6, 2001

This GOES class X5.6 flare in NOAA AR 9415 was observed by SST, at 212 GHz only (because the atmosphere was practically opaque at 405 GHz). The event was well observed at microwaves by OVSA (New Jersey Institute of Technology Owens Valley Solar Array) and by hard X-ray and and gamma-ray experiments onboard of Yohkoh (Japan/US/UK) and Shenzhou-2 (China) satellites (Kaufmann et al., 2002b). The SST beams are shown projected on a SOHO MDI

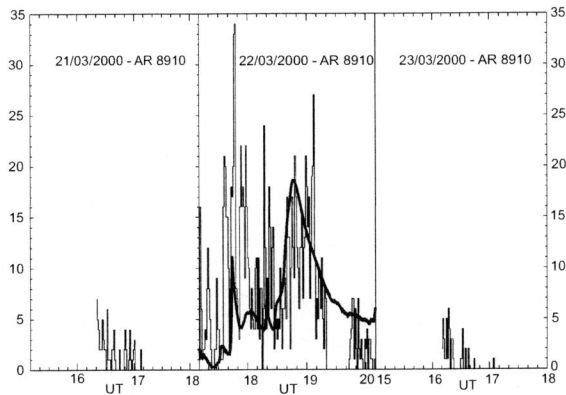

Fig. 11. Submillimeter 212 GHz pulse occurrence rates with time (derived from radiometer 2 corrected antenna temperature output) for March 21, 22 and 23, 2000, showing the dramatic increase together with the large flare in NOAA AR 8910 on March 22. GOES 0.5-4 Å X-rays are plotted together.

magnetogram, in Fig. 8(b). Figure 12(a) shows the time profiles at 18 GHz (OVSA), 212 GHz and 0.324-12.5 MeV hard-rays (Yohkoh/GRS). Figure 12(b) shows the close time correspondence of time profiles at 212 GHz, 0.2-0.6 MeV hard x-rays (from Shenzhou-2 and Yohkoh experiments) at the onset of the impulsive phase. The large flux observed at 212 GHz suggests a turnover frequency somewhere between 18-212 GHz, or a nearly flat spectrum from microwaves to submm-wavelengths.

10 seconds 212 GHz samples, labeled A, B, C and D in Fig. 12(a), are displayed in Fig. 13 with 5 ms time resolution. Rapid spikes are observed as the main flux increases - similarly to the pulses found in the March 22, 2000 event. The rate of pulse occurrence with time has an extraordinarily good correspondence with the time profile at ≥ 1 MeV gamma-rays, and with the flux onset of the 212 GHz emissions, as shown in Fig. 14.

The angular positions of the 212 GHz spikes relative to the main emission source position have been determined for the sections B and C of Fig. 13, by correlating the outputs from beams 2,3 and 4 (the multiple beam technique described by Giménez de Castro et al., 1999). Their spatial positions are shown in Fig. 15, suggesting the pulses are produced at different sites in the flaring source, separated by tens of arcseconds. Similar results have been obtained at a mm-wavelength (48 GHz), for impulsive components in complex bursts (Correia et al., 1995; Raulin et al., 2000).

4 The Onset of Pulses at Submm-Wavelengths and CMEs

The possible association of Coronal Mass Ejections (CMEs) to the onset of the rapid spikes observed at submm-waves has been suggested (Kaufmann et al.,

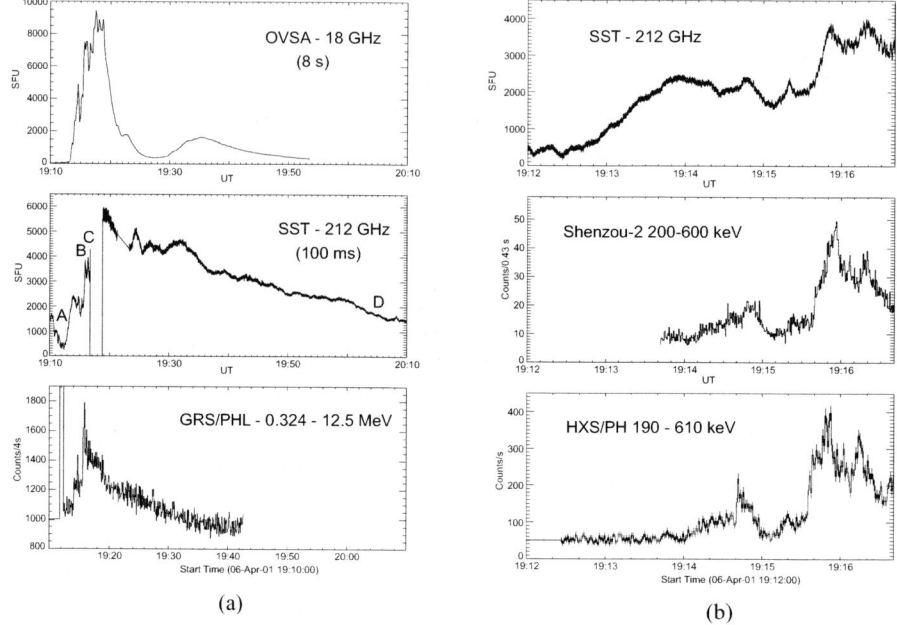

(a)   (b)

Fig. 12. (a) time profiles for the April 6, 2001 solar burst, at microwaves (top), 212 GHz (middle) and high energy hard X-rays (bottom). Sections labeled A,B.C, and D are shown expanded in Fig. 13. (b) Expanded 212 GHz burst time profile with ≥ 0.2 MeV hard X-rays obtained by two different satellite experiments (after Kaufmann et al., 2002b).

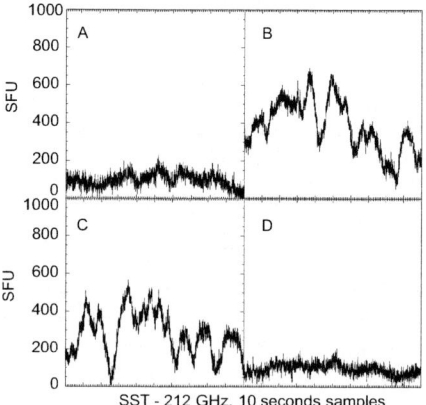

Fig. 13. 10 seconds expanded sections, labeled in Fig. 12(a), show the rapid spikes at the maximum phase of the event, B and C, in solar flux units above the main emission level (after Kaufmann et al., 2002b).

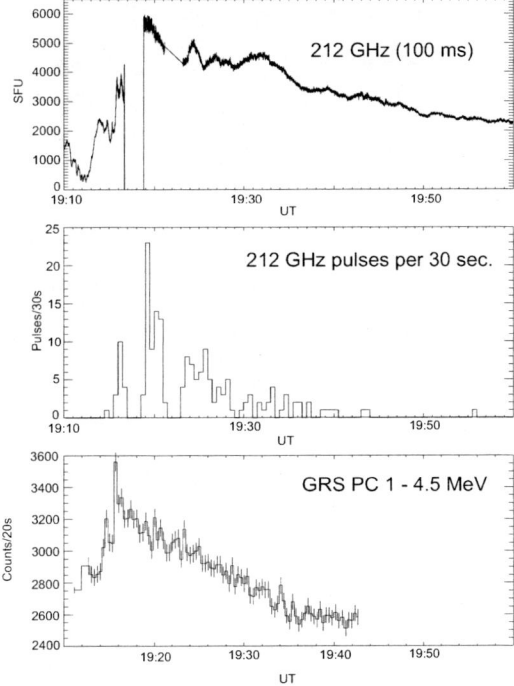

Fig. 14. The frequency of 212 GHz pulse occurrence (middle) correlates very well with the ≥ 1 MeV gamma-ray time profile (bottom) and the 212 GHz main emission fluxes (top) at the onset of the impulsive phase (after Kaufmann et al., 2002b).

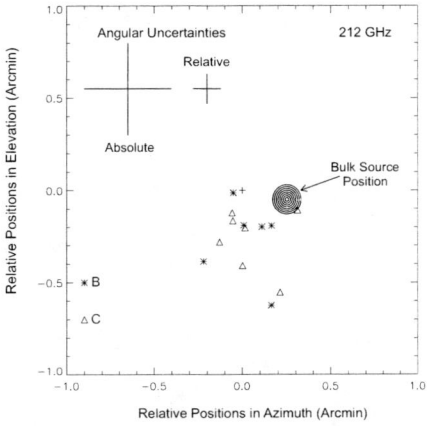

Fig. 15. The angular positions of the larger 212 GHz spikes in sections B and C of the burst (Figs. 12 and 13), with respect to the main emission source angular position (after Kaufmann et al., 2002b).

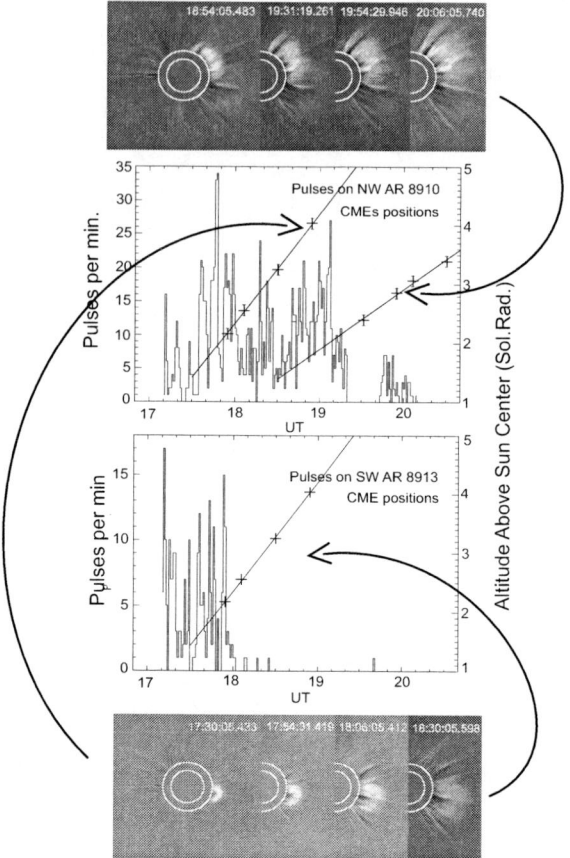

Fig. 16. The March 22, 2000 CMEs: difference image frames obtained by the LASCO C2 coronagraph on SOHO, compared to 212 GHz pulse count rates (see Fig. 11), ordinates at left. The first CME, related to the SW AR, is shown at the bottom. The second CME, related to the NW AR, is shown at the top. The CME altitudes above the AR are plotted as a function of time, with ordinates to the right. Arrows indicate the CMEs corresponding to the upper plot for beam 2, on AR 8910, and below for beam 1, on AR 8913 (after Kaufmann et al., 2001b).

2001b). Two CMEs were observed by LASCO coronagraphs during the March 22, 2000 large solar flare. The LASCO C2 images between 18 and 21 UT are shown in Fig. 16, together with the 212 GHz pulse occurrence rates with time, measured at beam 2, near AR 8910, in the NW quadrant, as well as by beam 1, pointed approximately on the SW quadrant, near AR 8913. Velocities of 450 km/s and 370 km/s have been computed for the first CME (SW, bottom white-light image differences in Fig. 16) and for the second CME (NW, top white-light image differences in Fig. 16), respectively. The positions of the ejected material were extrapolated back to the solar surface for both CMEs, strongly suggesting a close association between the initial phase of CMEs, near the solar surface,

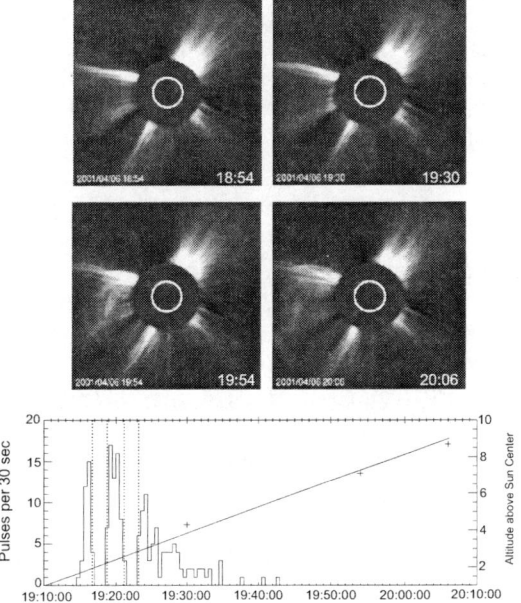

Fig. 17. The 212 GHz pulse count rates, bottom, for the April 6, 2001 flare. The "halo" class CME is shown in LASCO C2 frames above. The CME positions relative to the flaring site at the solar surface are plotted below, with altitude ordinates at right. Within the time uncertainty, the position extrapolation suggests that the start of the coronal event is close to or just before the onset of the submm-wavelength pulses (after Kaufmann et al., 2001b).

and the onset of the enhanced rate of occurrence of submm-wavelengths pulses. The first CME, ejected in the SW direction, corresponded to the first onset of submm-wave pulses at about 1730 UT in the approximate direction of the SW active region - where no Hα brightening has been reported - and observed by both beams 2 (on AR 8910 at NW) and 1 (on AR 8913 at SW). The second CME, in the NW direction, corresponded to another enhancement of submm-wave pulses starting at about 1820 UT, in the direction of the NW AR 8910, the site of the main Hα flare. Moreover, a large scale coronal connection is suggested between ARs in both hemispheres for the first CME. One "halo" class CME was associated to the April 6, 2001 flare. The LASCO C2 images are shown in Fig. 17, together with the 212 GHz pulse rate of occurrence with time. A rough estimate of the CME velocity of 1450 km/s, extrapolated to the SE AR 9415 position at the solar surface, suggests a starting time at about 1908 UT, just before the onset of the submm-wave pulses.

The submm-wave pulses appear to be present in certain AR independently from detectable flares, flashing at a lower occurrence rate, as mentioned in Sect. 3.1, and Fig. 12. One suggested possibility is that emissions at other energy ranges (visible, X- and gamma-rays) associated to single discrete pulses might be too weak, and too fast, for the sensitivity and time resolution of the usual detectors. As the active region magnetic complexity grows, conditions are reached for a major large scale instability to set in, inducing multiple smaller scale and faster instabilities (Sturrock, 1986, and references therein) which might be associated to the observed pulses at higher occurrence rates. Time integrated fluxes at other ranges of energy will increase, becoming measurable within the sensitivities of the usual detectors. Further investigations are required to understand why the combination of the large and small scale instabilities might allow the acceleration of large masses of ionized gas outwards, as observed in the solar corona and interplanetary space.

5 Final Remarks: New Questions and Prospects

The observations of flares at submillimeter waves bring promising new diagnostics on the flaring processes and a number of challenging difficulties for interpretation. No flare exhibiting a bulk emission flux density spectrum rising from mm to submm waves (such as suggested by examples in Figs. 3 and 4) has been analyzed yet. Rapid subsecond submm pulses have been detected whose rate of occurrence varies with the main burst fluxes at X- and gamma-rays. This result might be compared to the proportionality between fluxes and repetition rates of superimposed fast structures found at cm-mm wavelengths (Kaufmann et al., 1980; Raulin et al., 1998).

For the few events studied the main (bulk) fluxes decrease with frequency in the microwave-submm range, as predicted for optically thin synchrotron emissions. However the flux densities of the submm spikes, defined as the excess above the underlying emission level, seem to increase with frequency in the range 200-400 GHz. The nature of the pulses raises several basic questions to be further investigated with more observations and theoretical explorations, such as: (a) have they a physical nature distinct from the bulk emission; (b) are they related to subsecond time structures observed at optical wavelengths (Wang et al., 2000); (c) are they associated to rapid modulation of emission by waves and quakes (Zharkova and Kosovichev, 2000); (d) are they thermal bombs with short cooling times in a dense ambient plasma ($\geq 10^{14}$ cm^{-3}); (e) are they non-thermal dense short-lived synchrotron sources in which inverse Compton action might be one effective loss mechanism (Kaufmann et al., 1986).

The first submm flare observations confirm the idea that, irrespective of the emission process, the flare main emissions at higher energy X- and gamma-rays appear as a response to numerous fast energetic injections whose signatures are represented by the submm pulses, discrete in time, at different spatial locations. Suggestive evidence has been found that submm spikes count rates might be one early signature of the CME's initial acceleration process. The CMEs' start times extrapolated to the solar surface are close to or just preceding the onset time of submm spikes. This result agrees with previous suggestions that CME onsets precede any clearly related flare activity, being sometimes associated to minor X-ray flare precursors (Harrison et al. 1990). The presently suggested association to the onset of submm-wavelength fast pulses might bring new important clues for the understanding of these highly energetic phenomena in the solar atmosphere, which are known to have a profound influence on space weather conditions.

Solar flare observations in the submm-IR range are still at their very beginning. The results which are now being obtained are highly suggestive on how much we still have to learn in this nearly unexplored range of wavelengths in order to get a better and full description of the flare process. SST is the only dedicated solar instrument available for submm observations.

For the present solar cycle the use of existing non-solar submm-wavelength telescopes in the world for solar flare observations should be highly stimulated. Some of those telescopes can be pointed to the Sun if properly protected for heat damages. The University of Köln 3-m radio telescope KOSMA, located in Gornergrat, Switzerland, was recently used to observe a solar flare in the submm-wavelengths range, whose main features were reported at the 2001 CESRA Workshop (cf. the contribution of Lüthi et al. to Kliem at al., *this volume*). For the next solar cycle, on a time scale of 5-10 years, we may expect new steps in diagnostics from a new generation of THz ground-based telescopes, with new technology bolometers and imagers, and perhaps using single-dish antenna technology of the ALMA project, with imaging focal plane arrays. A new space experiment MIRAGES (Trottet, 2001, pers. comm.) has been proposed for far-IR solar flare detection at 2 and 8 THz, combined with a gamma-ray experiment, to the French space agency CNES for a flight during the next solar cycle maximum.

Acknowledgements

The author expresses his thanks to CESRA organizers for the invitation, and partial support of travel expenses to attend the meeting. Corrections and suggestions given by an anonymous referee and by the Editor have improved considerably the clarity of the text. Thanks are due to Dr. J.-P. Raulin, Mr. V. M. S. Pereira and Ms. N.G. Escolano for their help in composing the manuscript edition. This research was partially supported by Brazilian Agencies FAPESP, contract 99/06126-7, and Argentina Agency CONICET.

References

1. M.A.F. Allaart, J. van Nieuwkoop, C. Slottje, L.H. Sondaar: Solar Phys. **130**, 183 (1990).
2. A.T. Altyntsev, H. Nakajima, T. Takano, G.V. Rudenko: Solar Phys. **195**, 401 (2000).
3. T.S. Bastian, A.O. Benz, D.E. Gary: Ann. Rev. Astron. Astrophys. **36**, 131 (1998).
4. J.E. Beckman: Nature **220**, 52 (1968).
5. A.O. Benz, H. Su, A. Magun, K. Stehling: A&A **93**, 539 (1992).
6. J.C. Brown: Trans. Roy. Soc. London **A291**, 473 (1976).
7. Chupp, E.L.: in "High Energy Solar Physics", ed. by R. Ramaty, N. Mandzhavidze and X.-M. Hua, AIP Conf. Proc. **374**, p. 3 (1996).

8. C.D. Clark, W.M. Park: Nature **219**, 922 (1968).

9. E. Correia, J.E.R. Costa, A. Magun, Herrmann, R.: Solar Phys. **159**, 143 (1995).

10. D.L. Croom: in "High Energy Phenomena on the Sun", ed. by R. Ramaty and R.G. Stone, NASA Publ. SP 342, p. 114 (1973).

11. C. de Jager: Space. Sci. Rev. **44**, 43 (1986).

12. C. de Jager, G. de Jonge: Solar Phys. **58**, 127 (1978).

13. F. Dröge: A&A **57**, 285 (1977).

14. G.A. Dulk, K.A. Marsh: ApJ **259**,350 (1982).

15. A.G. Emslie, S. Mehta, and M.E. Machado: AGU-EOS Trans. 75, 295 (1994).

16. D.J. Forrest, E.L. Chupp: Nature **305**, 291 (1983).

17. K.J. Frost: ApJ **158**, L159 (1969).

18. V. Gaizauskas, K.F. Tapping,: ApJ **241**, 804 (1980).

19. C.G. Giménez de Castro, J.-P. Raulin, et al.: A&A **140**, 373 (1999).

20. V.I. Ginzburg, S.I. Syrovatskii: Ann. Rev. Astron. Astrophys. **3**, 297 (1965).

21. O. Hachenberg, G. Wallis: Z. Astrophys. **52**, 42 (1961).

22. R. A. Harrison, E. Hildner et al.: J. Geophys. Res. **95**, 917 (1990).

23. H.S. Hudson: Solar Phys. **45**, 69 (1975).

24. K. Hurley, M. Niel et al.: ApJ **265**, 1076 (1983).

25. P. Kaufmann: Solar Phys. **102**, 97 (1985).

26. P. Kaufmann: in "Infrared Tools for Solar Astrophysics: What's Next?", ed. by J.R. Kuhn and M.J.Penn, World Scientific, Singapore, p. 127 (1995).

27. P. Kaufmann: Solar Phys. **169**, 377 (1996).

28. P. Kaufmann, L.R. Piazza, J.C. Raffaelli: Solar Phys. **54**, 179 (1977).

29. P. Kaufmann, F.M. Strauss, R. Opher, C. Laporte: A&A **87**, 58 (1980).

30. P. Kaufmann, E. Correia et al.: Nature **313**, 380 (1985).

31. P. Kaufmann, E. Correia, J.E.R. Costa, A.M. Zodi Vaz, A&A 157, 11, 1986.

32. P. Kaufmann, N.J. Parada et al.: in Proc. Kofu Symp. "New Look at the Sun", NRO Report No. 360, p. 323 (1994).

33. P. Kaufmann, G. Trottet et al. : Solar Phys. **197**, 361 (2000).

34. P. Kaufmann, J.-P. Raulin et al.: ApJ **548**, L95 (2001a).

35. P. Kaufmann, J.-P. Raulin et al.: in SCOSTEP ISCS 2001 Workshop, "2001-A Space Odyssey", CEDAR Workshop & 10th Quadrennial STP Symposium, Longmont, CO, USA, June 12-22 (2001b).

36. P. Kaufmann, J.E.R. Costa et al.: in Proc. SMBO/IEEE MTT-S International Microwave and Optoelectronics Conference, August 6-10, 2001, Belém, PA, Brazil, 439, reproduced with IEEE copyright permission in: Telecomunicações **4**, 18 (2002a).

37. P. Kaufmann, J.-P. Raulin et al.: ApJ **574**, 1059 (2002b).

38. A.L. Kiplinger, B.R. Dennis et al.: ApJ **265**, L99 (1983).

39. K.-L. Klein: A&A **183**, 341 (1987).

40. M.R. Kundu: "Solar Radio Astronomy", John Wiley & Sons, NY (1965)

41. M.E. Machado, K.K. Ong et al.: Adv.Space Res. **13(9)**, 175 (1993).

42. K.G. McClements, J.C. Brown, A&A **165**, 235 (1986).

43. A.M. Melo, C.G. Giménez de Castro et al.: Rev. Mackenzie de Engenharia, submitted (2002).

44. H. Nakajima: in "High Energy Solar Physics - Anticipating HESSI", ed. by R. Ramaty and N. Mandzhavidze, ASP Conf. Ser. **206**, p. 313 (2000).

45. K. Ohki, H. S. Hudson, Solar Phys. **43**, 405 (1975).

46. G.K. Parks, J.R. Winckler: ApJ **155**, L117 (1969).

47. R. Ramaty: in "High Energy Phenomena on the Sun", ed. by R. Ramaty and R.G. Stone, NASA Publ. SP 342, p. 188 (1973).

48. R. Ramaty, N. Mandzhavidze: in "High-Energy Solar Phonomena - a New Era of Spacecraft Measurements", ed. by J. Ryan and W.T. Vestrand, AIP Conference Proc. **294**, p. 26 (1994).
49. R.Ramaty, V. Petrosian: ApJ **178**, 241 (1972).
50. J.-P. Raulin, P. Kaufmann et al.: ApJ **498**, L173 (1998).
51. J.-P. Raulin, N. Vilmer et al.: A&A **355**, 355 (2000).
52. J.I. Sakai, R. Zaidman et al.: in "Rapid Fluctuations in Solar Flares", ed. by B.R. Dennis, L.E. Orwig and A.L. Kiplinger, NASA Conf.Publ. 2449, p. 393 (1986).
53. J. Shklovsky: Nature **202**, 275 (1964).
54. C. Slottje: Nature **275**, 520 (1978).
55. W.A. Stein, E.P. Ney: J. Geophys. Res. **68**, 65 (1963).
56. P. A. Sturrock: in "Rapid Fluctuations in Solar Flares", ed. by B.R. Dennis, L.E. Orwig and A.L. Kiplinger, NASA Conf.Publ. 2449, p. 1 (1986).
57. P.A Sturrock, Y.Uchida: ApJ **246**, 331 (1981).
58. P.A. Sturrock, P. Kaufmann, R.L. Moore, D.F. Smith: Solar Phys. **94**, 341 (1984).
59. T.Takakura: Solar Phys. **1**, 304 (1967).
60. T. Takakura, P Kaufmann et al.: Nature **302**, 317 (1983).
61. G. Trottet, N. Vilmer et al.: A&A Suppl. **97**, 337 (1993).
62. G. Trottet, J.-P. Raulin et al.: A&A **381**, 694 (2002).
63. H.F van Beek, L.D. de Feiter, C. de Jager: Space Res. XIV, 447 (1974).
64. H. Wang, J. Qiu et al.: ApJ **542**, 1080 (2000).
65. J.P. Wild, S.F Smerd, A.A. Weiss: Ann. Rev. Astron. Astrophys. **1**, 291 (1963).
66. V.V. Zharkova, A.G. Kosovichev: in "High Energy Solar Physics - Anticipating HESSI", ed. by R. Ramaty and N. Mandzhavidze, ASP Conf. Ser. 206, p. 77 (2000).
67. A.M. Zodi Vaz, P. Kaufmann et al.: in "Rapid Fluctuations in Solar Flares", ed. by B.R. Dennis, L.E. Orwig and A.L. Kiplinger, NASA Conf. Publ. 2449, p. 171 (1985).

Index